01-CTS-3

# FACTORS INFLUENCING THE RELIABILITY OF ENHANCED BIOLOGICAL PHOSPHORUS REMOVAL

by:

**J.B. Neethling**
**Brian Bakke**
**Mario Benisch**
**April Gu**
**Heather Stephens**
HDR

**H. David Stensel**
University of Washington

**Rebecca Moore**
Moore Engineering Consultants

*2005*

WERF

The Water Environment Research Foundation, a not-for-profit organization, funds and manages water quality research for its subscribers through a diverse public-private partnership between municipal utilities, corporations, academia, industry, and the federal government. WERF subscribers include municipal and regional water and wastewater utilities, industrial corporations, environmental engineering firms, and others that share a commitment to cost-effective water quality solutions. WERF is dedicated to advancing science and technology addressing water quality issues as they impact water resources, the atmosphere, the lands, and quality of life.

For more information, contact:
Water Environment Research Foundation
635 Slaters Lane, Suite 300
Alexandria, VA 22314-1177
Tel: (703) 684-2470
Fax: (703) 299-0742
www.werf.org
werf@werf.org

This report was co-published by the following organizations. For non-subscriber sales information, contact:

IWA Publishing
Alliance House, 12 Caxton Street
London SW1H 0QS, United Kingdom
Tel: +44 (0) 20 7654 5500
Fax: +44 (0) 20 7654 5555
www.iwapublishing.com
publications@iwap.co.uk


Library of Congress Catalog Card Number: 2005921964

IWAP ISBN: 978-1-84339-740-3



HDR

# ACKNOWLEDGMENTS

The project team wishes to recognize the contribution of the many wastewater treatment plant and utility staff who contributed their time to provide information in support of this project, and colleagues overseas who solicited participation from wastewater treatment facilities outside of North America.

## Report Preparation

**Principal Investigator:**
J.B. Neethling, Ph.D., P.E., *HDR*

**Project Team:**
Brian Bakke, E.I.T., *HDR*
Mario Benisch, *HDR*
April Gu, Ph.D., *HDR*
Heather Stephens, P.E., *HDR*
H. David Stensel, Ph.D., P.E., D.E.E, *University of Washington*
Rebecca Moore, P.E., *Moore Engineering Consultants*

## Project Subcommittee

Tyler Richards, *Gwinnett County Department of Public Utilities*
James Barnard, Ph.D., *Black and Veatch*
David Jenkins, Ph.D., *University of California at Berkeley*
David Stevens, Ph.D., P.E., *Utah State University*
David Waltrip, *Hampton Roads Sanitation District*
Steve Williams, *RWE Thames Water*

## Water Environment Research Foundation Staff

**Director of Research:** Daniel M. Woltering, Ph.D.
**Project Manager:** Mary Strawn

# ABSTRACT AND BENEFITS

**Abstract:**

Enhanced biological phosphorus removal (EBPR) has been used for decades to remove phosphorus from municipal wastewater because it allows facilities to meet water quality goals while minimizing chemical consumption and sludge production. However, there is still substantial variability in both the practices applied to achieve EBPR and the level of soluble phosphorus removal achieved.

The objective of this research project was to develop information that can be used to help municipal wastewater treatment plants more efficiently and cost effectively remove phosphorus through EBPR processes. This project included detailed analysis of routine water quality and operating data, field testing observations, and special studies conducted over the course of the project to evaluate the variability of EBPR, factors influencing EBPR performance, and the relationship between EBPR and the presence of glycogen accumulating organisms (GAOs).

The study showed that EBPR is capable of achieving very low effluent phosphate concentrations. When operating well, it is possible to achieve soluble phosphate concentrations under 0.1 mg/L. However, many facilities struggle to routinely achieve soluble phosphate concentrations under 1.0 mg/L. For this reason, most facilities experience significant variability in EBPR performance, requiring augmentation of EBPR through the use of substrate addition or polishing with iron- or aluminum-containing chemicals.

**Benefits:**

- Documents the performance of full-scale EBPR systems.
- Evaluates reliability of facilities with history of EBPR based on key influent characteristics and operational parameters.
- Documents phosphorus release and uptake for facilities with a history of EBPR and shows the lack of relationship between phosphorus release and performance.
- Documents the difficulty associated with predicting successful EBPR in full-scale plants based on a single parameter.
- Identifies the need for chemical addition for long-term, stable phosphorus removal to high levels.
- Quantifies the importance of recycle streams and controlling return loads to maintain good EBPR performance.
- Evaluates presence of glycogen accumulating organisms in full-scale facilities and summarizes research on conditions that may lead to the proliferation of glycogen accumulating organisms.

**Keywords:** Phosphorus, enhanced biological phosphorus removal, biological nutrient removal

# TABLE OF CONTENTS

# LIST OF TABLES

# LIST OF FIGURES

# LIST OF ACRONYMS

| | |
|---|---|
| A2O | Anaerobic-Anoxic-Oxic |
| ARCY | Anoxic Recycle |
| BNR | Biological Nutrient Removal |
| BOD | Biochemical Oxygen Demand |
| CLS | Chemical Sludge |
| COD | Chemical Oxygen Demand |
| DO | Dissolved Oxygen |
| EBPR | Enhanced Biological Phosphorus Removal |
| ffCOD | Flocculated Filtered Chemical Oxygen Demand |
| FISH | Fluorescence In-Situ Hybridization |
| GAB | Glycogen Accumulating Bacteria |
| GAO | Glycogen Accumulating Organisms |
| GBT | Gravity Belt Thickener |
| GPD | Gallons per Day |
| HRT | Hydraulic Retention Time |
| IWA | International Water Association |
| K | Potassium |
| Mg | Magnesium |
| MLSS | Mixed Liquor Suspended Solids |
| MLVSS | Mixed Liquor Volatile Suspended Solids |
| NRCY | Nitrified Recycle |
| P | Phosphorus |
| PAO | Phosphorus Accumulating Organism |
| PE | Primary Effluent |
| PHA | Polyhydroxyalkanoates |
| PHB | Polyhydroxybutyrate |
| PSL | Primary Sludge |
| RAS | Return Activated Sludge |
| RBC | Rotating Biological Contactor |
| RBCOD | Readily Biodegradable COD |
| SBOD | Soluble BOD |
| SBR | Sequencing Batch Reactor |
| SCOD | Soluble COD |
| SE | Secondary Effluent |
| SRT | Solids Retention Time |
| SVI | Sludge Volume Index |
| TCA | Tricarboxylic Acid Cycle |
| TP | Total Phosphorus |
| TSCOD | Truly Soluble COD |
| TSS | Total Suspended Solids |
| UCT | University of Cape Town |
| URT | Uptake and Release Tests |
| VFA | Volatile Fatty Acid |
| VIP | Virginia Initiative Plant |

| | |
|---|---|
| VSS | Volatile Suspended Solids |
| WAS | Waste Activated Sludge |
| WERF | Water Environment Research Foundation |
| WWTP | Wastewater Treatment Plant |

# Executive Summary

## ES.1 Project Objectives

Enhanced biological phosphorus removal (EBPR) has been used for decades to remove phosphorus from municipal wastewater. The anaerobic /aerobic activated sludge process train associated with EBPR is attractive to treatment plant operators and utilities because it allows facilities to meet water quality goals while minimizing chemical consumption and sludge production. Despite its growing use, there is still substantial variability in both the practices applied to achieve EBPR and the level of phosphate removal achieved. There is also a lack of understanding of the impact on EBPR of other plant processes, such as chemical addition and anaerobic digestion, used at municipal wastewater treatment plants.

This research project was designed to develop information that can be used to help municipal wastewater treatment plants more efficiently and cost-effectively remove phosphorus through EBPR processes. The main objectives of this research were to:

- Develop a database of phosphorus-removal performance in plants that have practiced EBPR for more than one year.
- Evaluate the reliability of these plants to meet phosphorus limits under normal operating conditions. Evaluate reliability without chemical addition.
- Identify factors that contribute to reliable EBPR performance. These may include factors such as:
  - Influent wastewater characteristics;
  - Nitrogen removal requirements and the presence of nitrate;
  - Role of fermenters or other processes used to enhance EBPR;
  - Management of return flows from anaerobic solids processing steps; and
  - Chemical addition for phosphorus polishing or EBPR backup.

In this study reliability is considered the ability of the process to produce the same effluent consistently. Reliability is expressed in terms of the percentage of time that various effluent phosphorus concentrations are achieved. The efficiency of the process is expressed in terms of effluent phosphorus concentration.

The study also sought to evaluate the presence of glycogen accumulating organisms (GAOs), which can compete with phosphorus accumulating organisms (PAOs) for substrate and thus impact EBPR performance. Recent research from laboratory studies and limited full-scale plant reports clearly shows that operation of an anaerobic/aerobic activated sludge process does not guarantee that the system will only select for the PAOs on which EBPR relies. A wide range of diverse bacteria can compete for substrate in the anaerobic zone and may out-compete the PAOs, seriously impairing EBPR. Environmental factors such as pH, temperature, influent characteristics, and operational factors such as solids retention time (SRT) have been shown to strongly influence the population of anaerobic/aerobic tolerant bacteria and, therefore, the effectiveness of EBPR.

## ES.2 Project Approach

A three-phased approach was used to gather data on plant performance, influent characteristics, and operating conditions. These phases were designed to provide increasingly more detailed evaluations of a decreasing number of facilities.

The first phase in the project was to publish a screening survey on the internet. This survey was designed to solicit sufficient information with which to characterize the treatment plant operation and performance, yet limit the time and data required to complete the survey. As such, the survey collected general operator interpretations of the facility rather than values supported by routine data analysis. In addition to providing general information about the range of processes used and performance achieved, the screening survey identified facilities for follow-up evaluation.

Thirteen facilities were identified for additional investigation in the second phase, with detailed studies conducted at five facilities in the third phase. These investigations included:

- Compiling three years of operating data to determine long term performance and the influence of conventional process parameters on EBPR performance. Operating data included effluent quality as well as influent flows and loads, and operating conditions (sludge age, mixed liquor suspended solids (MLSS), temperature, etc.).
- Intensive sample collection and analysis over two 2-week periods. Sample collection included characterization of influent, primary effluent (PE), secondary effluent (SE), final effluent, and recycle streams. Samples were analyzed for various measures of EBPR substrate, as well as various forms of suspended solids, nutrients, and metals.
- Measurements of aeration-basin phosphate profiles through the secondary treatment train to provide a snapshot of process operation and determine the phosphate release and uptake. Where possible, in addition to phosphate, samples were analyzed for ammonia-nitrogen, nitrate, nitrite, pH, and dissolved oxygen (DO).
- Bench-scale phosphate uptake and release tests (URT) conducted at the University of Washington using mixed liquor collected during the intensive sampling periods. During the URT, a mixed-liquor sample was spiked with a solution of sodium acetate, and the phosphate release measured for the first 45 or 60 minutes. After the anaerobic period, the sample was aerated for 2 to 3 hours during which time the phosphate uptake was measured.
- Microbial examination at the University of Washington to observe the presence of PAOs and GAOs using mixed liquor collected during the intensive sampling periods.
- Microbial examination to detect some of the known GAO bacteria using fluorescent in-situ hybridization (FISH) tests by Dr. Linda Blackall's laboratory at the University of Queensland in Australia using mixed liquor collected during the intensive sampling periods.
- Development of a phosphorus-mass balance to determine the phosphorus recycle load and aeration basin phosphorus inventory.

This study makes extensive use of plant operating data and measurements reported by process laboratories. Since the plant laboratories used different analytical techniques, phosphorus species are often reported differently and are used as such in the report. In some cases, one plant may measure only total phosphorus (TP) in the effluent, while another reports orthophosphate. In this report, the following conventions are used:

- All concentrations are as reported by field laboratories. Analytical methods may vary.
- TP refers to all phosphorus species measured.
- Phosphate (or orthophosphate) refers to measurements recorded as orthophosphate or $PO_4$. It is possible that particulate phosphate (from some precipitant) may be included in the measure if the precipitant phosphate is detected in the measure.
- The term *phosphorus* is used in this report to denote the general phosphorus species. It may refer to TP or phosphate, depending on the context of the measure.
- All phosphorus measurements in the activated sludge basins were first filtered and then measured, typically as phosphate.
- All phosphorus concentrations are expressed as P.

## ES.3 Screening Survey

Screening survey inputs were received from facilities representing a wide range of sizes, geographic locations, operating processes, and performance. Over 70 facilities were contacted directly to solicit participation; others learned of the survey through word of mouth and through a posting on WERF's website. Completed surveys were received from 47 facilities in the United States, Canada, Spain, Germany, and South Africa.

The following general observations are based on screening survey results:

- Plants achieving the highest percentage of influent phosphorus removal supplement EBPR by adding volatile fatty acids (VFAs), or by polishing with iron- or aluminum-containing chemicals, or both.
- Two-thirds of the facilities meeting effluent TP levels lower than 0.5 mg/L do not have solids recycle streams from the dewatering of anaerobically digested sludge.
- A wide range of SRTs (3 to more than 20 days) is represented in all categories of removal performance (effluent TP under 0.5 mg/L, 0.5–1.0 mg/L, and over 1.0 mg/L).
- The maximum TP removal for facilities reporting no VFA or chemical addition is 95%. The effluent TP concentrations of facilities reporting no VFA or iron- or aluminum-containing chemical addition ranges from 0.2 mg/L to over 2.0 mg/L.

Facilities were ranked by effluent TP achieved and summarized in a table outlining other key operating parameters (e.g., SRT, VFA addition, chemical addition). This information is shown in Table ES-1.

**Table ES-1. Phosphorus Removal Summary Ranked by Effluent TP Based on Screening Survey.**

| Facility[1] | Location | Influent TP (mg/L) | BOD: TP [2] | SRT Range (days) | Second. Effl. (SE) TP [3] (mg/L) | Effluent TP (mg/L) | Secondary Removal [4] | Total Removal % | Lime Addition[5] | Solids. Recycle | VFA and/or Chemical Addition [6] | | |
|---|---|---|---|---|---|---|---|---|---|---|---|---|---|
| | | | | | | | | | | | None | VFA | Chemical |
| **FACILITIES WITH EFFLUENT PHOSPHORUS < 0.5 mg/L** | | | | | | | | | | | | | |
| Luepinghausen | Luepinghausen, Ger | 5.4 | 21 | 10 - 20 | | 0.01 | | 98 | | X | | | Ferric-AB |
| Water Reclamation Facility | McMinnville, OR | 5 | 40 | 10 - 20 | 0.15 | 0.05 | 97% | 99 | | | | | Alum-TCL |
| Durham AWWTP | Tigard, OR | 7.5 | 19 | 10 - 20 | 0.572 | 0.06 | 92% | 99 | X | X | | X | Alum-SCL, TCL |
| City of Ft. Myers Central AWWT Facility | Fort Myers, FL | | | 10 - 20 | | 0.1 | | | | | | | Alum-SCL |
| McDowell Creek WWTP[9] | Huntersville, NC | 7 | 35 | 10 - 20 | 0.1 | 0.1 | | 99 | X | X | | X | Alum-PCL,SCL |
| Kalispell AWT & BNR Facility | Kalispell, MT | 4.73 | 56 | | | 0.11 | | 98 | | | | X | |
| Back River Wastewater Treatment Plant | Baltimore, MD | 3.6 | 34 | 5 - 10 | 0.17 | 0.12 | | 97 | | X | | | Ferrous-AB |
| City of Kelowna Wastewater Treatment Facility | Kelowna, BC | 6.63 | 18 | 5 - 10 | 0.21 | 0.15 | | 98 | | | | X | Alum-SCL |
| Iron Bridge Regional WRF | Oviedo, FL | 8 | 19 | 5 - 10 | 0.3 | 0.2 | | 98 | | | | | Alum-SCL |
| Green Bay Metropolitan Sewerage District | Green Bay, WI | | | 10 - 20 | | 0.21 | | | | | X | | |
| Reedy Creek Improvement District WRF | Lake Buena Vista, FL | 6.7 | 22 | 5 - 10 | -- | 0.26 | | 96 | | | | X | Ferric-AB,SCL |
| Brownstown Wastewater Utility [8] | Brownstown, IN | 5.4 | 25 | 3 - 5 | | 0.26 | | 95 | | | X | | |
| Hildesheim | Hildesheim, Germany | 7.85 | 26 | 10-20 | | 0.34 | | 96 | | X | | X | Ferric/Ferrous - SCL, Centr. |
| American-Bath WWTP [8] | Lima, OH | 8 | 88 | 10-20 | | 0.35 | | 96 | | | | | Alum-SCL |
| City of Cheney WWTP & Reclamation Plant [7] | Cheney, WA | 4.75 | 45 | > 20 | | 0.4 | | 92 | | | X | | |
| **FACILITIES WITH EFFLUENT PHOSPHORUS 0.5 - 1.0 mg/L** | | | | | | | | | | | | | |
| Nine Springs Wastewater Treatment Plant | Madison, WI | 6 | 24 | 5 - 10 | | 0.5 | | 92 | | X | | | Ferric-Centr. |
| Stevens Point Wastewater Treatment Plant | Stevens Point, WI | | | < 3 | | 0.5 | | | | X | | X | Ferric-PCL, AB |
| Waiblingen | Waiblingen, Germany | 8.4 | 24 | 5-10 | | 0.5 | | 94 | | X | | X | Ferric-SCL |
| Sheboygan Regional Wastewater Treat. Facil. | Sheboygan, WI | 5.14 | 23 | 5 - 10 | | 0.56 | | 89 | | X | | | Ferric-AB, Centr. |
| City of Grande Prairie WWTP | Grand Prairie, Alberta | 7.19 | 36 | | | 0.58 | | 92 | | | X | | |
| Kamen-Koernerback | Kamen, Germany | | | 10 - 20 | | 0.6 | | | | X | | | Ferric-SCL |
| Duelmen | Duelmen, Germany | | | 10 - 20 | >0.8 | 0.6 | | | | X | | | Ferric-SCL |
| Fergus Falls WWTP | Fergus Falls, MN | 6 | 19 | > 20 | | 0.6 | | 90 | | X | | | Alum-SCL |
| Virginia Initiative Plant (VIP) | Norfolk, VA | 5.4 | 26 | 5 - 10 | | 0.6 | | 89 | | | | X | |
| Bonnybrook Wastewater Treatment Plant | Calgary, Alberta | 6.3 | 19 | 5 - 10 | | 0.75 | | 88 | | X | | X | Alum-PCL,SCL |
| Nansemond Treatment Plant | Suffolk, VA | 10 | 15 | 5 - 10 | | 0.8 | | 92 | | X | | X | Ferric-PCL, AB, SCL |

Table ES-1 (continued). Phosphorus Removal Summary Ranked by Effluent TP Based on Screening Survey.

| Facility[1] | Location | Influent TP (mg/L) | BOD: TP [2] | SRT Range (days) | Second. Effl. TP [3] (mg/L) | Effluent TP (mg/L) | Secondary Removal [4] | Total Removal % | Lime Addition[5] | Solids. Recycle | VFA and/or Chemical Addition [6] None | VFA | Chemical |
|---|---|---|---|---|---|---|---|---|---|---|---|---|---|
| **FACILITIES WITH EFFLUENT PHOSPHORUS 0.5 - 1.0 mg/L (continued)** | | | | | | | | | | | | | |
| Laughlin Water Reclamation Facility | Laughlin, NV | 6.75 | 41 | > 20 | -- | 0.85 | | 87 | | | | | Alum-SCL |
| University of Florida, Water Rec Facility | Gainesville, FL | | | > 20 | 0.86 | 0.86 | | | | | | | Alum-NS |
| Marshall Wastewater Treatment Facility | Marshall, WI | | | | | 1 | | | | | X | | |
| Rogers Pollution Control Facility | Rogers, AR | 11 | 25 | | | 1 | | 91 | | | | | Ferrous-NS |
| Sod Run Wastewater Treatment Plant | Perryman, MD | 8 | 24 | 10 - 20 | | 1 | | 88 | | X | | | Ferric-PCL,SCL |
| Pocatello Water Pollution Control Dept. [7] | Pocatello, ID | 8 | 23 | 3 - 5 | | 1 | | 88 | | X | X | | |
| Mauldin Road WWTP | Greenville, SC | 7 | 23 | 10 - 20 | 1 | 1 | | 86 | | | | X | |
| Village of Minoa Wastewater Treat. Facil. | Minoa, NY | 6 | 33 | | | 1 | | 83 | | | X | | |
| **FACILITIES WITH EFFLUENT PHOSPHORUS > 1.0 mg/L** | | | | | | | | | | | | | |
| Olifantsfontein WCW | Johannesburg, SA | 17.1 | | 10 - 20 | | 1.15 | | 93 | | X | | X | |
| Zeekoegat Wastewater Treatment Works | Pretoria, South Africa | 6 | | 10 - 20 | | 1.2 | | 80 | | | | X | Ferric-Filter |
| Eugene/Springfield WPCF [8] | Eugene, OR | 8.5 | 10 | 5 - 10 | | 1.4 | | 84 | | X | X | | |
| Proctors Creek Wastewater Plant | Chester, VA | 7 | 21 | 10 - 20 | -- | 1.4 | | 80 | | X | | | Ferric-PCL |
| Ephrata Regional WWTP No. 2 | Ephrata, PA | 6.8 | 42 | > 20 | | 1.5 | | 78 | | | | | Ferrous-AB |
| Lower Reedy | Greenville, SC | 6.5 | 42 | 10 - 20 | | 1.5 | | 77 | X | X | X | | |
| Flip Human Water Care Works | Krugersdorp, SA | 12 | | < 3 | | 2 | | 83 | | | X | | |
| Missoula Wastewater Division [8] | Missoula, MT | 4.98 | 37 | 3 - 5 | | 2.02 | | 59 | | X | | X | |
| Rooiwal Wastewater Treatment Works | Pretoria, South Africa | 9.79 | | 10 - 20 | | 2.42 | | 75 | | | X | | |
| Percy Stewart Water Care Works | Krugersdorp, SA | 12 | | < 3 | 4.52 | 4.04 | 62% | 66 | | | | | Ferric-SCL |
| ERAR Viveros de la Villa | Madrid, Spain | 6.1 | 19 | 5 - 10 | 4.7 | 4.7 | | 23 | | X | X | | |

Notes:

1. Blank spaces in the table indicate that no data were given or the value was unknown.
2. BOD:TP ratio entering anaerobic tank/zone. Thirty-five percent reduction in influent BOD was assumed for facilities with primary clarifiers.
3. Facility has effluent filter. Dash indicates that facility has effluent filter, but no secondary effluent TP reported.
4. Facility adds alum to tertiary clarifier
5. Lime addition for alkalinity control.
6. Chemical addition for phosphorus removal: PCL – Primary Clarifier, SCL – Secondary Clarifier; TCL – Tertiary Clarifier; AB – Aeration Basin, Centr – Dewatering Centrate, NS – Not Specified.
7. Facility only nitrifies.
8. Facility nitrifies and denitrifies.
9. No values below 0.1 mg/L were reported

## ES.4 Field Testing and Detailed Data Analysis Results

Three years of plant operating records and two 2-week intensive sampling periods provided the data for a detailed analysis of EBPR performance.

### ES.4.1 Influent Characteristics and Effluent Performance

Five facilities were selected for in-depth study. The locations and general characteristics of these facilities are shown in Table ES-2.

**Table ES-2. Summary of Wastewater Treatment Plants Participating in Detailed Study.**

| | Durham | VIP | Nansemond | McDowell Creek | Lower Reedy |
|---|---|---|---|---|---|
| State | Oregon | Virginia | Virginia | North Carolina | Florida |
| Process | A2O† | VIP* | VIP | Modified UCT** | Modified UCT |
| Design Capacity, MGD | 25 | 40 | 30 | 6 | 9 |
| Notable feature | Stable process with low effluent phosphorus | Stable process, with apparent annual upset in August | Struggled to achieve EBPR until 2003 | Add sugar waste. | Consistently meets permit limit. Long HRT system |
| Permit Limit (mg/L P) | 0.07 (seasonal) | 2.0 | 2.0 | 1.0 | 2.0 |

*Virginia Initiative Plant

**University of Cape Town

†Anaerobic/anoxic/oxic

The five facilities selected for in-depth study had similar influent characteristics. Influent biochemical oxygen demand (BOD) concentrations ranged from approximately 150–200 mg/L, ammonia concentrations ranged from 20–25 mg/L, and TP ranged from 5–8 mg/L. The exception was the Nansemond facility, which had higher influent ammonia and TP concentrations (35 mg/L and 14 mg/L, respectively), likely due to industrial discharges. Table ES-3 summarizes average aeration basin influent characteristics at the facilities during the intensive sampling periods. The McDowell Creek plant adds a high BOD waste from a soft drink manufacturing plant directly to the anaerobic zone. The BOD concentration in this wastewater is 40,000–130,000 mg/L with an average concentration of approximately 80,000 mg/L (Goins, et al., 2003). The Durham plant adds fermenter overflow directly to the anaerobic zone.

**Table ES-3. Summary Comparison of Aeration Basin Influent Parameters from Intensive Testing.**

| Parameter | Unit | Durham | VIP | Nansemond | McDowell Creek | Lower Reedy |
|---|---|---|---|---|---|---|
| BOD | mg/L | 148 | 148 | 196 | 186 | 199 |
| TKN | mg/L | NA | 25 | 152 | 31 | 34 |
| $NH_4$-N | mg/L | 26 | 19 | 35 | 24 | 25 |
| TP | mg/L | 6.1 | 5.3 | 14 | 6.5 | 8.5 |
| $PO_4$-P | mg/L | 3.4 | 4.1 | 12 | -- | 5.8 |
| VFA:P Ratio | -- | 8.1 | 12 | 1.6 | 9.1 | 8.8 |
| ffCOD:P Ratio | -- | 20 | 31 | 6.7 | 35 | 15 |
| COD:P Ratio | -- | 40 | 59 | 29 | 75 | 34 |
| BOD:P Ratio | -- | 28 | 33 | 12 | 26 | 20 |
| ffCOD:COD Ratio | -- | 0.48 | 0.51 | 0.23 | 0.46 | 0.46 |
| VFA:RBCOD Ratio | -- | 0.41 | 0.38 | 0.24 | 0.26 | 0.29 |

On average all of the plants achieved high levels of EBPR. Table ES-4 compares the average secondary effluent water quality at the facilities using operating data collected between January 2001 and the fall of 2003 (datasets from various facilities end during different months). Not all of the facilities collected secondary effluent phosphate data as part of their routine sampling.

**Table ES-4. Summary Comparison of Average Secondary Effluent (SE) Parameters (3-year data set).**

| Parameter | Unit | Durham | VIP | Nansemond | McDowell Creek | Lower Reedy |
|---|---|---|---|---|---|---|
| $NH_4$-N | mg/L | 0.65 | 2.4 | 12 | 0.25 | 0.62 |
| $NO_3$-N | mg/L | 8.4 | 6.5 | 6.4 | 4.1 | 4.7 |
| TN | mg/L | 10 | 9.4 | 19 | NA | NA |
| TP | mg/L | 0.15 | 0.5 | 1.2 | 0.15 | 0.67 |
| $PO_4$-P | mg/L | 0.1 | 0.6 | NA | NA | 0.5 |
| BOD | mg/L | 1.6 | 6.2 | 6.1 | 3.4 | 4.1 |
| TSS | mg/L | 1.0 | 3.9 | 7.7 | 3.1 | 4.7 |
| Alkalinity | mg/L | 113 | 76 | 169 | 86 | 111 |

## ES.4.2 Phosphate Profile Measurements

Samples were collected at several intervals during the study to determine the phosphate concentration in each compartment of the activated-sludge aeration basin. Because of internal recycles, reductions in measured phosphate concentrations from one compartment to the next may not accurately represent the increase or decrease in phosphorus mass. Therefore, profile concentrations were normalized to account for dilution/concentration effects of the internal recycle streams to produce a flow-corrected phosphate concentrations based on the aeration-basin influent flow rate. Figure ES-1 shows average flow-corrected phosphate profiles from the six facilities included in the detailed study. (Mauldin Road facility was initially included in field testing but lacked adequate historical data to complete a full analysis.) The phosphate profiles showed that the phosphorus release in the anaerobic zone occurred to a highly variable extent, and neither correlated with the effluent phosphate concentration nor the net phosphorus removal.

## ES.4.3 Uptake and Release Tests

URTs provide a qualitative assessment of the acetate-removing or PAO activity in the activated sludge. By adding an excess of VFA (sodium acetate) to mixed liquor, the measured phosphate release is limited by the PAO activity in the biomass and by the amount of stored phosphorus. URTs can be used for:

- Determining the amount of PAO activity from the phosphate release rate. The release rate is indicative of PAO activity.
- Determining the potential competition for VFA in the anaerobic zone. The ratio of phosphate released to VFA consumed indicates PAO activity. Literature values for this ratio range from 0.02 g P/g acetate to 0.8 g P/g acetate (Liu, et al., 1997; Mino, et al., 1987; Wentzel, et al., 1988; Smolders, et al., 1994), with lower ratios indicative of competition from other VFA consuming organisms such as GAOs.
- Determining kinetic parameters for process simulation model calibration, such as the ratio of phosphate released to acetate consumed and magnesium and potassium uptake and release rates.
- Providing information on the phosphate uptake kinetics that can be used to determine the time required to remove phosphate in the aerobic zone.

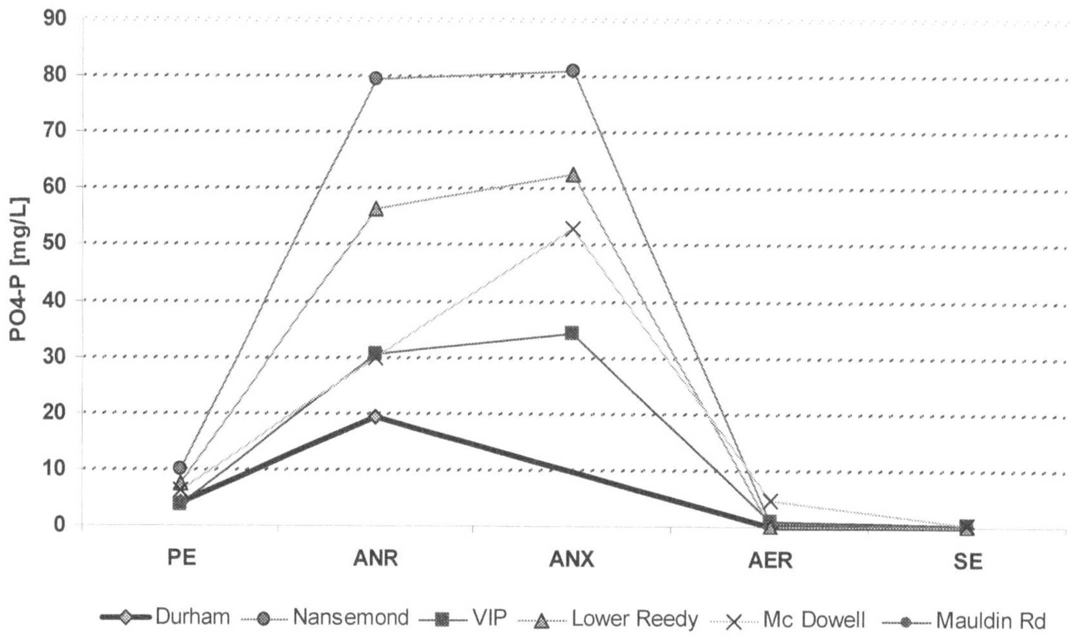


**Figure ES-1: Average Flow-Corrected Phosphate Profiles.**

Results of the uptake and release tests are shown in Table ES-5 below. Definitions of test parameters are provided in Appendix P.

**Table ES-5. Summary of Uptake and Release Test Results.**

| Test Parameter | Unit | Durham | VIP | Nansemond | McDowell Creek | Lower Reedy |
|---|---|---|---|---|---|---|
| Phosphate Release Rate | mg P /g VSS /hr | 7.4 | 12.0 | 6.4 | 11.9 | 24.6 |
| Phosphate Uptake Rate | mg P /g VSS /hr | 2.4 | 5.3 | 3.0 | 3.6 | 5.9 |
| Release Rate:Uptake Rate | ratio | 3.1 | 2.7 | 2.3 | 3.6 | 4.0 |
| ΔP/Acetic acid | mg $PO_4$-P/ mg COD | 0.28 | 0.34 | 0.22 | 0.27 | 0.51 |
| ΔP/ΔMg - Release | mg $PO_4$-P / mg Mg | 4 | 6 | 14 | 6 | 4 |
| ΔP/ΔMg - Uptake | mg $PO_4$-P / mg Mg | 3.5 | 4.6 | 5.7 | 5.1 | 3.6 |
| MLVSS | mg/L | 2600 | 2070 | 2010 | 2200 | 1850 |

## ES.5 Discussion of Findings

### ES.5.1 General Process Performance

The following conditions have been found to favor EBPR and allow the achievement of consistently low effluent phosphate concentrations. These conditions, and the subsequent evaluation which follows, are described in detail in Chapter 6.0.

- Influent composition — high influent BOD:TP
- Exclusion of oxidants (nitrate and dissolved oxygen) from the anaerobic zone
- Excluding or minimizing recycle flows and loads fluctuations from solids processing
- Favorable operating conditions (low SRT, modest temperature, sufficient DO in aeration basin, balanced anaerobic and aerobic HRTs).

Based on these four conditions, the expected performance of each facility in achieving EBPR was rated on a scale of one to 10 in which 10 means that the plant meets all four factors favorably (See Section 6.6.6.5). The observed plant performance and reliability was also rated on a scale of one to 10 (See Table 6-23). Results of this analysis are shown in Table ES-6. The results show that the expected performance rating for the facilities closely correlates to the observed performance rating.

The ratio of the uptake HRT to release HRT also closely correlates to the observed performance rating. The ratio between the uptake rate and the release rate in Tables ES-5 (and ES-8) are remarkably consistent — typically between three and four based on release rate:uptake rate. This implies that in full-scale treatment facilities, one may expect that the aerobic (and some anoxic) contact time should be three to four times as long as the release time (anaerobic plus excess-anoxic time). Table ES-6 shows that the observed performance rating very closely matches the ratio of uptake to release times available in the plants.

### ES.5.2 Impact of Influent Characterization

The influent BOD:TP ratio is often used as a measure of EBPR potential. The data collected during this study did not indicate a direct correlation between the influent BOD:TP, COD:TP, VFA:TP or ffCOD:TP (flocculated-filtered COD) ratios and effluent TP or phosphate concentrations. The facility with the lowest ratios (Nansemond) had the highest average effluent TP concentrations, but one of the facilities with the highest influent ratios (Lower Reedy) had the highest effluent phosphate concentrations. Facilities with aeration basin influent BOD:TP ratios greater than 25:1 generally had more stable performance than those with lower BOD:TP ratios, but some facilities performed well with BOD:TP ratios as low as 15:1.

In almost all cases, the *variability* of primary effluent BOD load had a greater impact on changes in effluent phosphorus concentrations than the absolute value of the primary effluent BOD:TP ratio. Figure ES-2 illustrates this point for the Nansemond facility (only two points are indicated but others can be identified in the trend plot). On almost every occasion when primary effluent BOD or the primary effluent BOD:TP ratio decreased, effluent phosphorus increased. The same correlation was not observed between effluent TP and the absolute value of the BOD:TP ratio or the primary effluent BOD load. The change in the BOD:TP ratio from either an increase in BOD or decrease in TP, appeared to cause a change in the steady operation of the biology. Because the biomass in the aeration basin retains a large mass of phosphorus, the change in BOD:TP could cause a high release of accumulated phosphorus without a corresponding increase in uptake to cause the change in effluent phosphorus.

Table ES-6. Summary of Expected and Actual Performance.

| Parameter | Durham | VIP | Nansemond | McDowell | Lower Reedy |
|---|---|---|---|---|---|
| Process | A2O | VIP | VIP | UCT | UCT |
| Lime addition | Yes | No | No | Yes | Yes |
| VFA addition | Fermenter | - | - | Soft drink waste | - |
| SRT, days | 10.9 | 9.5 | 8.5 | 6.7 | 9.4 |
| % Anaerobic volume | 9 | 12 | 15 | 7 | 13 |
| Ratio Uptake:Release HRT[4] | 3.9 | 3.1 | 2.3 | 4.9 | 3.3 |
| Infl BOD/P[1] | 17-28 | 29-33 | 7-13 | 31-35 | 15-25 |
| Infl BOD/P[2] | 21.7 | 31.9 | 12.6 | 31.3 | 21.5 |
| ffCOD/P[1] | 19.5 | 30.9 | 6.7 | 34.9 | 15.6 |
| Avg sec effl NO3-N[2], mg/L | 8.4 | 6.5 | 6.4 | 4.1 | 4.7 |
| RAS, % | 61 | 31 | 84 | 87 | 52 |
| % Internal recycle rate | 300 | 73 / 77 | 91 / 99 | 70/ 320 | 116 |
| Avg effl sec effl ortho P[2], mg/L | 0.1 | 0.6 | NA | NA | 0.5 |
| Percent of SE ortho P data <0.50 mg/L[2] | 50 | 70 | 24 | 95 | 76 |
| Percent of SE ortho P data <1.0 mg/L[2] | 80 | 82 | 64 | 99 | 81 |
| Percent of SE ortho P data <2.0 mg/L[2] | 98 | 92 | 85 | 100 | 91 |
| Phosphorus Recycle Load[3], % | 24 | 9 | 86 | 26 | 54 |
| Expected Performance Rating | 7 | 10 | 4 | 9 | 7 |
| Observed Performance Rating | 7 | 6 | 4 | 9 | 5 |

1. From intense sample periods (Table 6-17) – exclude fermenter or sugar additive
2. From routine plant operating data (2000 – 2003)
3. From mass balance
4. Release HRT = Anaerobic plus ½ Anoxic HRT; Uptake HRT = Aerobic + ½ Anoxic HRT

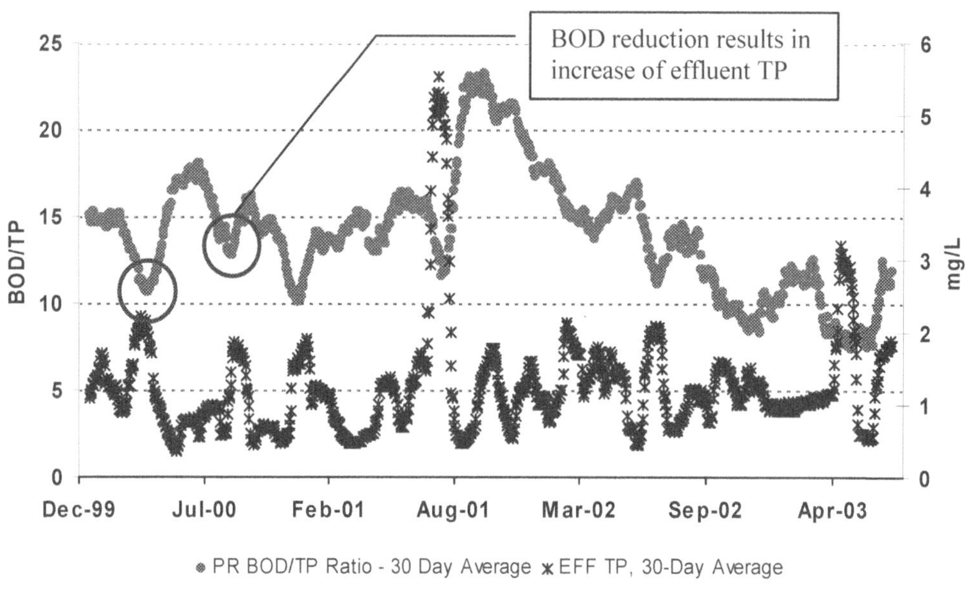


Figure ES-2: 30 Day Averages of Primary Effluent BOD:TP Ratio and Effluent Total Phosphorus.

### ES.5.3 Achievable Effluent Phosphate Concentrations

Table ES-7 shows the lowest achievable phosphate concentrations from the operating data. These concentrations represent the lower 10% of the operating data and are indicative of the level of phosphorus removal that appears feasible with a well functioning EBPR plant. These effluent phosphate concentrations are quite low—0.04 mg/L for Durham, and <0.2 mg/L for three of the other facilities. Optimizing and improving the reliability of phosphate removal through VFA addition is therefore a viable method to meet very low effluent phosphorus concentrations.

Table ES-7 also shows the variability of the data, calculated as the ratio between the 90$^{th}$ percentile and the 50$^{th}$ percentile. This measure provides an indication of the variability in the data. The ratio is quite large, typically ranging from about 3 to 6. This means that 90% of the data is below three to six times the average value. This is quite a large spread of observed values.

**Table ES-7. Achievable Secondary Effluent Phosphorus Concentrations (3-year data set).**

| Parameter | Unit | Durham | VIP | Nansemond | McDowell Creek | Lower Reedy |
|---|---|---|---|---|---|---|
| Phosphorus species | | OrthoP | TP | TP | OrthoP | OrthoP |
| Lower 10% | mg/L | 0.04 | <0.2 | 0.4 | <0.1 | 0.1 |
| Average 50% | mg/L | 0.06 | 0.29 | 0.76 | <0.1 | 0.46 |
| Upper 90% | mg/L | 0.99 | 1.68 | 2.53 | 0.27 | 2.16 |
| Ratio 90$^{th}$/50$^{th}$ | ratio | 3.4 | 5.8 | 3.3 | >2.7 | 4.8 |

### ES.5.4 Uptake and Release Tests and GAO Investigations

Table ES-8 below compares the PAO and GAO observations with phosphate release and uptake rates (from URT tests) and effluent phosphate concentrations at the time samples were collected for URT and PAO/GAO analysis. Table ES-8 shows that the results of some of these experiments were surprising. Some samples in which PAOs were readily observed (Durham, VIP, Nansemond) showed low release ratios of between 0.14 and 0.40 mg P/mg acetic acid (as COD) while Lower Reedy (which also showed dominant PAOs) had a high release ratio of 0.78 mg P/mg acetic acid (COD). The one plant with dominant GAOs (McDowell Creek) has a low release ratio (0.28 mg P/mg acetic acid).

**Table ES-8. Comparison of PAO/GAO Observations and Uptake and Release Test Results.**

| Sample | Summary of Microbial Investigation | Phosphate Release Mg P/mg acetic acid | Phosphate Release Rate (mg P/g VSS/hr) | Phosphate Uptake Rate (mg P/g VSS/hr) | RR:UR ratio[2] |
|---|---|---|---|---|---|
| Durham, Side 1 | PAOs dominant , some GAOs were easily observed | 0.4 | 10 | 3.6 | 2.8 |
| Durham, Side 2 | PAOs dominant , some GAOs were easily observed | 0.14 | 4.2 | 1.5 | 2.8 |
| VIP, July 2003 | PAO dominant, both tetrapod-shaped and *Competibacter*-type GAOs observed | 0.32 | 11.6[1] | 3.0[1] | 3.9 |
| VIP, September 3, 2003 | PAO dominant, both tetrapod-shaped and *Competibacter*-type GAOs easily observed | — | 11.1 | 2.2 | 5.0 |
| VIP, September 12, 2003 | Abundant GAOs, almost as abundant as PAOs. | — | None | None | — |
| Nansemond, July 2003 | PAO dominant, no tetrapod-shaped GAOs were observed, but *Competibacter*-type GAOs were observed. | — | None | None | — |
| Nansemond, September 2003 | PAO dominant, no GAOs was observed | 0.30 | 7.7 | 2.6 | 3.0 |
| McDowell | GAOs seemed to be dominant, more abundant than PAOs. | 0.28 | 5.5 | 1.4 | 3.9 |
| Lower Reedy | PAOs dominant in big flocs, only a few GAOs were found | 0.78 | 24.3 | 4.6 | 5.2 |

1. URT run in August 2003
2. RR:UR = Ratio of Release Rate to Uptake Rate
3. Rates are shown normalized to VSS for illustration. PAO/VSS composition will vary from plant to plant.

The uptake and release rates shed light on EBPR process dynamics. The phosphate release rates were significantly higher than the phosphate uptake rates. Release rates varied from 4.2 mg P/g VSS/h in Train 2 of the Durham facility to 11.1 mg P/g VSS/h at the VIP plant. In fact, the release and uptake rates in Train 2 of the Durham plant were lower than even the McDowell facility, where GAOs appeared to dominate PAOs in the mixed liquor. Phosphate release was much faster than phosphate uptake. Uptake rates were generally quite low (as demonstrated in other URT tests) and indicate the importance of providing sufficient aerobic time to complete the phosphate uptake in full-scale facilities.

## ES.6 Conclusions

The results of this study led to following conclusions.

- EBPR is capable of achieving very low effluent phosphate concentrations. When operating well, phosphate concentrations <0.1 mg/L can be achieved for extended periods (more than a month), 0.03 mg/L for a week, and even below 0.02 mg/L for several sequential days. Excursions above these levels are common.
- EBPR performance can be estimated by examining the plant influent BOD and TP loading, the level of oxidants in the anaerobic zone, the degree of recycle phosphorus control, and operating parameters (SRT, temperature, DO, anaerobic and aerobic HRT).
- To reliably achieve effluent phosphorus concentrations of 0.5 mg/L at $50^{th}$ percentile, chemical backup or BOD augmentation (through fermentation or adding readily biodegradable substrate source) are necessary.
- The performance of phosphorus removal improves when the BOD:TP ratio is above 25:1. Absolute values of the BOD:TP ratio or primary effluent BOD load did not correlate directly with phosphorus removal as long as the BOD:TP ratio was high.
- Changes in activated sludge influent load (as indicated by the BOD:TP ratio) correlated to changes in effluent phosphorus concentration. This variation appears to be caused by a change in the steady state operation of the PAOs, resulting in excessive release of the large accumulated phosphorus mass in the biomass.
- Phosphate release varied widely (from under 20 mg/L to over 70 mg/L) between plants and even under different loading conditions at one plant. A high anaerobic phosphate release did not necessarily correlate with low effluent phosphate concentrations. Other factors such as chemical precipitation with calcium in the anaerobic zone or excess ("secondary") phosphorus release could impact the measured phosphate concentration in the anaerobic zone.
- A range of phosphate release to VFA uptake ratios can be associated with functioning EBPR and low effluent phosphorus concentrations.
- Recycle streams play a significant role in adding variability to EBPR performance. Eliminating or attenuating the recycle load fluctuations greatly improves the process reliability.
- The presence of GAOs in full-scale facilities does not necessarily preclude the plant from achieving EBPR provided there is sufficient carbon in the influent.
- Chemical addition (such as ferric, alum, or lime) for phosphate precipitation or to meet other operational needs such as alkalinity or odor control can be recycled to the aeration basin and react with phosphate.
- Optimizing EBPR by adding of a carbon source (VFA) is a viable method for reliably achieving very low (<0.02 mg/L) effluent phosphate concentrations.

# CHAPTER 1.0

# PROJECT BACKGROUND AND OBJECTIVES

## 1.1 Project Background

In this time of increasingly stringent effluent discharge limits and pressure from elected officials to avoid increases in sewer rates, many wastewater treatment plant operators and managers are seeking ways to provide enhanced treatment while minimizing operating costs. Enhanced biological phosphorus removal (EBPR) is in many cases an efficient method to meet these goals. This phenomenon was observed in wastewater treatment plants as early as the 1960s (Levin and Shapiro, 1965), but a comprehensive biological model describing the process was not introduced until the 1980s (Wentzel, et al., 1986). Although the use of EBPR has increased over the past decades and the understanding of the biological mechanisms has continued to advance, there is still substantial variability in both the practices applied to achieve EBPR and the results of these practices in terms of soluble phosphorus removal. There is also a lack of understanding of the interaction of EBPR and other processes such as chemical precipitation and anaerobic digestion used at many municipal wastewater treatment plants.

The main goals of this research project are two-fold: first to evaluate and document EBPR practices and facilities, and second to conduct a detailed evaluation of the plant design and performance data and to identify factors contributing to the production of very low effluent phosphorus concentrations (<1.0 mg/L). Expected results of this research are a greater understanding of the range of operating conditions, operating practices, and treatment efficiency seen throughout the United States and abroad, and a better understanding of a specific process or processes that could lead to more reliable phosphorus removal.

### 1.1.1 Definition

EBPR is a biological process designed to cultivate bacteria that are able to take up and store phosphorus at levels greater than required for normal heterotrophic metabolic activity in the activated sludge process of wastewater treatment plants. Removal of phosphorus from wastewater by means of EBPR occurs through two major processes: 1) uptake by phosphorus-accumulating bacteria; and 2) removal, processing, and disposal or reuse of the phosphorus-enriched solids. Any evaluation of the performance of biological phosphorus-removal processes needs to address both components.

#### 1.1.1.1 Enhanced Phosphorus Uptake

Through the study and practice of EBPR, researchers and wastewater treatment plant operators discovered that the following conditions were required to support excess biological phosphorus uptake and storage (Barnard, 1976; Stensel, 1991):

- Exposing the activated sludge bacteria to influent wastewater in an anaerobic contacting zone, followed by an aerobic (or anoxic) zone;
- Minimizing the amount of nitrate or oxygen fed to the anaerobic zone; and

- Availability of volatile fatty acids (VFAs) or a source of readily biodegradable organic substrate in the anaerobic zone.

These conditions support the EBPR mechanism, which incorporates some or all of the following biochemical reactions (Wentzel, et al., 1986; Comeau, 1989; Mino, et al., 1987):

***Anaerobic Phase***

- Fermentation of complex and readily-biodegradable chemical oxygen demand (COD) to acetate and other VFAs under anaerobic conditions
- Acetate and propionate assimilation and conversion to intracellular polyalkanoates, such as polyhydroxybutyrate (PHB), under anaerobic conditions
- Degradation of stored polyphosphate and glycogen to provide energy for PHB formation, and subsequent release of phosphorus (and stored cations such as magnesium, potassium, and calcium).

***Aerobic Phase***

- PHB oxidation for new cell synthesis and glycogen formation
- Phosphate and cation uptake and reformation of polyphosphate granules

Other bacteria exist that can assimilate organic substrates in the anaerobic zone without phosphorus release. Anaerobic/aerobic processes have been operated in which acetate and other soluble substrates have been assimilated in the anaerobic phase without phosphorus release, and without polyphosphate storage and enhanced phosphorus removal in the aerobic phase. This type of metabolism was first reported by Cech and Hartman (1990, 1993) as a result of laboratory studies on EBPR. These non-phosphorus-accumulating organisms can deplete the VFA available for phosphorus-accumulating organisms (PAOs) and impair process efficiency. Therefore, it is important to provide conditions in wastewater treatment plants that favor the growth of PAOs and protect against the growth of organisms that would compete for substrate in the anaerobic zone.

### 1.1.2 Phosphorus Removal

Because soluble phosphorus becomes bound in the intracellular polyphosphates through EBPR, phosphorus is actually removed from the liquid stream through removal of the PAOs. Removal of PAOs from the liquid stream treatment process occurs through two avenues: as waste activated sludge (WAS) or as waste sludge from a tertiary polishing step such as effluent filtration. Both of these sludge streams typically find their way to solids-handling processes for thickening, stabilization, and dewatering. An anoxic/anaerobic environment often exists in solids handling processes such as thickening or anaerobic digestion. PAOs can release the stored phosphorus and associated cations under these anaerobic conditions.

## 1.2 Factors that Influence EBPR

The various conditions and parameters that impact EBPR efficiency can be grouped into three major categories: substrate requirements, environmental factors, and design/operating parameters.

### 1.2.1 Substrate Requirements

PAOs rely on a supply of VFAs, primarily acetate, for the biochemical reactions described above. The following conditions impact the availability of VFAs for EBPR:

- Supply of readily biodegradable COD in the influent wastewater;
- Anaerobic contact time to allow fermentation of readily biodegradable COD to VFAs; and
- Absence of electron acceptors such as oxygen and nitrate, which would deplete the VFAs available for phosphorus removal.

The VFA supply can be augmented. In many communities, especially those with combined sewer systems or short travel times in the sewer, the influent readily-biodegradable COD concentrations can be relatively low. To increase the readily biodegradable COD, some treatment plants add VFAs chemically or operate primary sludge (PSL) fermenters to produce the VFAs needed for EBPR.

The second two conditions can be addressed through careful design. The anaerobic zone must be sized to allow for both oxygen and nitrate reduction and VFA generation. Since VFAs are typically taken up by PAOs almost as quickly as they are generated in the anaerobic zone, uptake is not a limiting step in this process (Stensel and Barnard, 1992). The process must also be designed to minimize the introduction of oxygen and nitrate into the anaerobic zone. This is done by minimizing weir overflows, screw pumps, or other equipment that aerates streams sent to the anaerobic zone—primary effluent (PE), return activated sludge (RAS), etc. Dissolved oxygen (DO) control of internal recycle streams will also limit the oxygen returned to the anaerobic zone.

### 1.2.2 Environmental Factors

Environmental factors such as temperature and pH have been shown to have some impact on EBPR efficiency. Recent research (Filipe, et al., 2001a; Filipe, et al., 2001b) has documented an optimum pH range of 7.0 to 7.5 for PAOs with decreasing activity at lower pH values. Temperature impacts the process as it relates to VFA generation in the anaerobic zone (Stensel, 1991). Temperature also influences the microbiology of the activated sludge culture as well as the viability of PAOs. Analysis by Rabinowitz, et al., (2004, in press) of data from three full-scale, warm-weather biological nutrient removal (BNR) facilities showed that EBPR was significantly compromised or failed at temperatures over 30°C. The researchers hypothesized that the high temperatures reduced the PAO growth rate, resulting in washout. This condition can have the secondary impact of allowing readily biodegradable substrate normally consumed by PAOs to enter the aerobic zone, promoting the growth of filamentous bacteria.

### 1.2.3 Design/Operating Parameters

Designers and operators have control over the following design and operating parameters:

- The *anaerobic/aerobic* sequence has been implemented in a variety of configurations, using multiple anaerobic, anoxic, and aerobic zones with multiple recycle streams to optimize a single activated sludge tank for both phosphorus and ammonia or nitrogen removal.
- *Anaerobic contact time* can determine the amount of VFAs available to support EBPR. However, some research has shown that if PAOs are held under anaerobic conditions

after the VFA supply has been depleted, phosphorus can be released without any PHB formation (Barnard, 1984). This phenomenon, termed secondary release, is detrimental to EBPR because the released phosphorus is not taken up under aerobic conditions.

- Typical diurnal fluctuations in hydraulic and organic loading rates can lead to wide fluctuations in anaerobic and aerobic contact times and the amount of VFA available for the PAOs. Bench-scale reactor research using synthetic wastewater (Merseth, 1995) showed that EBPR performance could be negatively affected by periods of low VFA availability.
- *Nitrification* can have an impact on EBPR by increasing the nitrate concentration in the RAS stream, which results in more VFA consumption in the anaerobic contact zone for nitrate reduction, leaving less VFA for the PAOs to cause a reduction in EBPR efficiency.
- *Sidestream processes* may help or hinder EBPR efficiency. Sludge-thickening processes that produce VFAs can produce intermittent loads that may benefit phosphorus-removal performance, while other sludge-processing operations, such as digestion, may return phosphorus to the liquid-treatment system to essentially increase the influent phosphorus loading. This may result in increased effluent phosphorus concentrations if sufficient readily-degradable COD is not available for the PAOs.
- The *solids retention time (SRT)* is usually dictated by requirements of biochemical oxygen demand (BOD) removal, ammonia removal, or denitrification. However, the design SRT could impact phosphorus-removal performance. At longer SRTs, a greater proportion of the PAOs that are produced are lost to endogenous decay, resulting in less PAOs in the waste sludge, and thus a lower phosphorus-removal efficiency (Stensel, 1991; Whang and Park, 2001 and 2002).

Some additional processes appear to have a significant impact on EBPR performance but are not well understood. These include the following:

- *Solids processing* components, such as anaerobic digestion, which can cause significant phosphorus release. Based on the known metabolic pathways, stored phosphorus would be released in anaerobic digesters. In fact, dewatering centrate accounts for half of the phosphorus load on the activated-sludge process at the Durham Facility in Tigard, OR (Benisch, et al., 2001).
- Most EBPR plants have the ability to use some level of *chemical phosphorus removal* for effluent polishing to prevent struvite deposits or to supplement EBPR during times of inadequate performance. It is difficult to separate chemical phosphorus removal from biological phosphorus removal, especially when recycle flows from dewatering processes are returned to the liquid-treatment stream. Chemical precipitation benefits EBPR by reducing the recycled soluble phosphorus load.

### 1.2.4 Competing Organisms

Results from bench-scale and pilot plant studies have shown that in some systems, VFAs can be captured in the anaerobic zone by bacteria that do not perform EBPR (Satoh, et al., 1992; and Liu, et al., 1996). When these bacteria dominate, there is little if any phosphorus release in the anaerobic zone and little phosphorus uptake in the aerobic zone. These bacteria also convert VFAs to polyhydroxyalkanoates (PHAs) and produce glycogen in the aerobic zone but with different glycogen to PHA ratios than PAOs. They have been referred to as glycogen-accumulating organisms (GAOs); this group includes G bacteria, which are distinguished by their tetrad, coccoid morphology. Chapter 5.0 provides a summary of prior research on GAOs and factors that favor their dominance in anaerobic/aerobic activated-sludge processes.

## 1.3 Research Objectives

The main objectives of this research are to:

- Develop a database of phosphorus-removal performance in plants that have practiced EBPR for more than 1 year.
- Evaluate the reliability of these plants to meet phosphorus limits without chemical addition.
- Identify factors that contribute to reliable EBPR performance. These may include factors such as:
  - Influent wastewater characteristics;
  - Nitrogen removal requirements and the presence of nitrate;
  - Role of fermenters or other processes used to enhance EBPR;
  - Management of return flows from anaerobic solids-processing steps; and
  - Impacts of chemical addition.

# CHAPTER 2.0

# PROJECT APPROACH

The first phase approach was designed to gather a comprehensive set of information about plants performing EBPR and develop a more detailed data set for a smaller number of plants willing to participate in extensive testing in Phase 2. Each of these steps will be described in detail in subsequent chapters.

## 2.1 Screening Survey

An online survey was used to develop a comprehensive inventory of full-scale EBPR plants in the United States and abroad, documenting general information regarding plant size, operating parameters, and system performance. These data were used to evaluate:

- Effluent phosphorus levels being achieved (soluble and total)
- Processes used to accomplish these levels
- Whether supplementary VFAs were being provided
- Whether chemicals were being added for phosphorus removal or for other treatment goals
- Types and impacts of solids processing

The survey (Appendix A) was posted on the Internet in February 2002 in both English and German. The survey questions were designed to solicit sufficient information with which to characterize treatment plant operation and performance yet limit the time required to complete the survey to approximately 30 minutes and to enable the respondent to answer the majority of the questions without having to gather additional data. The data presented in the screening survey portion did not require calculated performance results but relied on the respondents' perceptions of plant performance. Typically, ranges of data were requested rather than asking for specific values (i.e., performance 0.5–1.0 mg/L and not 0.78 mg/L).

Concurrent with the screening survey, the team identified treatment plants with which to request further participation. A list of plants was developed by the project team members and from literature and technical publications. These inquiries resulted in a list of 80 facilities that had low effluent phosphorus limits and/or practiced EBPR. Team members solicited participation by phone. Team members in Germany and South Africa solicited participation from plants in these countries. These efforts resulted in 47 screening survey responses (Chapter 3.0).

## 2.2 Detailed Survey

Ten plants provided plant operational data for detailed evaluation and review. These facilities provided three years of daily performance data.

All plants provided effluent phosphorus data. Most plants provided operating data (e.g., flow, sludge age, mixed liquor suspended solids [MLSS]). Sometimes, conventional water quality parameters at various process streams were available—P, total suspended solids (TSS), total phosphorus (TP), BOD, and so forth. Some facilities provided data on soluble and particulate measurements.

The data records were summarized to determine the plant loading and performance using trend and probability plots. Chapter 4.0 summarizes the detailed survey results. Appendices B through F include reports summarizing detailed data for several facilities.

This study makes extensive use of plant operating data and measurements reported by process laboratories. Since the plant laboratories used different analytical techniques, phosphorus species are often reported differently and are used as such in the report. In some cases, one plant may measure only TP in the effluent, while another reports orthophosphate. In this report, the following convention is used:

- All concentrations are as reported by field laboratories. Analytical methods may vary.
- TP refers to all phosphorus species measured.
- Phosphate (or orthophosphate) refers to measurements recorded as orthophosphate or $PO_4$. It is possible that particulate phosphate (from some precipitant) may be included in the measure if the precipitant phosphate is detected in the measure.
- The term *phosphorus* is used in this report to denote the general phosphorus species. It may refer to TP or phosphate, depending on the context of the measure.
- All phosphorus measurements in the activated-sludge basins were first filtered and then measured, typically as phosphate.
- All phosphorus concentrations are expressed as P.

## 2.3 Field Testing

Site visits were conducted at nine facilities. The site visits were conducted to:

- Collect additional information to gain a better understanding of plant operations
- Conduct phosphate and ammonia profile sampling of the aeration basin
- Collect samples for GAO testing
- Determine phosphorus and chemical precipitant content of the MLSS
- Gather data for process simulation with the BioWin model.

Methods used to meet these objectives are described below. The results of the field testing are described in Chapter 4.0. Detailed field testing reports are included in Appendices B through F and Appendix H.

### 2.3.1 Plant Operation

Informal discussions were held with operators to determine the process operation and to identify operating problems and strategies. The discussions were generally conducted in conjunction with a walk-through of the plant to identify sampling locations and to photograph the facilities.

### 2.3.2 Process Stream Characterization

Samples were collected to characterize the wastewater and recycle streams and to define process performance. Plant staff collected composite samples for activated sludge influent (usually PE), secondary effluent (SE), and return flows (dewatering return flow) for the 24-hour period preceding the site visit. Samples were refrigerated at 4°C. During the site visits, grab samples were collected for WAS, dewatered solids, and anaerobic zone mixed liquor.

Samples were analyzed for organics, nutrients, and metals. Detailed results are given in chapters 3.0 and 5.0 and in the plant reports included in the appendix. For soluble COD and BOD (SCOD and SBOD) analyses, sample filtration was done using a 0.45 µm membrane filtrate. Influent and secondary clarifier effluent flocculated filtered COD (ffCOD) was used as a measure of the readily biodegradable COD (RBCOD) using the method of Mamais, et al. (1993).

### 2.3.3 Phosphate and Ammonia Profiles

Phosphate and ammonia profiles were measured twice during each site visit, generally once in the morning and once in the afternoon. Nitrate was initially measured in grab samples, but the analysis did not produce reliable results and was abandoned. Samples were collected over a 20- to 30-minute period from each basin (zone) starting at the effluent end and working forward. Samples were collected with a scoop, immediately poured into a large beaker, and allowed to settle while at the basin (approximately 2 to 4 minutes). When a clear supernatant appeared, the sample was decanted through a glass fiber filter into a sample bottle. Samples were taken to the laboratory and analyzed for phosphate and ammonia using Chemetrics cuvette colorimetric methods (see Appendix I).

### 2.3.4 Phosphate Uptake and Release Testing (URT)

Phosphate uptake and release was determined by a method similar to that of Janssen, et al. (2002). During Phase 1 the tests were conducted in the field. During Phase 2 the first set of tests was conducted in the field while the second set of testing was conducted at the University of Washington. For that, an aeration basin mixed-liquor sample was collected and shipped overnight to the University of Washington environmental engineering laboratory. The next day the mixed-liquor sample was aerated and then added to six test containers. The containers were sealed tight and mixed gently using nitrogen gas for 20 minutes. At that time, sodium acetate was added to the samples at doses of 0, 20, 40, 60, 80, and 100 mg/L as COD. The bottles were gently stirred for 45 minutes then 100 mL of clear supernatant was decanted and analyzed for phosphate. The bottle was then aerated for 90 minutes and sampled again for phosphate analysis. pH was controlled during the test by adding sodium hydroxide or hydrochloric acid. The pH and temperature were recorded at the end of the test. The complete test procedure is described in Appendix J.

The intent of the test was to detect the presence of GAO bacteria (or limitation of PAO bacteria) by observing the amount of phosphorus release relative to the amount of acetate COD added. When PAOs dominate the population, a rapid P release and uptake of phosphorus will occur under the anaerobic/aerobic test conditions. Theoretically, about 0.5 mg P are released per mg VFA consumed. If all VFA were consumed in the 45-minute anaerobic contact time, a plot of P release versus VFA consumed should have a slope of 0.5 mg P/mg VFA as COD.

A second test was conducted to determine the P uptake rate of biomass (PAOs) under subsequent aerobic conditions. This is similar to the URT test with the following modifications: (1) All samples received a VFA (acetic acid) dose of 80 mg COD/L, and (2) the aeration time was 5, 10, 20, 30, 60, and 90 minutes to generate a phosphate uptake measurement versus time. This test provides information on the P uptake rate under aerobic conditions.

## 2.4 Metals Sampling

Metals participate in the uptake and release of phosphate and can form chemical precipitates under certain conditions. The following metals were measured in the influent, effluent, and solids streams:

- Magnesium, calcium, and potassium—these ions are released with P release under anaerobic conditions and taken up with P uptake under aerobic conditions.
- Aluminum and iron—these metals are frequently added to wastewater streams to precipitate P. At some EBPR plants, small amounts of alum or a ferric salt are added to produce lower effluent P concentrations.

## 2.5 Process Modeling

The BioWin model was used to simulate the performance of field-tested plants. The basis of the BioWin model is the IWA Activated Sludge Model Number 2, with some additional enhancements including:

- Evaluation of additional state variables such as pH and metals; and
- The possibility of simulating anaerobic digestion (including struvite precipitation).

The goal of the simulation work was to determine how well this model reproduced the field test results and to determine which model parameters required modification to improve simulation.

The models were set up to closely match the plant operating characteristics as determined from the site visits. Where data were missing, assumptions were made to obtain realistic results from the models. Modeling results are described in Chapter 4.0 and in appendices K through N.

## 2.6 GAO Investigation

The goal of the first year of the project was to review existing literature on GAOs, evaluate methods for detecting the presence of GAOs in activated sludge, and to develop a protocol for measuring GAOs. A protocol was developed and used in the field testing to evaluate the presence of GAOs and their impact on EBPR. The presence of GAO was judged by microscopic and molecular probing observations from mixed-liquor samples and the results of the URT test. The GAO investigation is discussed in Chapter 5.0.

# CHAPTER 3.0

# SCREENING SURVEY RESULTS

This summary of the survey results is organized by survey sections. The chapter concludes with a section titled, "Phosphorus Removal Summary." A copy of the screening survey is included in Appendix A.

Survey results may be based on operator experience and recollection of typical performance rather than on data analysis because the intent of the survey was to provide an approximate performance overview for as many plants as possible with as little effort as possible. Calculated values are, at best, estimates roughly indicative of influent characteristics or process performance.

## 3.1 General Information

This survey section requested information on EBPR treatment capacity conditions, treatment goals including nutrient levels, and other general information related to EBPR. Tables 3-1 and 3-2 present the plant capacities, treatment goals, permit status, and nutrient removal targets. The Green Bay Metropolitan Sewerage District facility does not have a P permit or P removal goals, but operates in an EBPR mode. All other facilities have either P permit limits or treatment targets.

Figures 3-1 through 3-5 summarize the general information survey results. Figure 3-1 shows the average plant flow during the phosphorus permit season. Only 43 of the 47 plants that completed surveys submitted flow information. Plants of all sizes responded—from under 1 mgd to over 50 mgd. The largest plant was the 180-mgd Back River Wastewater Treatment Plant, Baltimore, Maryland.

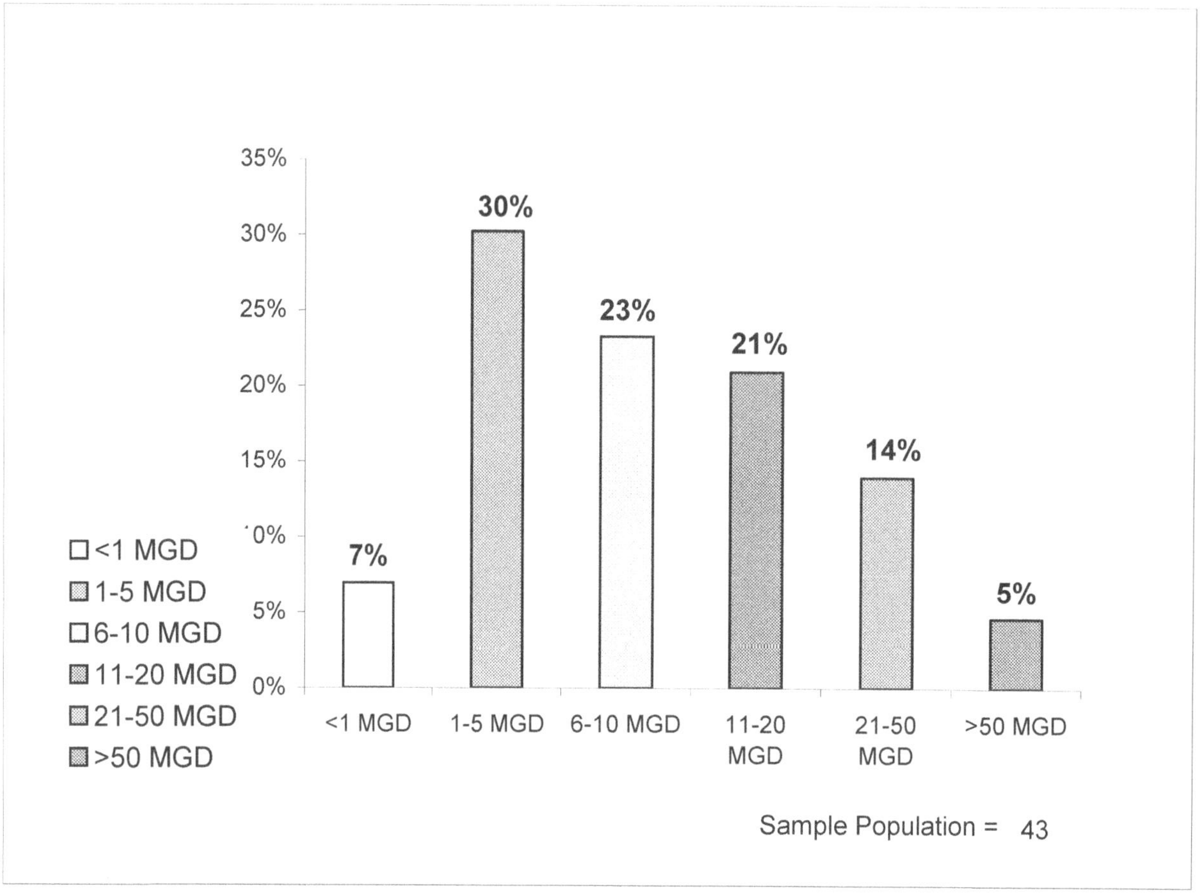


**Figure 3-1. Average Plant Flow.**

Table 3-1. General Facility and Permit Information.

| No. | Facility | Location | Facility Size (mgd) Design Flow | Avg. Flow | Treatment Goal Phosphorus Removal | Nitrification Only | Nit / Denit | Years of Bio-P |
|---|---|---|---|---|---|---|---|---|
| 1 | Green Bay Metropolitan Sewerage District | Green Bay, WI | 49.0 | 32 | X | | X | 5-10 |
| 2 | Kalispell AWT & BNR Facility | Kalispell, MT | 3.1 | 2.4 | X | | X | 5-10 |
| 3 | Durham AWWTP | Tigard, OR | 28.0 | 20 | X | | X | 2-5 |
| 4 | Bonnybrook Wastewater Treatment Plant | Calgary, Alberta | 132.1 | 108.3 | X | | X | 5-10 |
| 5 | City of Cheney WWTP and Reclamation Plant | Cheney, WA | 2.7 | 1.19 | X | X | | 5-10 |
| 6 | Laughlin Water Reclamation Facility | Laughlin NV | 8.0 | 3 | X | | X | 2-5 |
| 7 | Reedy Creek Improvement District WRF | Lake Buena Vista, FL | 15.0 | 9.9 | X | | X | > 10 |
| 8 | Proctors Creek Wastewater Plant | Chester, VA | 27.0 | 16.5 | | | X | |
| 9 | Sod Run Wastewater Treatment Plant | Perryman, MD | 20.0 | 12 | X | | X | 1-2 |
| 10 | City of Grande Prairie WWTP | Grande Prairie, Alberta | 5.8 | 3.2 | X | | X | |
| 11 | Fergus Falls WWTP | Fergus Falls, MN | 2.8 | 2.2 | X | | X | 2-5 |
| 12 | Eugene/Springfield WPCF | Eugene, OR | 49.0 | 24 | X | X | | > 10 |
| 13 | McDowell Creek WWTP | Huntersville, NC | 6.0 | 5 | X | | X | 2-5 |
| 14 | Mauldin Road WWTP | Greenville, SC | 70.0 | 20 | X | | X | > 10 |
| 15 | Lower Reedy | Greenville, SC | 10.0 | 6.5 | X | | X | 1-2 |
| 16 | City of Ft. Myers Central AWWT Facility | Fort Myers, FL | 11.0 | 11.0 | X | | X | > 10 |
| 17 | Nine Springs Wastewater Treatment Plant | Madison, WI | 57.0 | 42 | X | X | | 2-5 |
| 18 | City of Kelowna Wastewater Treat. Facility | Kelowna, BC | 10.4 | 7.2 | X | | X | > 10 |
| 19 | Missoula Wastewater Division | Missoula, MT | 9.0 | 8 | X | | | 5-10 |
| 20 | Nansemond Treatment Plant | Suffolk, VA | 30.0 | 18 | X | | X | 5-10 |
| 21 | Back River Wastewater Treatment Plant | Baltimore, MD | 180.0 | 145 | X | | X | 5-10 |
| 22 | Water Reclamation Facility | McMinnville, OR | 5.6 | 3.1 | X | | X | 5-10 |
| 23 | Rogers Pollution Control Facility | Rogers, AR | 6.7 | 5.2 | X | | X | > 10 |
| 24 | University of Florida, Water Rec Facility | Gainesville, FL | 3.0 | 3.0 | X | | X | 5-10 |
| 25 | Pocatello Water Pollution Control Dept. | Pocatello, ID | 12.0 | 7.4 | X | | No | > 10 |
| 26 | Sheboygan Regional Wastewater Treat. Facil. | Sheboygan, WI | 18.6 | 12.3 | X | | X | |
| 27 | Village of Minoa Wastewater Treat. Facil | Minoa, NY | 0.9 | 0.5 | X | | X | 2-5 |

Table 3-1. General Facility and Permit Information (Continued).

| No. | Facility | Location | Facility Size (MGD) Design Flow | Avg. Flow | Treatment Goal Phosphorus Removal | Nitrification Only | Nit / Denit | Years of Bio-P |
|---|---|---|---|---|---|---|---|---|
| 28 | Marshall Wastewater Treatment Facility | Marshall, WI | 0.6 | 0.35 | X | | X | 2-5 |
| 29 | Ephrata Regional WWTP No. 2 | Ephrata, PA | 2.3 | 1.274 | X . | | X | 2-5 |
| 30 | American-Bath WWTP | Lima, OH | 2.8 | 1.2 | X | | | 1-2 |
| 31 | Brownstown Wastewater Utility | Brownstown, IN | 0.7 | 0.3 | X | | | |
| 32 | Iron Bridge Regional WRF | Oviedo, FL | 40.0 | 24.6 | X | | X | > 10 |
| 33 | Virginia Initiative Plant | Norfolk,VA | 40.0 | | X | | X | > 10 |
| 34 | Kamen-Koernerback | Kamen, Germany | | | X | | X | 1-2 |
| 35 | Duelmen | Duelmen, Germany | | | X | | X | 5-10 |
| 36 | Luepinghausen | Luepinghausen, Ger. | | | X | | X | 5-10 |
| 37 | ERAR Viveros de la Villa | Madrid, Spain | | | X | | X | 1-2 |
| 38 | Olifantsfontein WCW | Johannesburg, SA | 27.7 | 16.4 | X | | X | > 10 |
| 39 | Stevens Point Wastewater Treatment Plant | Stevens Point, WI | 5.2 | 3 | X | | X | 2-5 |
| 40 | Hildesheim | Hildesheim, Germany | 9.9 | 8.5 | X | | X | 5-10 |
| 41 | Waiblingen | Waiblingen, Germany | | 2 | X | | X | > 10 |
| 42 | Flip Human Water Care Works | Krugersdorp, SA | 13.2 | 6.1 | X | | X | > 10 |
| 43 | Percy Stewart Water Care Works | Krugersdorp, SA | 7.1 | 4.3 | X | | X | > 10 |
| 44 | Daspoort Wastewater Treatment Works | Pretoria, South Africa | 11.1 | 11.6 | X | | X | > 10 |
| 45 | Zeekoegat Wastewater Treatment Works | Pretoria, South Africa | 7.9 | 8.7 | X | | X | > 10 |
| 46 | Sunderlandridge Wastewater Treat. Works | Pretoria, South Africa | 10 | 10.4 | X | | X | > 10 |
| 47 | Rooiwal Wastewater Treatment Works | Pretoria, South Africa | 39.6 | 31.7 | X | | X | > 10 |

Note: Proctors Creek facility (Facility No. 8) indicates it does not operate for phosphorus removal, but Table 3-2 indicates year-round P limit, with TP target of 0.5 – 2.0 mg/L

**Table 3-2. General Facility Nutrient Removal Information.**

| No. | Facility | Phosphorus Permit | Years of Bio-P | Nutrient Removal Targets | | | |
|---|---|---|---|---|---|---|---|
| | | | | PO4-P (mg/L) | TP (mg/L) | NH4-N (mg/L) | TN (mg/L) |
| 1 | Green Bay Metropolitan Sewerage District | No Permit | 5-10 | | | No target | No target |
| 2 | Kalispell AWT & BNR Facility | Year-round | 5-10 | | 0.1 - 0.5 | 0.1 - 0.5 | > 10.0 |
| 3 | Durham AWWTP | Seasonal | 2-5 | 0.1 - 0.5 | | 0.1 - 0.5 | No target |
| 4 | Bonnybrook Wastewater Treatment Plant | Year-round | 5-10 | 0.5 - 2.0 | 0.5 - 2.0 | 5.0 - 10.0 | No target |
| 5 | City of Cheney WWTP and Reclamation Plant | Year-round | 5-10 | 2.0 - 5.0 | 2.0 - 5.0 | 2.0 - 5.0 | No target |
| 6 | Laughlin Water Reclamation Facility | Year-round | 2-5 | 0.1 - 0.5 | 0.1 - 0.5 | 0.1 - 0.5 | 5.0 - 10.0 |
| 7 | Reedy Creek Improvement District WRF | Year-round | > 10 | 0.1 - 0.5 | 0.5 - 2.0 | | |
| 8 | Proctors Creek Wastewater Plant | Year-round | | | 0.5 - 2.0 | 2.0 - 5.0 | 5.0 - 10.0 |
| 9 | Sod Run Wastewater Treatment Plant | Year-round | 1-2 | 0.5 - 2.0 | 0.5 - 2.0 | | 5.0 - 10.0 |
| 10 | City of Grande Prairie WWTP | No Permit | | | 0.5 - 2.0 | No target | No target |
| 11 | Fergus Falls WWTP | Year-round | 2-5 | 0.1 - 0.5 | 0.1 - 0.5 | 0.1 - 0.5 | 2.0 - 5.0 |
| 12 | Eugene/Springfield WPCF | No Permit | > 10 | 0.5 - 2.0 | 0.5 - 2.0 | 5.0 - 10.0 | 5.0 - 10.0 |
| 13 | McDowell Creek WWTP | Year-round | 2-5 | 0.1 - 0.5 | 0.1 - 0.5 | 0.1 - 0.5 | 5.0 - 10.0 |
| 14 | Mauldin Road WWTP | Year-round | > 10 | 0.1 - 0.5 | 0.5 - 2.0 | 0.5 - 2.0 | No target |
| 15 | Lower Reedy | Year-round | 1-2 | 0.5 - 2.0 | 0.5 - 2.0 | 0.5 - 2.0 | No target |
| 16 | City of Ft. Myers Central AWWT Facility | Year-round | > 10 | 0.1 - 0.5 | 0.1 - 0.5 | | 2.0 - 5.0 |
| 17 | Nine Springs Wastewater Treatment Plant | Seasonal | 2-5 | 0.1 - 0.5 | 0.5 - 2.0 | 0.5 - 2.0 | > 10.0 |
| 18 | City of Kelowna Wastewater Treat. Facility | Year-round | > 10 | | 0.1 - 0.5 | 0.5 - 2.0 | 2.0 - 5.0 |
| 19 | Missoula Wastewater Division | Year-round | 5-10 | 0.5 - 2.0 | 0.5 - 2.0 | No target | No target |
| 20 | Nansemond Treatment Plant | Year-round | 5-10 | 0.1 - 0.5 | 0.5 - 2.0 | 0.5 - 2.0 | 5.0 - 10.0 |
| 21 | Back River Wastewater Treatment Plant | Year-round | 5-10 | | | 0.5 - 2.0 | 2.0 - 5.0 |
| 22 | Water Reclamation Facility | Seasonal | 5-10 | | 0.1 - 0.5 | | 5.0 - 10.0 |
| 23 | Rogers Pollution Control Facility | No Permit | > 10 | 0.5 - 2.0 | 0.5 - 2.0 | 0.5 - 2.0 | |
| 24 | University of Florida, Water Rec Facility | No Permit | 5-10 | 0.5 - 2.0 | 0.5 - 2.0 | No target | No target |
| 25 | Pocatello Water Pollution Control Dept. | No Permit | > 10 | | 0.5 - 2.0 | No target | No target |
| 26 | Sheboygan Regional Wastewater Treat. Facil. | Year-round | | 0.1 - 0.5 | 0.5 - 2.0 | 2.0 - 5.0 | No target |
| 27 | Village of Minoa Wastewater Treat. Facil | Year-round | 2-5 | 0.5 - 2.0 | 0.5 - 2.0 | 0.5 - 2.0 | |
| 28 | Marshall Wastewater Treatment Facility | Year-round | 2-5 | 0.5 - 2.0 | 0.5 - 2.0 | 0.1 - 0.5 | No target |
| 29 | Ephrata Regional WWTP No. 2 | Year-round | 2-5 | | 0.5 - 2.0 | 2.0 - 5.0 | > 10.0 |
| 30 | American-Bath WWTP | Year-round | 1-2 | 0.1 - 0.5 | 0.1 - 0.5 | 0.1 - 0.5 | No target |
| 31 | Brownstown Wastewater Utility | Year-round | | | 0.1 - 0.5 | No target | No target |
| 32 | Iron Bridge Regional WRF | Year-round | > 10 | 0.1 - 0.5 | 0.1 - 0.5 | 0.1 - 0.5 | |
| 33 | Virginia Initiative Plant | Year-round | > 10 | 0.5 - 2.0 | 0.1 - 0.5 | 0.1 - 0.5 | 5.0 - 10.0 |
| 34 | Kamen-Koernerback | Year-round | 1-2 | 0.1 - 0.5 | 0.1 - 0.5 | 5.0 - 10.0 | > 10.0 |
| 35 | Duelmen | Year-round | 5-10 | 0.1 - 0.5 | 0.1 - 0.5 | 2.0 - 5.0 | > 10.0 |
| 36 | Luepinghausen | Seasonal | 5-10 | 0.1 - 0.5 | 2.0 - 5.0 | > 10.0 | |
| 37 | ERAR Viveros de la Villa | Year-round | 1-2 | 0.5 - 2.0 | 0.5 - 2.0 | 0.5 - 2.0 | 5.0 - 10.0 |
| 38 | Olifantsfontein WCW | Year-round | > 10 | 0.5 - 2.0 | 0.5 - 2.0 | 5.0 - 10.0 | No target |
| 39 | Stevens Point Wastewater Treatment Plant | Year-round | 2-5 | 0.1 - 0.5 | 0.1 - 0.5 | No target | No target |
| 40 | Hildesheim | Year-round | 5-10 | 0.1 - 0.5 | 0.5 - 2.0 | 0.5 - 2.0 | 5.0 - 10.0 |

| No. | Facility | Phosphorus Permit | Years of Bio-P | Nutrient Removal Targets PO4-P (mg/L) | TP (mg/L) | NH4-N (mg/L) | TN (mg/L) |
|---|---|---|---|---|---|---|---|
| 41 | Waiblingen | Year-round | > 10 | 0.5 - 2.0 | 0.5 - 2.0 | 0.5 - 5.0 | 18 |
| 42 | Flip Human Water Care Works | Year-round | > 10 | > 5.0 | > 5.0 | 5.0 - 10.0 | No target |
| 43 | Percy Stewart Water Care Works | Year-round | > 10 | 0.5 - 2.0 | 0.5 - 2.0 | 5.0 - 10.0 | No target |
| 44 | Daspoort Wastewater Treatment Works | Year-round | > 10 | 0.5 - 2.0 | No target | 0.5 - 2.0 | 5.0 - 10.0 |
| 45 | Zeekoegat Wastewater Treatment Works | Year-round | > 10 | 0.5 - 2.0 | No target | 0.5 - 2.0 | 5.0 - 10.0 |
| 46 | Sunderlandridge Wastewater Treat. Works | Year-round | > 10 | 0.5 - 2.0 | 0.5 - 2.0 | 0.5 - 2.0 | 5.0 - 10.0 |
| 47 | Rooiwal Wastewater Treatment Works | Year-round | > 10 | 0.5 - 2.0 | 0.5 - 2.0 | 0.5 - 2.0 | 5.0 - 10.0 |

Figure 3-2 shows the nutrient removal processes implemented by the surveyed plants. EBPR, nitrification, and denitrification existed at most facilities (85%); 9% operated for P removal only, and 6% for phosphorus removal and nitrification.

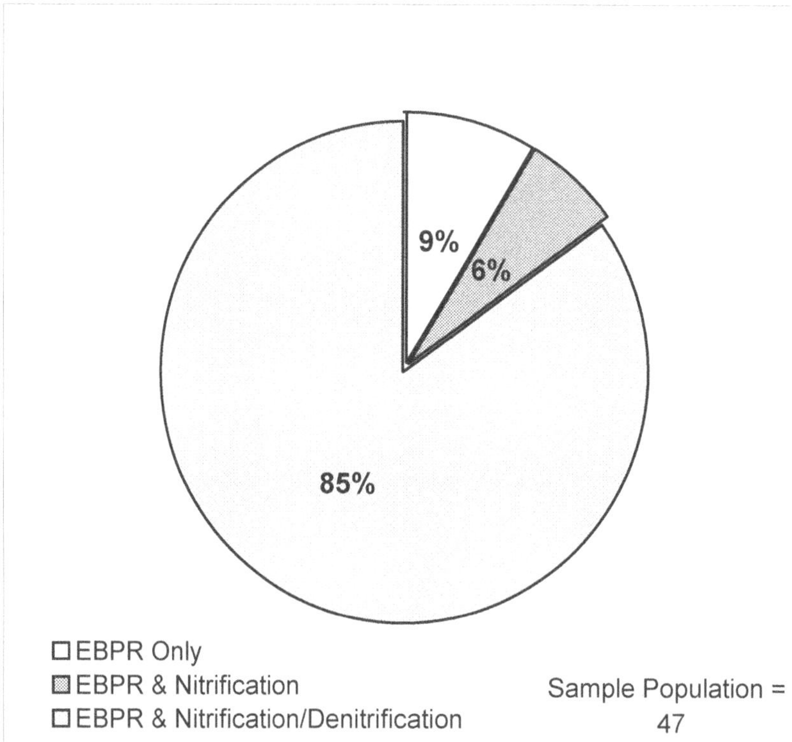


**Figure 3-2. Facility Nutrient Removal Processes.**

Figure 3-3 shows that most of the treatment plants have year-round effluent P permits. A few survey participants reported achieving EBPR even though phosphorus removal is not required by permit. These facilities either operate EBPR voluntarily (in some cases anticipating future phosphorus removal requirements) or accomplish EBPR as a side effect of operating an anaerobic selector. Phosphorus removal in these plants is incidental.

Figure 3-4 shows the number of years of EBPR operation. Two-thirds of the survey participants have practiced EBPR over 5 years.

Figure 3-5 shows the SRT operating range of the plants during the phosphorus permit season. The SRTs range from 3 to >20 d. Most of the treatment plants reported SRTs of 5 d or more. It is therefore likely that most of the plants nitrify, which is confirmed by the performance summary in Figure 3-2.

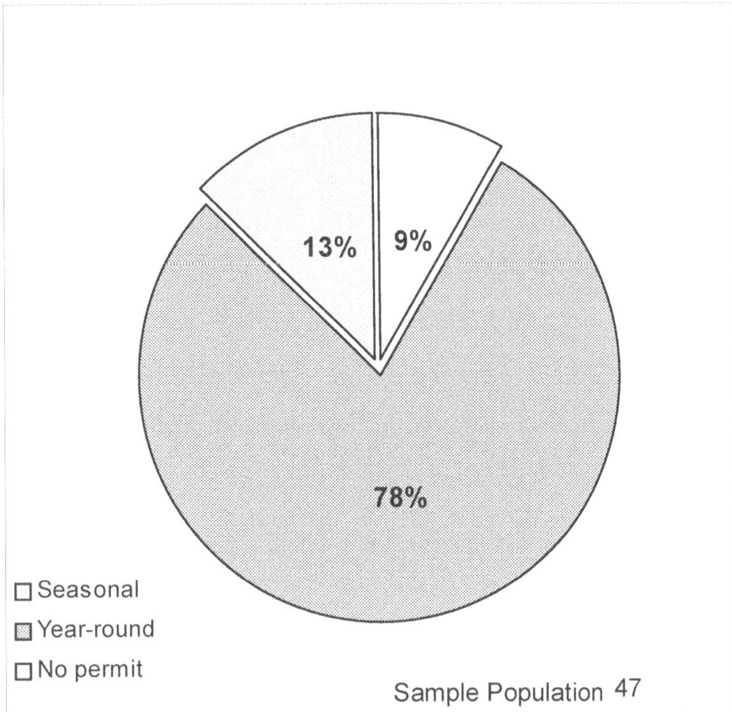


**Figure 3-3. Phosphorus Removal Permits.**

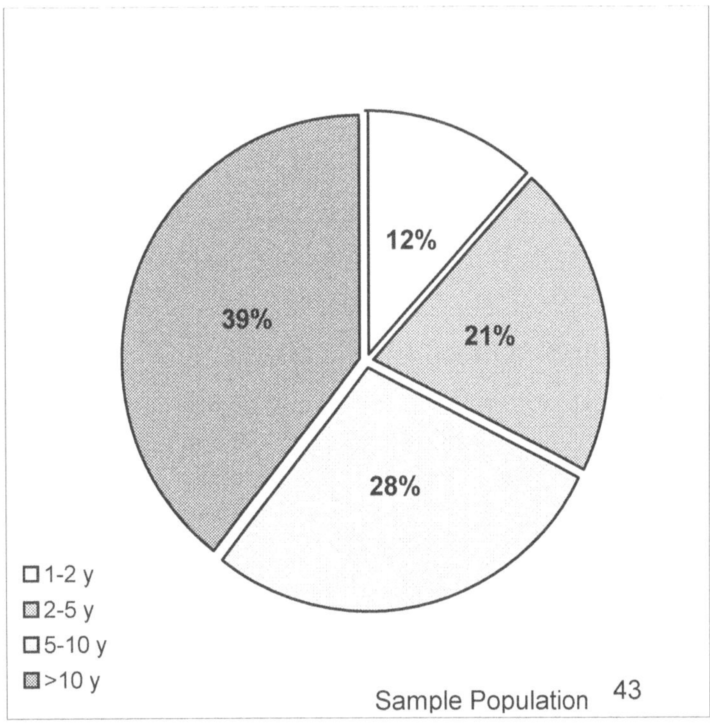


**Figure 3-4. Years of EBPR Operation.**

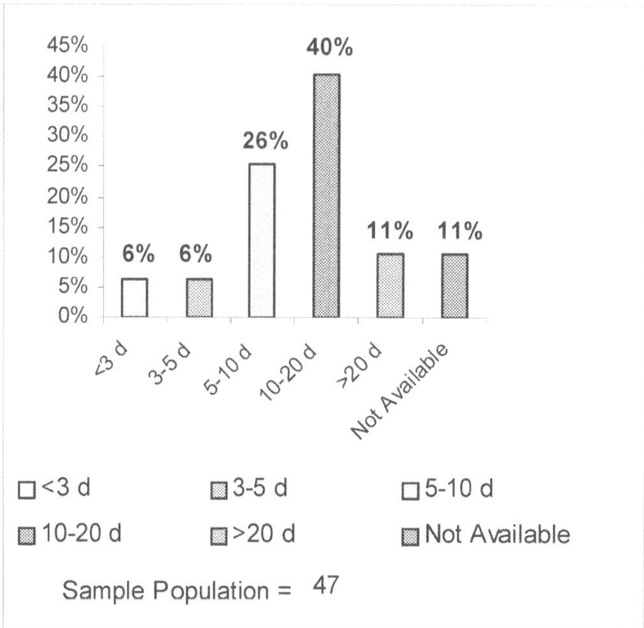


**Figure 3-5. EBPR Solids Retention Time (Not Available indicates data not provided).**

## 3.2 Influent Information

This survey section requested information on plant influent characteristics and metals monitored at the treatment plants. Table 3-3 presents the influent wastewater characteristics and a BOD:P ratio calculated from survey data[1]. Table 3-4 presents influent wastewater temperatures and documents whether the utility measures incoming metals. The table shows that most plants have influent BOD:P ratios higher than 20:1. This would support some level of EBPR (Stensel, 1991). Influent soluble BOD (SBOD) and VFA data are limited, so it was not possible to relate EBPR performance with the amount of RBCOD in the influent wastewaters. No correlation could be found between the effluent phosphorus concentration and the reported influent BOD:P concentration ratio during Phase 1. Many other factors such as the plant design, chemicals addition, sludge handling, or recycle streams play major roles with regard to phosphorus-removal efficiency.

## 3.3 Secondary Treatment Process Information

This survey section requested information on secondary treatment processes, VFA addition, and supplementary alkalinity addition. Table 3-5 presents the secondary treatment process components of the plants. About 75% of the surveyed plants utilized primary clarifiers and activated sludge; 20% used oxidation ditches; one used rotating biological contactors (RBCs), and one used a sequencing batch reactor (SBR). Nearly all of the plants used secondary clarification[2] and about 25% of the plants used effluent filtration.

[1] Calculation is based on values that may be only estimates of influent characteristics.

[2] Two plants reported oxidation ditch or SBR activated-sludge processes with no secondary clarifier. Three facilities reported conventional activated-sludge aeration basins with no secondary clarifiers.

**Table 3-3. Influent Wastewater Characteristics.**

| No. | Facility | COD (mg/L) | BOD (mg/L) | sBOD (mg/L) | VFA (mg/L) | TSS (mg/L) | TKN (mg/L) | NH4-N (mg/L) | NO3-N (mg/L) | PO4-P (mg/L) | TP (mg/L) | Alkalinity (mg/L) | BOD: P |
|---|---|---|---|---|---|---|---|---|---|---|---|---|---|
| 1 | Green Bay Metro. Sewerage District | 400 | 190 | | | 230 | 30 | 18 | | | | | |
| 2 | Kalispell AWT & BNR Facility | | 270 | | | 330 | 32 | 28 | 0.13 | | 4.7 | | 56 |
| 3 | Durham AWWTP | | 220 | | | 260 | | 26 | | | 7.5 | | 29 |
| 4 | Bonnybrook Wastewater Treatment Plant | | 190 | | 15 | 190 | | 21 | | | 6.3 | | 29 |
| 5 | City of Cheney WWTP and Reclamation Plant | 420 | 210 | | | 230 | 28 | 8 | 2 | | 4.8 | 200 | 45 |
| 6 | Laughlin Water Reclamation Facility | | 280 | | | 290 | | 23 | 1.1 | | 6.8 | 270 | 41 |
| 7 | Reedy Creek Improvement District WRF | | 230 | | | 220 | 40 | 35 | | | 6.7 | | 34 |
| 8 | Proctors Creek Wastewater Plant | | 230 | | | 220 | 42 | 28.6 | | | 7 | | 32 |
| 9 | Sod Run Wastewater Treatment Plant | | 300 | | | 350 | 40 | 35 | | | 8 | 200 | 38 |
| 10 | City of Grande Prairie WWTP | | 390 | | | 670 | | 33 | | | 7.2 | | 55 |
| 11 | Fergus Falls WWTP | | 180 | | | 230 | | 17 | | | 6 | | 29 |
| 12 | Eugene/Springfield WPCF | | 140 | | | 200 | 36 | 24 | | | 8.5 | | 16 |
| 13 | McDowell Creek WWTP | 700 | 380 | | | 400 | 40 | 26 | 0.5 | | 7 | 140 | 54 |
| 14 | Mauldin Road WWTP | 400 | 250 | | 25 | 250 | 40 | 25 | | | 7 | 250+ | 36 |
| 15 | Lower Reedy | | 420 | | 40 | 300 | 40 | 20 | | | 6.5 | 180 | 65 |
| 16 | City of Ft. Myers Central AWWT Facility | | 180 | | | 220 | 32 | | | | | | |
| 17 | Nine Springs WWTP | | 220 | | 20 | 230 | 31 | 23 | | | 6 | 240 | 37 |
| 18 | City of Kelowna Wastewater Treat. Facility | 1000 | 190 | | | 930 | 33 | 18 | 0.7 | 3.61 | 6.6 | 250 | 28 |
| 19 | Missoula Wastewater Division | | 290 | | | 240 | 33 | 19 | 0.37 | 2.81 | 5.0 | | 58 |
| 20 | Nansemond Treatment Plant | | 230 | | | 180 | 45 | 32 | | | 10 | 250 | 23 |
| 21 | Back River WWTP | 480 | 190 | | | 150 | 28 | 15 | 0.1 | | 3.6 | 200 | 53 |
| 22 | Water Reclamation Facility | 450 | 200 | | | 240 | 40 | 17 | | 2.5 | 5 | 150 | 40 |
| 23 | Rogers Pollution Control Facility | | 280 | | | 400 | | 17 | 0.09 | 5.4 | 11 | | 25 |
| 24 | University of Florida, Water Rec Facility | | 170 | | | 150 | | | | | | | |

Table 3-3. Influent Wastewater Characteristics (Continued).

| No. | Facility | COD (mg/L) | BOD (mg/L) | sBOD (mg/L) | VFA (mg/L) | TSS (mg/L) | TKN (mg/L) | NH4-N (mg/L) | NO3-N (mg/L) | PO4-P (mg/L) | TP (mg/L) | Alkalinity (mg/L) | BOD: P |
|---|---|---|---|---|---|---|---|---|---|---|---|---|---|
| 25 | Pocatello Water Pollution Control Dept. | | 290 | | | 202 | 36 | 28 | | | 8 | | 36 |
| 26 | Sheboygan Regional Wastewater Treat. Facil. | | 180 | | | 200 | | | | | 5.1 | | 36 |
| 27 | Village of Minoa WWTF | | 200 | | | 200 | 40 | 30 | | | 6 | | 33 |
| 28 | Marshall Wastewater Treatment Facility | | 190 | | | 180 | | | | 5 | | | |
| 29 | Ephrata Regional WWTP No. 2 | | 290 | | | 240 | | 32 | | | 6.8 | | 42 |
| 30 | American-Bath WWTP | | 700 | | | 350 | | | | | 8 | | 88 |
| 31 | Brownstown Wastewater Utility | | 140 | | | 180 | | | | | 5.4 | | 25 |
| 32 | Iron Bridge Regional WRF | | 150 | 85 | | 150 | 40 | 25 | 1 | 6.5 | 8 | 240 | 19 |
| 33 | Virginia Initiative Plant | | 220 | 100 | | 150 | 30 | 22 | <0.2 | | 5.4 | 190 | 41 |
| 34 | Kamen-Koernerback | 340 | 91 | | | | | 43 | | 5.3 | | | |
| 35 | Duelmen | 320 | 130 | | | | | 31 | | 6.1 | | | |
| 36 | Luepinghausen | 350 | 180 | | | | | 24 | | 5 | 5.4 | | 33 |
| 37 | ERAR Viveros de la Villa | 370 | 180 | | | 100 | 47 | 35 | 0.68 | 5.4 | 6.1 | 240 | 29 |
| 38 | Olifantsfontein WCW | 1000 | | | | 530 | 69 | 43 | | 1 | 17 | 170 | |
| 39 | Stevens Point Wastewater Treatment Plant | | 300 | | | 280 | | | | | | | |
| 40 | Hildesheim | 400 | 200 | | | | 52 | | | | 7.9 | 400 | 26 |
| 41 | Waiblingen | 430 | 200 | | | | | 40 | 2.1 | | 8.4 | 410 | 24 |
| 42 | Flip Human Water Care Works | 1700 | | | | 610 | 70 | 32 | 0.26 | 10 | 12 | 270 | |
| 43 | Percy Stewart Water Care Works | 1700 | | | | 740 | 100 | 50 | 1.4 | 8 | 12 | 370 | |
| 44 | Daspoort Wastewater Treatment Works | 400 | | | | 180 | 36 | 20 | <1 | 4 | 6 | 220 | |
| 45 | Zeekoegat Wastewater Treatment Works | 420 | | | | 220 | 45 | 23 | <1 | 5.5 | 6 | 250 | |
| 46 | Sunderlandridge Wastewater Treat. Works | 680 | | | 75 | | | 32 | | 8.2 | | 270 | |
| 47 | Rooiwal Wastewater Treatment Works | 640 | | | | 580 | 46 | 23 | 0 | 6 | 9.8 | 330 | |

### Table 3-4. Influent Wastewater Temperatures and Metals Analysis.

| No. | Facility | Temperatures (°C) | | | Routine Metal Analysis | | | | |
|---|---|---|---|---|---|---|---|---|---|
| | | Minimum | Average | Maximum | Ca | K | Mg | Fe | Al |
| 1 | Green Bay Metropolitan Sewerage District | 16 | 24 | 32 | | | | | |
| 2 | Kalispell AWT & BNR Facility | 9 | 15 | 21 | | | | | |
| 3 | Durham AWWTP | 15 | 20 | 25 | X | X | X | X | X |
| 4 | Bonnybrook Wastewater Treatment Plant | 9 | 12 | 19 | X | X | X | X | X |
| 5 | City of Cheney WWTP | 9 | 16 | 20 | | | | | |
| 6 | Laughlin Water Reclamation Facility | 18 | 23 | 28 | | | | | |
| 7 | Reedy Creek Improvement District WRF | 26 | 28 | 30 | | | | | |
| 8 | Proctors Creek Wastewater Plant | 13 | 20 | 27 | | | | | |
| 9 | Sod Run Wastewater Treatment Plant | 12 | 16 | 21 | | | | | |
| 10 | City of Grande Prairie WWTP | | 15 | | | | | | |
| 11 | Fergus Falls WWTP | 10 | 16 | 19 | | | | | |
| 12 | Eugene/Springfield WPCF | 17 | 20 | 22 | | | | | |
| 13 | McDowell Creek WWTP | 14 | 20 | 22 | | | | | |
| 14 | Mauldin Road WWTP | 15 | 19 | 22 | X | X | X | X | X |
| 15 | Lower Reedy | 15 | 18 | 22 | X | X | X | X | X |
| 16 | City of Ft. Myers Central AWWT Facility | 25 | 27 | 29 | | | | | |
| 17 | Nine Springs Wastewater Treatment Plant | 10 | 16 | 22 | | | | | |
| 18 | City of Kelowna Wastewater Treat. Facility | 12 | | 23 | | | | | |
| 19 | Missoula Wastewater Division | 8 | 15 | 21 | | | | | |
| 20 | Nansemond Treatment Plant | 13 | 18 | 23 | X | X | X | X | X |
| 21 | Back River Wastewater Treatment Plant | 15 | 20 | 24 | | X | X | X | X |
| 22 | Water Reclamation Facility | 16 | 20 | 22 | | | | | |
| 23 | Rogers Pollution Control Facility | 9 | 20 | 30 | | | | | |
| 24 | University of Florida, Water Rec Facility | | | | | | | | X |
| 25 | Pocatello Water Pollution Control Dept. | 11 | 16 | 22 | | | | | |
| 26 | Sheboygan Regional WWTF | 10 | 14 | 18 | | | | | |
| 27 | Village of Minoa Wastewater Treat. Facil | 10 | 15 | 20 | | | | | |
| 28 | Marshall Wastewater Treatment Facility | | | | | | | | |
| 29 | Ephrata Regional WWTP No. 2 | 13 | 18 | 23 | | | | | |
| 30 | American-Bath WWTP | 10 | 18 | 26 | | | | | |
| 31 | Brownstown Wastewater Utility | | | | | | | | |
| 32 | Iron Bridge Regional WRF | 19 | 26 | 31 | | | X | X | X |
| 33 | Virginia Initiative Plant | 13 | 22 | 29 | | | | | |
| 34 | Kamen-Koernerback | 8 | 15 | 18 | | | | | |
| 35 | Duelmen | 8 | 14 | 20 | | | | | |
| 36 | Luepinghausen | 8 | 15 | 19 | | | | | |
| 37 | ERAR Viveros de la Villa | 17 | 19 | 24 | | | | | |
| 38 | Olifantsfontein WCW | 15 | 18 | 21 | X | X | X | X | X |
| 39 | Stevens Point Wastewater Treatment Plant | 4.7 | | 10 | | | | | |
| 40 | Hildesheim | 11 | 15 | 21 | | | | | |
| 41 | Waiblingen | 9.8 | 18 | 24 | X | X | X | X | X |
| 42 | Flip Human Water Care Works | 12 | 18 | 23 | X | | | X | |
| 43 | Percy Stewart Water Care Works | 12 | 18 | 22 | X | | | X | |
| 44 | Daspoort Wastewater Treatment Works | 16 | 22 | 25 | | | | X | |
| 45 | Zeekoegat Wastewater Treatment Works | 16 | 22 | 25 | | | | X | |
| 46 | Sunderlandridge Wastewater Treat. Works | 16 | 23 | 28 | | | | | |
| 47 | Rooiwal Wastewater Treatment Works | 17 | 22 | 28 | | | | X | |

Table 3-5. Secondary Treatment Process Components.

| No. | Facility | Primary Clarifiers | Oxidation Ditch | Aeration Basin | SBR | Trickling Filter | RBC | Secondary Clarifiers | Effluent Filtration | Chlorine Disinfect. | Other Disinfect. | Dechlor. |
|---|---|---|---|---|---|---|---|---|---|---|---|---|
| 1 | Green Bay Metropolitan Sewerage Dist. | X | | X | | | | | | | X | X |
| 2 | Kalispell AWT & BNR Facility | X | | X | | | | X | | | | |
| 3 | Durham AWWTP | X | | X | | | | X | X | X | | X |
| 4 | Bonnybrook Wastewater Treatment Plant | X | | X | | | | X | | | X | |
| 5 | City of Cheney WWTP | | | X | | | | X | | X | | X |
| 6 | Laughlin Water Reclamation Facility | | X | X | | | | X | X | X | | X |
| 7 | Reedy Creek Improvement District WRF | X | | X | | | | X | X | X | | |
| 8 | Proctors Creek Wastewater Plant | X | | X | | | | X | X | X | | X |
| 9 | Sod Run Wastewater Treatment Plant | X | | X | | | | X | | X | | X |
| 10 | City of Grande Prairie WWTP | X | | | | | X | X | | | | |
| 11 | Fergus Falls WWTP | X | | X | | | | X | | X | | X |
| 12 | Eugene/Springfield WPCF | X | | X | | | | X | | X | | X |
| 13 | McDowell Creek WWTP | X | | X | | | | X | X | | | |
| 14 | Mauldin Road WWTP | X | | X | | | | X | X | X | | X |
| 15 | Lower Reedy | X | | X | | | | X | | X | | X |
| 16 | City of Ft. Myers Central AWWT Facility | | | X | | | | X | | X | | X |
| 17 | Nine Springs WWTP | X | | X | | | | | | | X | |
| 18 | City of Kelowna WWTF | X | | X | | | | X | X | | | |
| 19 | Missoula Wastewater Division | X | | X | | | | X | | X | | |
| 20 | Nansemond Treatment Plant | X | | X | | | | X | | X | | X |
| 21 | Back River Wastewater Treatment Plant | X | | X | | | | X | X | X | | X |
| 22 | Water Reclamation Facility | | X | | | | | X | X | | X | |
| 23 | Rogers Pollution Control Facility | | X | | | | | X | | X | | X |
| 24 | University of Florida, Water Rec Facility | | X | | | | | | X | X | | |
| 25 | Pocatello Water Pollution Control Dept. | X | | X | | | | X | | X | | X |
| 26 | Sheboygan Regional WWTF | X | | X | | | | X | | X | | X |
| 27 | Village of Minoa Wastewater Treat. Facil | | | | X | | | | | | | |
| 28 | Marshall Wastewater Treatment Facility | | X | | | | | X | | | X | |
| 29 | Ephrata Regional WWTP No. 2 | | X | | | | | X | | X | | X |
| 30 | American-Bath WWTP | | X | | | | | X | | | | |
| 31 | Brownstown Wastewater Utility | | X | | | | | X | | | X | |
| 32 | Iron Bridge Regional WRF | | X | | | | | X | X | X | | X |
| 33 | Virginia Initiative Plant | X | | X | | | | X | | X | | X |
| 34 | Kamen-Koernerback | X | | X | | | | X | | | | |
| 35 | Duelmen | X | | X | | | | X | X | | | |
| 36 | Luepinghausen | X | | X | | | | X | | | | |
| 37 | ERAR Viveros de la Villa | X | | X | | | | X | X | | | |
| 38 | Olifantsfontein WCW | X | | X | | X | | X | | X | | |
| 39 | Stevens Point WWTP | X | | X | | | | X | | | X | |
| 40 | Hildesheim | X | | X | | | | X | | X | | |
| 41 | Waiblingen | X | | X | | | | X | | X | | |
| 42 | Flip Human Water Care Works | X | | X | | | | X | | X | | |
| 43 | Percy Stewart Water Care Works | X | | X | | X | | X | | X | | |
| 44 | Daspoort Wastewater Treatment Works | X | | X | | | | X | | X | | |
| 45 | Zeekoegat Wastewater Treatment Works | X | | X | | | | X | | X | | |
| 46 | Sunderlandridge Wastewater Treat. Works | X | | X | | X | | X | | X | | |
| 47 | Rooiwal Wastewater Treatment Works | X | | X | | | | X | | X | | |



WERF

Table 3-6 shows the reported reactor zone(s), mixed-liquor recycle configuration(s), and other process features utilized at each plant. Figure 3-6 and Figure 3-7 show VFA and alkalinity addition, respectively.

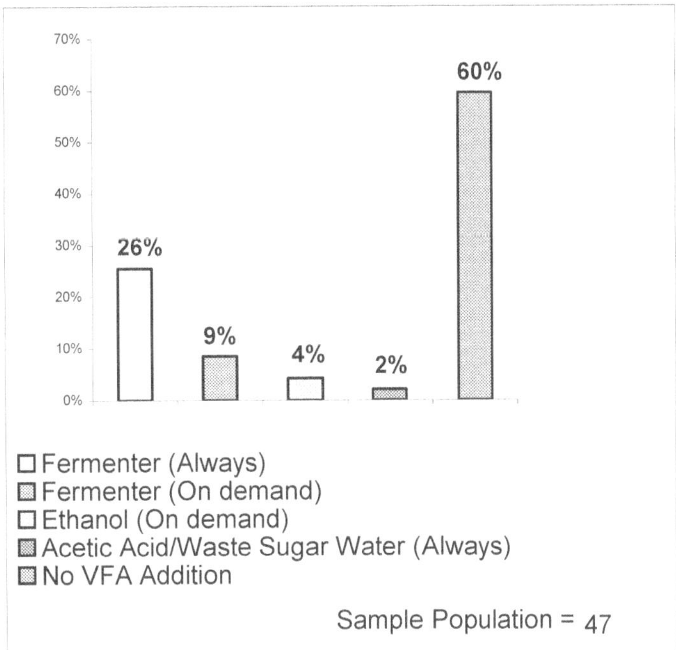


**Figure 3-6. Volatile Fatty Acid Addition.**

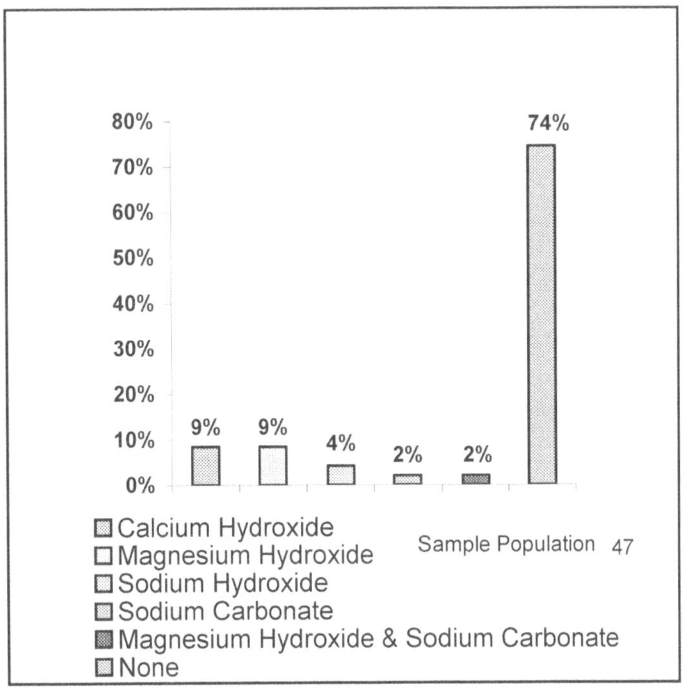


**Figure 3-7. Alkalinity Addition.**

Table 3-6. Activated Sludge Process Elements.

| No. | Facility | EBPR Reactor Zones | | | Mixed-Liquor Recycle | | | Other Process Features | | | |
|---|---|---|---|---|---|---|---|---|---|---|---|
| | | Aer. | Anox. | Anaer. | Aer. to Anox. | Anox. to Anaer. | Aer. to Anaer. | P-Stripper | SBR | Step Feed | Simult. N/DN |
| 1 | Green Bay Metropolitan Sewerage District | X | X | | | | | | | | X |
| 2 | Kalispell AWT & BNR Facility | X | X | X | | X | | | | | X |
| 3 | Durham AWWTP | X | X | X | X | | | | | | |
| 4 | Bonnybrook Wastewater Treatment Plant | X | X | X | X | | | | | | X |
| 5 | City of Cheney WWTP and Reclamation Plant | X | X | | X | | | | | | |
| 6 | Laughlin Water Reclamation Facility | X | X | X | | | | | | | X |
| 7 | Reedy Creek Improvement District WRF | X | X | X | X | | X | | | | |
| 8 | Proctors Creek Wastewater Plant | X | X | | X | | | | | | X |
| 9 | Sod Run Wastewater Treatment Plant | X | X | X | X | | | | | | |
| 10 | City of Grande Prairie WWTP | X | X | X | | | | | | X | X |
| 11 | Fergus Falls WWTP | X | X | X | | | | | | | X |
| 12 | Eugene/Springfield WPCF | X | | X | | | | | | | X |
| 13 | McDowell Creek WWTP | X | X | X | X | X | | X | | | |
| 14 | Mauldin Road WWTP | X | X | X | X | | | | | | |
| 15 | Lower Reedy | X | X | X | X | | | | | | |
| 16 | City of Ft. Myers Central AWWT Facility | X | X | X | X | | | | | | |
| 17 | Nine Springs Wastewater Treatment Plant | X | X | X | | X | | | | | |
| 18 | City of Kelowna Wastewater Treat. Facility | X | X | X | X | | | | | | X |
| 19 | Missoula Wastewater Division | X | X | X | | | | | | | |
| 20 | Nansemond Treatment Plant | X | X | X | X | X | | | | | X |
| 21 | Back River Wastewater Treatment Plant | X | X | | X | | | | | | X |
| 22 | Water Reclamation Facility | X | X | X | | | | | | | X |
| 23 | Rogers Pollution Control Facility | X | X | X | X | | | | | | |
| 24 | University of Florida, Water Rec Facility | X | X | X | X | X | X | | | | X |
| 25 | Pocatello Water Pollution Control Dept. | X | | X | | | | | | | |
| 26 | Sheboygan Regional Wastewater Treat. Facil. | X | | X | | X | | | | | |
| 27 | Village of Minoa Wastewater Treat. Facil | | | | | | X | | X | | |
| 28 | Marshall Wastewater Treatment Facility | | | | X | | | | | | X |
| 29 | Ephrata Regional WWTP 2 | X | X | X | | | | | | | |
| 30 | American-Bath WWTP | X | X | | X | | | | | | |
| 31 | Brownstown Wastewater Utility | X | | X | | | | | | | |

Table 3-6. Activated Sludge Process Elements (Continued).

| No. | Facility | EBPR Reactor Zones | | | Mixed-Liquor Recycle | | | Other Process Features | | | |
|---|---|---|---|---|---|---|---|---|---|---|---|
| | | Aer. | Anox. | Anaer. | Aer. to Anox. | Anox. to Anaer. | Aer. to Anaer. | P-Stripper | SBR | Step Feed | Simult. N/DN |
| 32 | Iron Bridge Regional WRF | X | X | X | X | | | | | | X |
| 33 | Virginia Initiative Plant | X | X | X | X | X | | | | | |
| 34 | Kamen-Koernerback | | | X | | | | | | X | |
| 35 | Duelmen | X | X | X | | X | | | | X | |
| 36 | Luepinghausen | | | X | | | | | | X | |
| 37 | ERAR Viveros de la Villa | X | X | X | X | | | | | | |
| 38 | Olifantsfontein WCW | X | X | X | X | | | | | X | |
| 39 | Stevens Point Wastewater Treatment Plant | X | X | | X | | | | | | X |
| 40 | Hildesheim | X | X | X | X | X | | | | | X |
| 41 | Waiblingen | X | | | | | | | | | |
| 42 | Flip Human Water Care Works | X | X | X | X | X | | | | | X |
| 43 | Percy Stewart Water Care Works | X | X | X | X | X | | | | | X |
| 44 | Daspoort Wastewater Treatment Works | X | X | X | X | | | | | | X |
| 45 | Zeekoegat Wastewater Treatment Works | X | X | X | X | | | | | | X |
| 46 | Sunderlandridge Wastewater Treat. Works | X | X | X | X | | | | | | |
| 47 | Rooiwal Wastewater Treatment Works | X | X | X | X | | | | | | |

Table 3-6 shows that nine facilities do not have an anaerobic zone. This implies that either the plants create anaerobic conditions in a portion of the process that they do not call an anaerobic zone (for instance, the Stevens Point WWTP has anaerobic/anoxic zones, some of which operate in an anaerobic mode), or that they failed to answer the question or answered incorrectly.

## 3.4 Supplementary Chemical Addition for Phosphorus Removal

This survey section requested information on the use, location, and dosage of chemical addition for phosphorus removal. This section includes a chart showing the percentage breakdown of surveyed plants adding chemical only, VFA only, both chemical and VFA, and neither.

Table 3-7 indicates which plants add chemicals, identifies the chemical and chemical addition location, and lists the reported dosage. Figure 3-8 shows a breakdown of the chemical added for phosphorus removal.

Figure 3-9 gives a percentage breakdown of the various chemical addition locations, and Figure 3-10 shows the percentage of plants adding VFA only, chemical only, both, or neither to augment phosphorus removal.

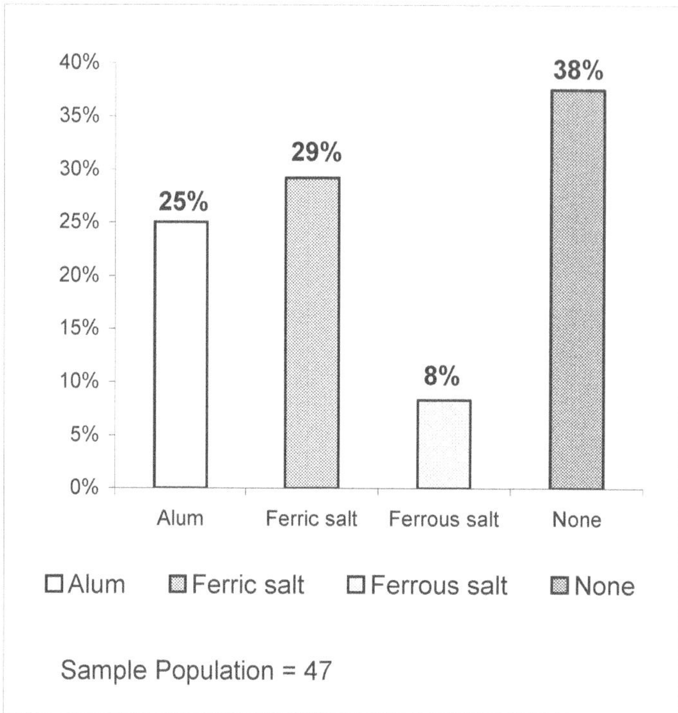


**Figure 3-8. Chemical Addition for Phosphorus Removal.**

**Table 3-7. Chemical Addition.**

| No. | Facility | | | | Addition Point for Supplementary P Removal | | | | | | |
|---|---|---|---|---|---|---|---|---|---|---|---|
| | | Avg. Flow (MGD) | Influent TP (mg/L) | Chemical Added | Primary Clarifier | Aeration Basin | Second. Clarifier | Tertiary Clarifier | Effluent Filter | Dewater. Centrate | Dosage (mg/L) |
| 1 | Green Bay Metropolitan Sewerage District | 32 | N/A | | | | | | | | |
| 2 | Kalispell AWT & BNR Facility | 2.4 | 4.73 | | | | | | | | |
| 3 | Durham AWWTP | 20 | 7.5 | Alum | | | X | X | | | 30 |
| 4 | Bonnybrook Wastewater Treatment Plant | 108 | 6.3 | Alum | X | | | | | | 10 |
| 5 | City of Cheney WWTP | 1.19 | 4.8 | | | | | | | | |
| 6 | Laughlin Water Reclamation Facility | 3 | 6.8 | Alum | | | X | | | | 50 |
| 7 | Reedy Creek Improvement District WRF | 9.9 | 6.7 | Ferric | | X | X | | X | | 8 |
| 8 | Proctors Creek Wastewater Plant | 16.5 | 7 | Ferric | X | | | | | | N/A |
| 9 | Sod Run WWTP | 12 | 8 | Ferric | | | X | | | | 18 |
| 10 | City of Grande Prairie WWTP | 3.2 | 7.12 | | | | | | | | |
| 11 | Fergus Falls WWTP | 2.2 | 6 | Alum | | | X | | | | N/A |
| 12 | Eugene/Springfield WPCF | 24 | 8.5 | | | | | | | | |
| 13 | McDowell Creek WWTP | 5 | 7 | Alum | X | | X | | | | 7 |
| 14 | Mauldin Road WWTP | 20 | 7 | | | | | | | | |
| 15 | Lower Reedy | 6.5 | 6.5 | | | | | | | | |
| 16 | City of Ft. Myers Central | 11.0 | N/A | Alum | | | X | | | | 60 |

| No. | Facility | Avg. Flow (MGD) | Influent TP (mg/L) | Chemical Added | Addition Point for Supplementary P Removal: Primary Clarifier | Aeration Basin | Second. Clarifier | Tertiary Clarifier | Effluent Filter | Dewater. Centrate | Dosage (mg/L) |
|---|---|---|---|---|---|---|---|---|---|---|---|
| | AWWTF | | | | | | | | | | |

Table 3-7. Chemical Addition (Continued).

| No. | Facility | Avg. Flow (MGD) | Influent TP (mg/L) | Chemical Added | Addition Point for Supplementary P Removal: Primary Clarifier | Aeration Basin | Second. Clarifier | Tertiary Clarifier | Effluent Filter | Dewater. Centrate | Dosage (mg/L) |
|---|---|---|---|---|---|---|---|---|---|---|---|
| 17 | Nine Springs Wastewater Treatment Plant | 42 | 6 | Ferric | | | | | | X | 1 |
| 18 | City of Kelowna Wastewater Treat. Facility | 7.2 | 6.6 | Alum | | | X | | | | N/A |
| 19 | Missoula Wastewater Division | 8 | 5.0 | | | | | | | | |
| 20 | Nansemond Treatment Plant | 18 | 10 | Ferric | X | X | X | | | | N/A |
| 21 | Back River Wastewater Treatment Plant | 145 | 3.6 | Ferrous | | X | | | | | 8 |
| 22 | Water Reclamation Facility | 3.1 | 5 | Alum | | | | X | | | 29 |
| 23 | Rogers Pollution Control Facility | 5.2 | 11 | Ferrous | | | | | | | N/A |
| 24 | University of Florida, Water Rec Facility | 3.0 | N/A | Alum | | | | | | | N/A |
| 25 | Pocatello Water Pollution Control Dept. | 7.4 | 8 | | | | | | | | |
| 26 | Sheboygan Regional Wastewater Treat. Facil. | 12.3 | 5.1 | Ferric | | X | | | | X | N/A |
| 27 | Village of Minoa Wastewater Treat. Facil | 0.5 | 6 | | | | | | | | |
| 28 | Marshall Wastewater Treatment Facility | 0.35 | N/A | | | | | | | | |
| 29 | Ephrata Regional WWTP No. 2 | 1.27 | 6.8 | Ferrous | | X | | | | | N/A |
| 30 | American-Bath WWTP | 1.2 | 8 | Alum | | | X | | | | N/A |
| 31 | Brownstown Wastewater Utility | 0.3 | 5.4 | | | | | | | | |
| 32 | Iron Bridge Regional WRF | 24.6 | 8 | Alum | | | X | | | | N/A |
| 33 | Virginia Initiative Plant | N/A | 5.4 | | | | | | | | |
| 34 | Kamen-Koernerback | N/A | N/A | Ferric | | | X | | | | N/A |
| 35 | Duelmen | N/A | N/A | Ferric | | | X | | | | N/A |
| 36 | Luepinghausen | N/A | 5.4 | Ferric | | X | | | | | N/A |
| 37 | ERAR Viveros de la Villa | N/A | 6.1 | | | | | | | | |
| 38 | Olifantsfontein WCW | 16.4 | 17 | | | | | | | | |
| 39 | Stevens Point Wastewater Treatment Plant | 3 | N/A | Ferric | X | X | | | | | N/A |
| 40 | Hildesheim | 8.5 | 7.9 | Ferric & Ferrous | | | X | | | X | 3.5 |
| 41 | Waiblingen | 2 | 8.4 | Ferric | | | X | | | | N/A |
| 42 | Flip Human Water Care Works | 6.1 | 12 | | | | | | | | |
| 43 | Percy Stewart Water Care Works | 4.3 | 12 | Ferric | | | X | | | | |
| 44 | Daspoort Wastewater Treatment Works | 11.6 | 6 | | | | | | | | |
| 45 | Zeekoegat Wastewater Treatment Works | 8.7 | 6 | Ferric | | | | | X | | N/A |

| No. | Facility | Avg. Flow (MGD) | Influent TP (mg/L) | Chemical Added | Addition Point for Supplementary P Removal | | | | | | Dosage (mg/L) |
|---|---|---|---|---|---|---|---|---|---|---|---|
| | | | | | Primary Clarifier | Aeration Basin | Second. Clarifier | Tertiary Clarifier | Effluent Filter | Dewater. Centrate | |
| 46 | Sunderlandridge Wastewater Treat. Works | 10.4 | N/A | Alum | | | | X | | X | 3-5 |
| 47 | Rooiwal Wastewater Treatment Works | 31.7 | 9.8 | | | | | | | | |

Note: N/A indicates not available or not provided

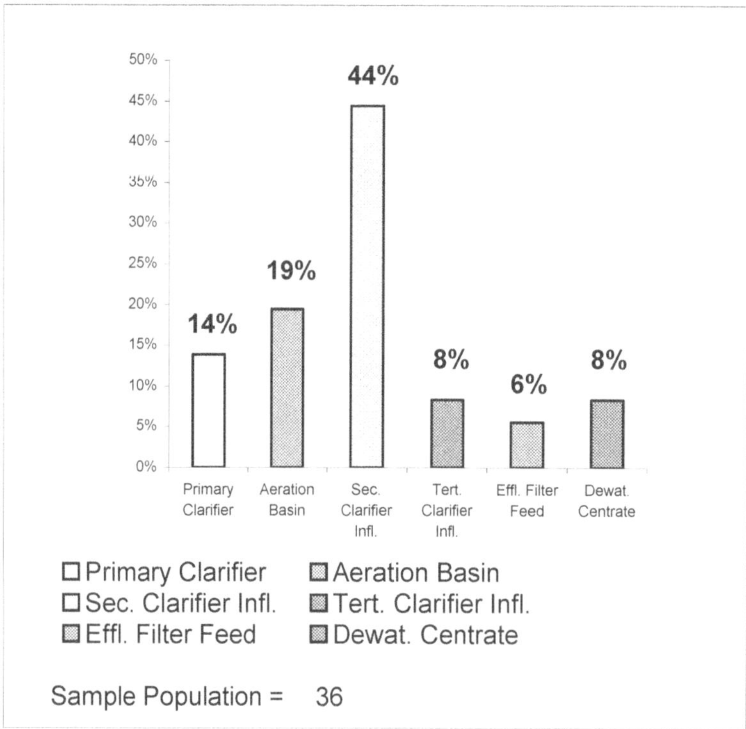


**Figure 3-8. Chemical Addition Locations.**



WERF

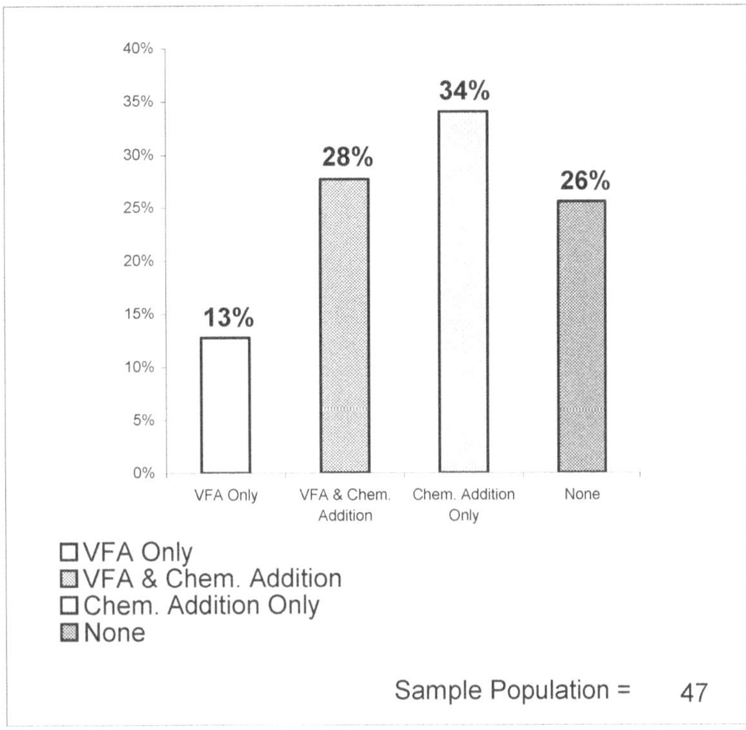


**Figure 3-9. Volatile Fatty Acid and Chemical Addition.**

## 3.5 Solids Handling Process Information

This survey section requested information on solids processing, including WAS thickening, solids digestion/processing, and solids dewatering. Figure 3-11 shows the percentage breakdown of WAS thickening methods. Over one-third of the plants utilized dissolved air flotation. This should keep the sludge aerobic, thereby minimizing phosphorus release. Another 30% used gravity belt thickeners. No information was requested on plants using other WAS thickening methods, but in at least one case, building a solids inventory in the secondary clarifier was the sole means for WAS thickening.

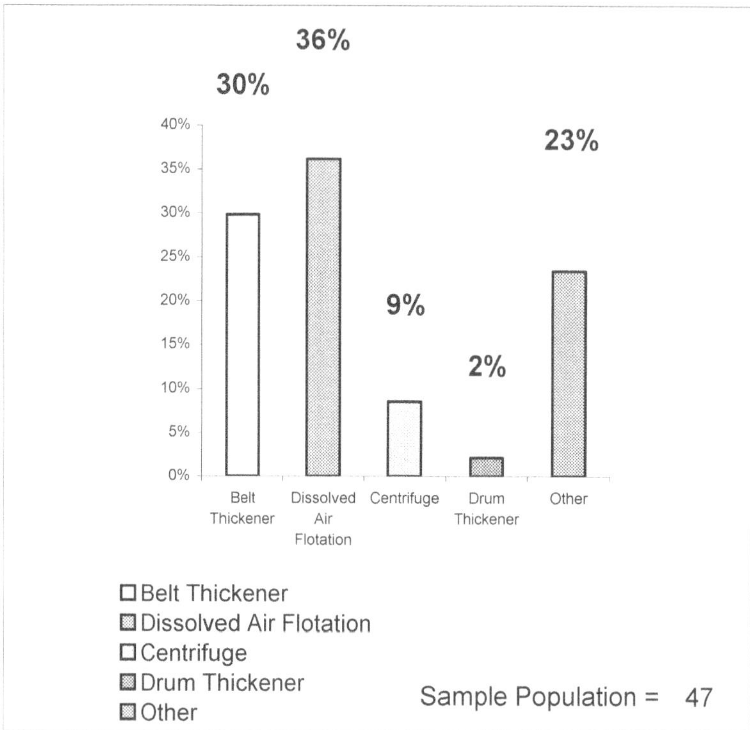


**Figure 3-10. Waste Activated Sludge Thickening Methods.**

Figure 3-12 shows the breakdown of WAS or solids treatment processes. The majority of the plants utilize anaerobic digesters. Several of the plants reporting "none" for solids processing either land-applied sludge or contracted out further processing of the sludge. One of the plants composted WAS. Almost 40% of the plants utilized belt filters to dewater treated solids, as shown in Figure 3-13. Another 23% used dewatering centrifuges. Those plants that did not report a biosolids dewatering method primarily used land application of liquid solids. Figure 3-14 shows the location of recycle flows returned to the mainstream.

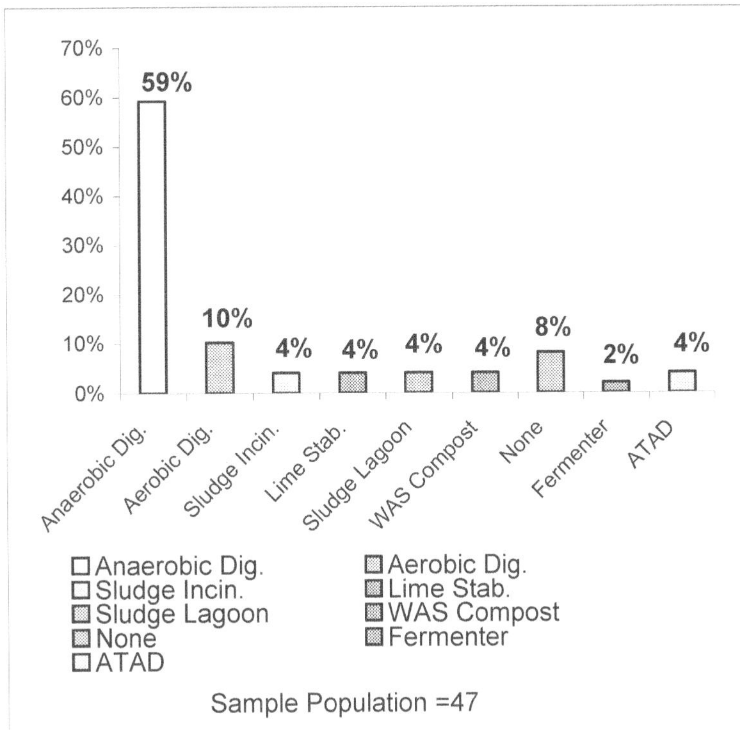


**Figure 3-11. Waste Activated Sludge/Biosolids Processing.**

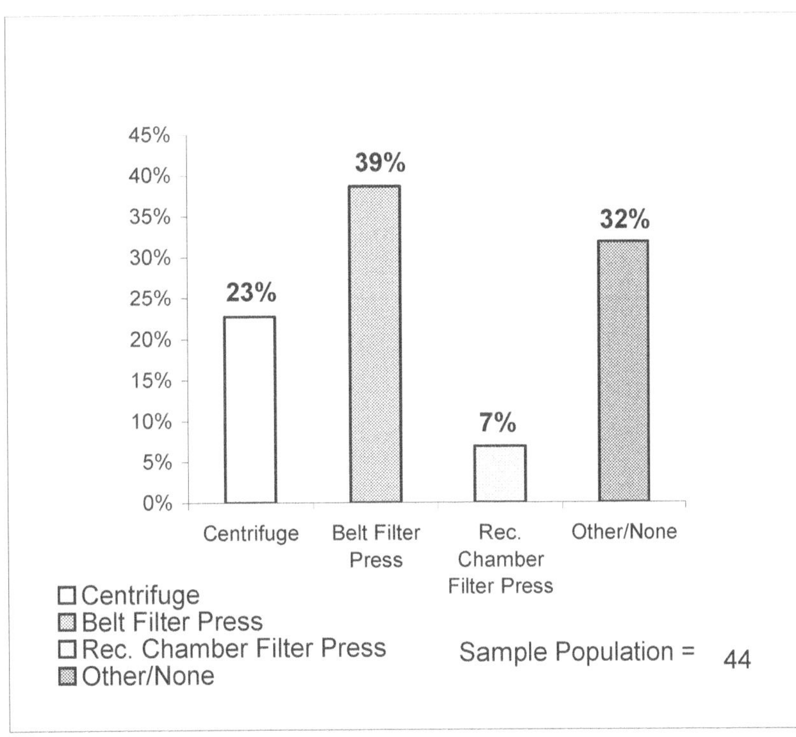


**Figure 3-12. Solids Dewatering.**

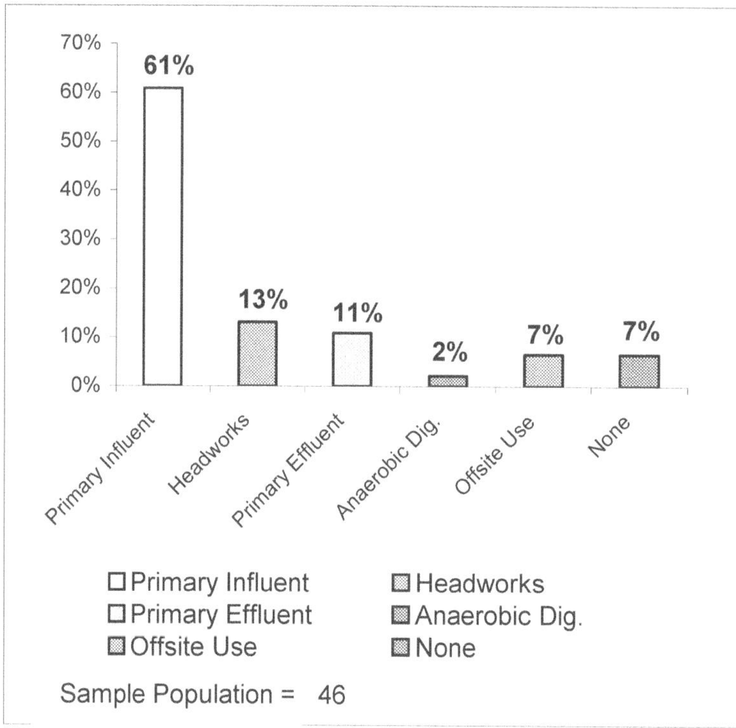


**Figure 3-13. Dewatering Recycle Return Location.**

## 3.6 Secondary and Plant Effluent

This survey section requested information on secondary and plant effluent. Tables 3-8 and 3-9 present the survey information for secondary effluent and plant effluent, respectively. Effluent nutrient targets were given in Table 3-2.

Table 3-9 shows that fewer than half of the facilities included in the survey provided effluent phosphate measurements, making it difficult to draw conclusions regarding EBPR performance. In cases where the effluent total phosphorus is relatively high (for example, the Ephrata Regional WWTP #2, with effluent TP of 1.5 mg/L), it is impossible to tell whether this is due to high phosphate or limitations in effluent clarification/filtration.

## 3.7 Phosphorus Removal Summary

This section groups the participating treatment plants based on VFA and/or chemical addition and summarizes their respective phosphorus-removal data. Other plant conditions, process configurations, and operating procedures impact phosphorus removal, perhaps even to a greater extent than VFA and chemical addition. Table 3-10 shows the effluent total phosphorus concentration and percent phosphorus removal for all facilities, grouped by type of EBPR augmentation. Figures 3-15 and 3-16 are provided to give a graphical overview of plant performance based on reported results in the survey but are not meant as exact representations of either average effluent concentrations or average phosphorus removal. Table 3-11 ranks all facilities based on their phosphorus-removal performance, and summarizes some of the key characteristics impacting phosphorus removal. The following general observations can be made:

- Plants achieving the highest percentage of influent phosphorus removal use chemical or VFA and chemical addition to supplement EBPR.
- Two-thirds of the facilities with effluent total phosphorus lower than 0.5 mg/L do not have solids recycle streams from dewatering of anaerobically digested sludge.
- The full range of SRTs is represented in all categories of P removal performance (under 0.5 mg/L, 0.5–1.0 mg/L, and more than 1.0 mg/L).
- The maximum total phosphorus removal for facilities reporting no VFA or chemical addition is 95%.

**Table 3-8. Secondary Effluent Characteristics.**

| No. | Facility | COD (mg/L) | TOC (mg/L) | BOD (mg/L) | TSS (mg/L) | TKN (mg/L) | NH4-N (mg/L) | NO3-N (mg/L) | $PO_4$-P (mg/L) | TP (mg/L) |
|---|---|---|---|---|---|---|---|---|---|---|
| 1 | Green Bay Metropolitan Sewerage District | NA | NA | 4 | 6.2 | 2.0 | NA | NA | 0.4 | 0.21 |
| 2 | Kalispell AWT & BNR Facility | | | | | | | | | |
| 3 | Durham AWWTP | | | | 5.3 | | 0.08 | 9.4 | 0.36 | 0.57 |
| 4 | Bonnybrook Wastewater Treatment Plant | | | 3.5 | 6 | 3.5 | 1.6 | 13 | | 0.75 |
| 5 | City of Cheney WWTP and Reclamation Plant | 11 | | 2.3 | 3.4 | | | 11 | | 0.4 |
| 6 | Laughlin Water Reclamation Facility | | | | | | | | | |
| 7 | Reedy Creek Improvement District WRF | | | | | 2 | | 0.5 | | |
| 8 | Proctors Creek Wastewater Plant | | | 3 | 4 | 0.96 | 2.7 | | | |
| 9 | Sod Run Wastewater Treatment Plant | | | 5 | 5 | 3 | 0.5 | 7 | 1 | 1 |
| 10 | City of Grande Prairie WWTP | | | 5.7 | 13 | | 0.5 | | | 0.43 |
| 11 | Fergus Falls WWTP | | | | | | | | | |
| 12 | Eugene/Springfield WPCF | | | | 6.7 | | | | | |
| 13 | McDowell Creek WWTP | <20 | | 2 | 2 | 1 | 0.2 | 7 | | 0.1 |
| 14 | Mauldin Road WWTP | | | >5 | >9 | >5 | >1 | >1 | 1 | 1 |
| 15 | Lower Reedy | | | | >10 | | >10 | >7 | | 1.5 |
| 16 | City of Ft. Myers Central AWWT Facility | | | | | | | | | |
| 17 | Nine Springs Wastewater Treatment Plant | | | | | | | | | |
| 18 | City of Kelowna Wastewater Treat. Facility | 32 | | | 2.7 | 1.8 | 0.33 | 1.7 | 0.03 | 0.21 |
| 19 | Missoula Wastewater Division | | | 160 | 90 | 39 | 30 | 0.14 | 6.2 | 7.8 |
| 20 | Nansemond Treatment Plant | | | | | | | | | |
| 21 | Back River Wastewater Treatment Plant | 78 | | 6.3 | 5.9 | 3.8 | 0.8 | 4.8 | | 0.17 |
| 22 | Water Reclamation Facility | | | 2 | 2 | | | | 0.05 | 0.15 |
| 23 | Rogers Pollution Control Facility | | | | | | | | | |
| 24 | University of Florida, Water Rec Facility | | | 2.4 | | | | | | 0.86 |
| 25 | Pocatello Water Pollution Control Dept. | | | | | | | | | |
| 26 | Sheboygan Regional Wastewater Treat. Facil. | | | 12 | 6 | | | | | 0.56 |
| 27 | Village of Minoa Wastewater Treat. Facil | | | <2 | <1 | 1 | <1 | 1.8 | | 1 |
| 28 | Marshall Wastewater Treatment Facility | | | | | | | | | |
| 29 | Ephrata Regional WWTP No. 2 | | | | | | | | | |
| 30 | American-Bath WWTP | | | 4 | 4 | | | | | 0.35 |
| 31 | Brownstown Wastewater Utility | | | | | | | | | |
| 32 | Iron Bridge Regional WRF | | 9 | 2 | 4 | 2 | 0.1 | 0.5 | 0.2 | 0.3 |
| 33 | Virginia Initiative Plant | | | | | | 1.3 | | | |
| 34 | Kamen-Koernerback | 28 | | | | 1.8 | 0.3 | 10 | | 0.6 |
| 35 | Duelmen | | | | | | 1.5 | 7.6 | 0.8 | |
| 36 | Luepinghausen | 23 | | 1.7 | | | 0.4 | 2.2 | 0.05 | 0.1 |
| 37 | ERAR Viveros de la Villa | 30 | | 5 | 5 | 8.1 | 0.81 | 8.0 | 4.4 | 4.7 |

**Table 3-8. Secondary Effluent Characteristics (Continued).**

| No. | Facility | COD (mg/L) | TOC (mg/L) | BOD (mg/L) | TSS (mg/L) | TKN (mg/L) | NH4-N (mg/L) | NO3-N (mg/L) | $PO_4$-P (mg/L) | TP (mg/L) |
|---|---|---|---|---|---|---|---|---|---|---|
| 38 | Olifantsfontein WCW | 29 | | | 11 | 9.8 | 8.9 | 4 | 0.25 | 1.15 |
| 39 | Stevens Point Wastewater Treatment Plant | | | 6 | 6 | | | | | 0.5 |
| 40 | Hildesheim | 38 | 13 | 3 | 11 | 3.8 | 1.4 | 2.8 | | 0.34 |
| 41 | Waiblingen | 22 | 4 | 22 | | 20 | 0.2 | 8.4 | 0.4 | 0.5 |
| 42 | Flip Human Water Care Works | 51 | | | 17 | | 0.42 | 0.92 | 1.0 | |
| 43 | Percy Stewart Water Care Works | 72 | | | 26 | 14 | 4.0 | 9.33 | 3.9 | 4.5 |
| 44 | Daspoort Wastewater Treatment Works | 34 | | | 9 | 3.5 | 3 | 5 | <0.1 | |
| 45 | Zeekoegat Wastewater Treatment Works | 45 | | | 6 | 10 | 5 | 7 | 1 | 1.2 |
| 46 | Sunderlandridge Wastewater Treat. Works | | | | | | | | | |
| 47 | Rooiwal Wastewater Treatment Works | 39 | | | 12 | | 0.83 | 2.7 | 1.6 | |

**Table 3-9. Plant Effluent Characteristics.**

| No. | Facility | COD (mg/L) | TOC (mg/L) | BOD (mg/L) | TSS (mg/L | TKN (mg/L | $NH_4$-N (mg/L) | $NO_3$-N (mg/L) | $PO_4$-P (mg/L) | TP (mg/L) | %P Removal |
|---|---|---|---|---|---|---|---|---|---|---|---|
| 1 | Green Bay Metropolitan Sewerage District | 34 | NA | 2.8 | 4.8 | 1.5 | 0.12 | NA | NA | | |
| 2 | Kalispell AWT & BNR Facility | | | <4.4 | 3 | 0.9 | <0.3 | 8.5 | | 0.11 | 98% |
| 3 | Durham AWWTP | | | 2 | 0.7 | 1.1 | 0.07 | 8.5 | | 0.06 | 99% |
| 4 | Bonnybrook Wastewater Treatment Plant | | | 3.5 | 6.5 | 3.5 | 1.6 | 13 | | 0.75 | 88% |
| 5 | City of Cheney WWTP and Reclamation Plant | | | | | | | | | | 92% |
| 6 | Laughlin Water Reclamation Facility | | | <2 | <4 | 0.65 | <.08 | 13 | 0.75 | 0.85 | 87% |
| 7 | Reedy Creek Improvement District WRF | | | 0.6 | 1.2 | 2 | | 0.5 | 0.2 | 0.26 | 96% |
| 8 | Proctors Creek Wastewater Plant | | | 2 | 1 | 1 | 0.37 | 3.3 | | 1.4 | 80% |
| 9 | Sod Run Wastewater Treatment Plant | | | 5 | 5 | 3 | 0.5 | 7 | 1 | 1 | 88% |
| 10 | City of Grande Prairie WWTP | | | 3.7 | 3.7 | | 2.6 | | | 0.58 | 92% |
| 11 | Fergus Falls WWTP | | | 3.8 | 8.5 | | 3.3 | 5 | | 0.6 | 90% |
| 12 | Eugene/Springfield WPCF | | | 2.7 | 5.8 | 12 | 11 | 6 | | 1.4 | 84% |
| 13 | McDowell Creek WWTP | <20 | | 2 | 1 | 1 | 0.2 | 0.7 | | 0.1 | 99% |
| 14 | Mauldin Road WWTP | | 25 | >5 | >5 | >5 | >1 | >1 | 1 | 1 | 86% |
| 15 | Lower Reedy | | >25 | >7 | >10 | >5 | >1 | >7 | | 1.5 | 77% |
| 16 | City of Ft. Myers Central AWWT Facility | | | 3 | 6.5 | 1.3 | | 0.45 | | 0.1 | |
| 17 | Nine Springs Wastewater Treatment Plant | | | 4 | 5 | | 0.5 | 11 | 0.35 | 0.5 | 92% |
| 18 | City of Kelowna Wastewater Treat. Facility | 31 | | 2.5 | 0.9 | 1.8 | 0.48 | 2.3 | 0.05 | 0.15 | 98% |
| 19 | Missoula Wastewater Division | | | 40 | 8.9 | 23 | 18 | 3.2 | 1.74 | 2.0 | 59% |
| 20 | Nansemond Treatment Plant | | | 8 | 10 | 2 | 0.6 | 10 | | 0.8 | 92% |
| 21 | Back River Wastewater Treatment Plant | 76 | | 4 | 2 | 4 | 1.4 | 4.1 | | 0.12 | 97% |

Table 3-9. Plant Effluent Characteristics (Continued).

| No. | Facility | COD (mg/L) | TOC (mg/L) | BOD (mg/L) | TSS (mg/L | TKN (mg/L | $NH_4$-N (mg/L) | $NO_3$-N (mg/L) | $PO_4$-P (mg/L) | TP (mg/L) | %P Removal |
|---|---|---|---|---|---|---|---|---|---|---|---|
| 22 | Water Reclamation Facility | | | 2 | 2 | 1.5 | 6 | 0.1 | 0.01 | 0.05 | 99% |
| 23 | Rogers Pollution Control Facility | | | 3 | 4 | | 0.3 | 2 | 0.6 | 1 | 91% |
| 24 | University of Florida, Water Rec Facility | | | 2.5 | 1.9 | | | | | 0.86 | |
| 25 | Pocatello Water Pollution Control Dept. | | | 10 | 10 | | 21 | | | 1 | 88% |
| 26 | Sheboygan Regional Wastewater Treat. Facil. | | | 12 | 6 | | | | | 0.56 | 89% |
| 27 | Village of Minoa Wastewater Treat. Facil | | | <2 | <1 | 1 | <1 | 1.8 | | 1 | 83% |
| 28 | Marshall Wastewater Treatment Facility | | | 4 | 4 | | 0.4 | | | 1 | |
| 29 | Ephrata Regional WWTP No. 2 | | | 2 | 4 | | 0.61 | 4 | | 1.5 | 78% |
| 30 | American-Bath WWTP | | | 4 | 4 | | | | | 0.35 | 96% |
| 31 | Brownstown Wastewater Utility | | | 3.4 | 13 | | | | | 0.26 | 95% |
| 32 | Iron Bridge Regional WRF | | 9 | 2 | 1 | 1 | 0.1 | 0.4 | 0.1 | 0.2 | 98% |
| 33 | Virginia Initiative Plant | | | 6 | 5 | 2.2 | | | | 0.6 | 89% |
| 34 | Kamen-Koernerback | | | | | | | | | | |
| 35 | Duelmen | 23 | | 5 | | 6 | 1.5 | 7.6 | 0.5 | 0.6 | |
| 36 | Luepinghausen | | | | | | | | | | 98% |
| 37 | ERAR Viveros de la Villa | 29 | | 4 | 4 | 7.9 | 0.77 | 8.0 | 4.4 | 4.7 | 23% |
| 38 | Olifantsfontein WCW | 29 | | | 11 | 9.8 | 8.9 | 4 | 0.25 | 1.2 | 93% |
| 39 | Stevens Point Wastewater Treatment Plant | | | 6 | 6 | | | | | 0.5 | |
| 40 | Hildesheim | 38 | 13 | 3 | 11 | 3.8 | 1.44 | 2.8 | | 0.34 | 96% |
| 41 | Waiblingen | 22 | 4 | 22 | | 2 | 0.2 | 8.4 | 0.4 | 0.5 | 94% |
| 42 | Flip Human Water Care Works | 68 | | | 32 | 6 | 0.29 | 1.1 | 1.4 | 2 | 83% |
| 43 | Percy Stewart Water Care Works | 57 | | | 23 | 12 | 4 | 4.7 | 3.7 | 4.0 | 66% |
| 44 | Daspoort Wastewater Treatment Works | 34 | | | 9 | 3.5 | 3 | 5 | <0.1 | | |
| 45 | Zeekoegat Wastewater Treatment Works | 45 | | | 6 | 10 | 5 | 7 | 1 | 1.2 | 80% |
| 46 | Sunderlandridge Wastewater Treat. Works | 45 | | | | | 3.1 | 2.9 | 0.94 | | |
| 47 | Rooiwal Wastewater Treatment Works | 44 | | | 13 | 4.4 | 1.0 | 2.7 | 1.9 | 2.4 | 75% |

Table 3-10. EBPR Performance and VFA/Chemical Addition.

| Facility No. | % P-Removal | | | | Effluent P, mg/L | | | |
|---|---|---|---|---|---|---|---|---|
| | No VFA or Chem. Addition | VFA Addition | VFA and Chem. Addition | Chemical Addition | No VFA or Chem. Addition | VFA Addition | VFA and Chem. Addition | Chemical Addition |
| 1 | | | | | 0.21 | | | |
| 2 | | 98% | | | | 0.11 | | |
| 3 | | | 99% | | | | 0.06 | |
| 4 | | | 88% | | | | 0.75 | |
| 5 | 92% | | | | 0.4 | | | |
| 6 | | | | 87% | | | | 0.85 |
| 7 | | | 96% | | | | 0.26 | |
| 8 | | | | 80% | | | | 1.4 |
| 9 | | | | 88% | | | | 1 |
| 10 | 92% | | | | 1 | | | |
| 11 | | | | | | | | 0.6 |
| 12 | 84% | | | | 1.4 | | | |
| 13 | | | 99% | | | | 0.1 | |
| 14 | | 86% | | | | 1 | | |
| 15 | 77% | | | | 1.5 | | | |
| 16 | | | | | | | | 0.1 |
| 17 | | | | 92% | | | | 0.5 |
| 18 | | | 98% | | | | 0.15 | |
| 19 | | 59% | | | | 2.02 | | |
| 20 | | | 92% | | | | 0.8 | |
| 21 | | | | 97% | | | | 0.12 |
| 22 | | | | 99% | | | | 0.05 |
| 23 | | | | 91% | | | | 1 |
| 24 | | | | | | | 0.86 | |
| 25 | 88% | | | | 1 | | | |
| 26 | | | | 89% | | | | 0.56 |
| 27 | 83% | | | | 1 | | | |
| 28 | | | | | 1 | | | |
| 29 | | | | 78% | | | | 1.5 |
| 30 | | | | 96% | | | | 0.35 |
| 31 | 95% | | | | 0.26 | | | |
| 32 | | | | 98% | | | | 0.2 |
| 33 | | 89% | | | | 0.6 | | |
| 34 | | | | | | | | 0.6 |
| 35 | | | | | | | 0.6 | |
| 36 | | | | 98% | | | | 0.1 |
| 37 | 23% | | | | 4.7 | | | |
| 38 | | 93% | | | | 1.15 | | |
| 39 | | | | | | | | |
| 40 | | | 96% | | | | 0.34 | |
| 41 | | | 94% | | | | 0.5 | |
| 42 | 83% | | | | 2 | | | |
| 43 | | | | 66% | | | | 4.04 |
| 44 | | | | | | | | |
| 45 | | | 80% | | | | 1.2 | |
| 46 | | | | | | | | |
| 47 | 75% | | | | 2.42 | | | |

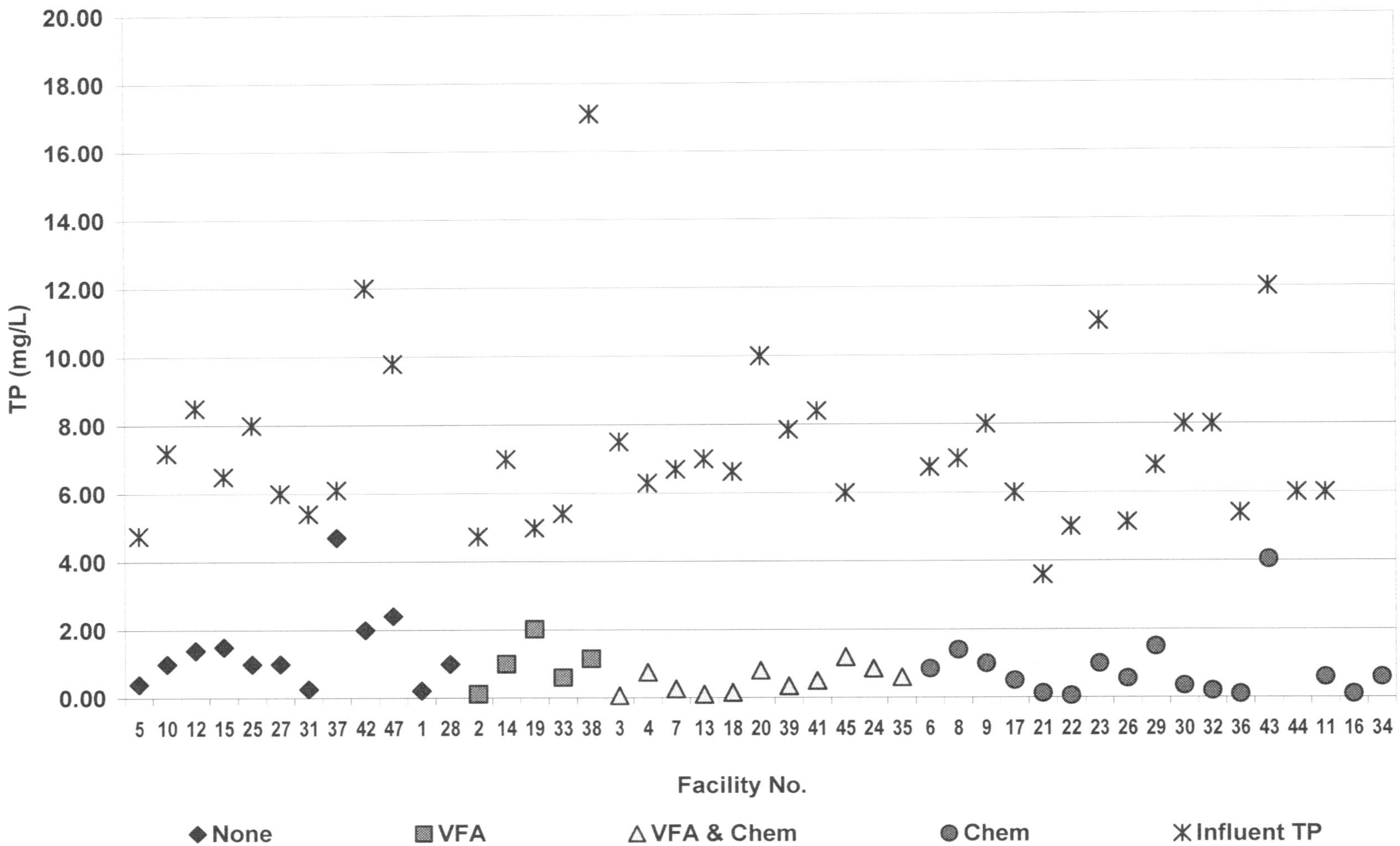


Figure 3-14. Influent and Effluent Total Phosphorus by VFA and/or Chemical Addition.

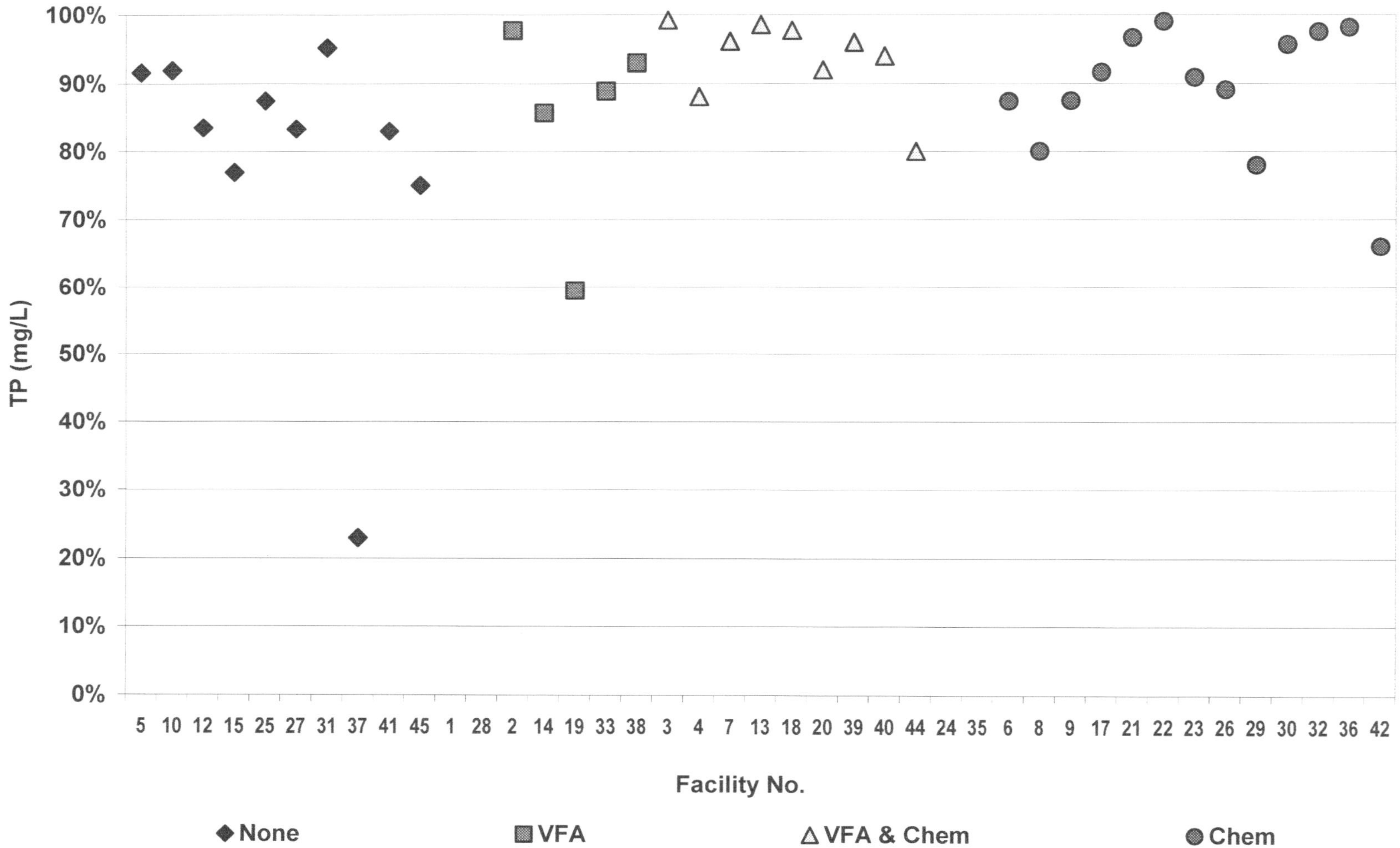


Figure 3-15. Percent Phosphorus Removal by VFA and/or Chemical Addition.

| No. | Facility[1] | Influent TP (mg/L) | BOD: TP [2] | SRT Range (days) | Second. Effl. TP [3] (mg/L) | Effluent TP (mg/L) | Secondary Removal [4] | Total Removal | Lime Addition [5] | Solids. Recycle | None | VFA Addition | Chemical Addition |
|---|---|---|---|---|---|---|---|---|---|---|---|---|---|
| **FACILITIES WITH EFFLUENT PHOSPHORUS < 0.5 mg/L** | | | | | | | | | | | | | |
| 36 | Luepinghausen | 5.4 | 21 | 10 - 20 | | 0.01 | | 98% | | X | | | Ferric-AB |
| 22 | Water Reclamation Facility | 5 | 40 | 10 - 20 | 0.15 | 0.05 | 97% | 99% | | | | | Alum-TCL |
| 3 | Durham AWWTP | 7.5 | 19 | 10 - 20 | 0.572 | 0.06 | 92% | 99% | X | X | | X | Alum-SCL, TCL |
| 16 | City of Ft. Myers Central AWWTF | | | 10 - 20 | | 0.1 | | | | | | | Alum-SCL |
| 13 | McDowell Creek WWTP[9] | 7 | 35 | 10 - 20 | 0.1 | 0.1 | | 99% | X | X | | X | Alum-PCL,SCL |
| 2 | Kalispell AWT & BNR Facility | 4.7 | 56 | | -- | 0.11 | | 98% | | | | X | |
| 21 | Back River WWTP | 3.6 | 34 | 5 - 10 | 0.17 | 0.12 | | 97% | | X | | | Ferrous-AB |
| 18 | City of Kelowna WWTF | 6.6 | 18 | 5 - 10 | 0.21 | 0.15 | | 98% | | | | X | Alum-SCL |
| 32 | Iron Bridge Regional WRF | 8 | 19 | 5 - 10 | 0.3 | 0.2 | | 98% | | | | | Alum-SCL |
| 1 | Green Bay Metropolitan Sewerage District | | | 10 - 20 | | 0.21 | | | | | X | | |
| 7 | Reedy Creek Improvement District WRF | 6.7 | 22 | 5 - 10 | -- | 0.26 | | 96% | | | | X | Ferric-AB,SCL |
| 31 | Brownstown Wastewater Utility [8] | 5.4 | 25 | 3 - 5 | | 0.26 | | 95% | | | X | | |
| 40 | Hildesheim | 7.9 | 26 | 10-20 | | 0.34 | | 96% | | X | | X | Ferric/Ferrous - SCL, Centr. |
| 30 | American-Bath WWTP [8] | 8 | 88 | 10-20 | | 0.35 | | 96% | | | | | Alum-SCL |
| 5 | City of Cheney WWTP & Reclamation Plant [7] | 4.8 | 45 | > 20 | | 0.4 | | 92% | | | X | | |
| **FACILITIES WITH EFFLUENT PHOSPHORUS 0.5 - 1.0 mg/L** | | | | | | | | | | | | | |
| 17 | Nine Springs WWTP | 6 | 24 | 5 - 10 | | 0.5 | | 92% | | X | | | Ferric-Centr. |
| 39 | Stevens Point WWTP | | | < 3 | | 0.5 | | | | X | | X | Ferric-PCL, AB |
| 41 | Waiblingen | 8.4 | 24 | 5-10 | | 0.5 | | 94% | | X | | X | Ferric-SCL |
| 26 | Sheboygan Regional WWTF | 5.1 | 23 | 5 - 10 | | 0.56 | | 89% | | X | | | Ferric-AB, Centr. |
| 10 | City of Grande Prairie WWTP | 7.2 | 36 | | | 0.58 | | 92% | | | X | | |
| 34 | Kamen-Koernerback | | | 10 - 20 | | 0.6 | | | | X | | | Ferric-SCL |
| 35 | Duelmen | | | 10 - 20 | >0.8 | 0.6 | | | | X | | | Ferric-SCL |
| 11 | Fergus Falls WWTP | 6 | 19 | > 20 | | 0.6 | | 90% | | X | | | Alum-SCL |
| 33 | Virginia Initiative Plant | 5.4 | 26 | 5 - 10 | | 0.6 | | 89% | | | | X | |
| 4 | Bonnybrook WWTP | 6.3 | 19 | 5 - 10 | | 0.75 | | 88% | | X | | X | Alum-PCL,SCL |
| 20 | Nansemond Treatment Plant | 10 | 15 | 5 - 10 | | 0.8 | | 92% | | X | | X | Ferric-PCL, AB, SCL |

Table 3-11. Phosphorus Removal Summary Ranked by Effluent TP (Continued).

| No. | Facility[1] | Influent TP (mg/L) | BOD: TP [2] | SRT Range (days) | Second. Effl. TP [3] (mg/L) | Effluent TP (mg/L) | Secondary Removal [4] | Total Removal | Lime Addition [5] | Solids. Recycle | VFA and/or Chemical Addition [6] None | VFA Addition | Chemical Addition |
|---|---|---|---|---|---|---|---|---|---|---|---|---|---|
| **FACILITIES WITH EFFLUENT PHOSPHORUS 0.5 - 1.0 mg/L (continued)** | | | | | | | | | | | | | |
| 6 | Laughlin WRF | 6.8 | 41 | > 20 | -- | 0.85 | | 87% | | | | | Alum-SCL |
| 24 | University of Florida, WRF | | | > 20 | 0.86 | 0.86 | | | | | | | Alum-NS |
| 28 | Marshall WWTF | | | | | 1 | | | | | X | | |
| 23 | Rogers Pollution Control Facility | 11 | 25 | | | 1 | | 91% | | | | | Ferrous-NS |
| 9 | Sod Run WWTP | 8 | 24 | 10 - 20 | | 1 | | 88% | | X | | | Ferric-PCL,SCL |
| 25 | Pocatello Water Pollution Control Dept. [7] | 8 | 23 | 3 - 5 | | 1 | | 88% | | X | X | | |
| 14 | Mauldin Road WWTP | 7 | 23 | 10 - 20 | 1 | 1 | | 86% | | | | X | |
| 27 | Village of Minoa WWTP | 6 | 33 | | | 1 | | 83% | | | X | | |
| **FACILITIES WITH EFFLUENT PHOSPHORUS > 1.0 mg/L** | | | | | | | | | | | | | |
| 38 | Olifantsfontein WCW | 17 | | 10 - 20 | | 1.15 | | 93% | | X | | X | |
| 45 | Zeekoegat Wastewater Treatment Works | 6 | | 10 - 20 | | 1.2 | | 80% | | | | X | Ferric-Filter |
| 12 | Eugene/Springfield WPCF [8] | 8.5 | 10 | 5 - 10 | | 1.4 | | 84% | | X | X | | |
| 8 | Proctors Creek Wastewater Plant | 7 | 21 | 10 - 20 | — | 1.4 | | 80% | | X | | | Ferric-PCL |
| 29 | Ephrata Regional WWTP No. 2 | 6.8 | 42 | > 20 | | 1.5 | | 78% | | | | | Ferrous-AB |
| 15 | Lower Reedy | 6.5 | 42 | 10 - 20 | | 1.5 | | 77% | X | X | X | | |
| 42 | Flip Human Water Care Works | 12 | | < 3 | | 2 | | 83% | | | X | | |
| 19 | Missoula Wastewater Division [8] | 5.0 | 37 | 3 - 5 | | 2.0 | | 59% | | X | | X | |
| 47 | Rooiwal Wastewater Treatment Works | 9.8 | | 10 - 20 | | 2.4 | | 75% | | | X | | |
| 43 | Percy Stewart Water Care Works | 12 | | < 3 | 4.5 | 4.0 | 62% | 66% | | | | | Ferric-SCL |
| 37 | ERAR Viveros de la Villa | 6.1 | 19 | 5 - 10 | 4.7 | 4.7 | | 23% | | X | X | | |

1. Blank spaces in the table indicate that no data were given, the value was unknown, or the data are not applicable.
2. BOD:TP ratio entering anaerobic tank/zone. Thirty-five percent reduction in influent BOD was assumed for facilities with primary clarifiers.
3. Facility has effluent filter. Dash indicates that facility has effluent filter, but no secondary effluent TP was reported.
4. Facility adds alum to tertiary clarifier.
5. Lime addition for alkalinity control.
6. Chemical addition for P-removal: PCL-Primary Clarifier, SCL-Secondary Clarifier, TCL-Tertiary Clarifier, AB-Aeration Basin, Centr.-Dewatering Centrate, NS-Not Specified.
7. Facility only nitrifies.
8. Facility nitrifies and denitrifies.
9. No values below 0.1 mg/L were reported

CHAPTER 4.0

# YEAR 1: DATA ANALYSIS AND FIELD TESTING

## 4.1 Introduction

This section presents the results of detailed survey and field testing from Phase 1. The initial survey provided detailed information from a selected group of plants. Field studies were conducted to fill in additional data gaps. These field studies were based on a one-day sampling event. The results from this initial work provide the baseline to select plants for in-depth study for an extended period. The results from the in-depth study is referred to as Phase 2 and reported in Chapter 6.0.

Using the results from the screening survey, a number of plants were identified for further investigation during the first year of the study. Selection criteria for these plants included EBPR history and performance, availability of data, and the level of interest and support from plant operations staff. Detailed information was collected from 12 facilities (Detailed Survey), nine of which were visited for field testing. Survey results and field testing were summarized in a variety of plant reports. Comprehensive results for detailed survey, field testing, and in-depth data collection at the five facilities investigated in both Year 1 and Year 2 are summarized in Appendices B through F. Appendix G describes the survey results for the Iron Bridge facility, and Appendix H describes the field test results for the Mauldin Road facility. Performance at four of the facilities was simulated using a biological process model (see Appendices K through N). This chapter presents the result from the detailed studies and one-day site visits at nine facilities. Chapter 6 presents the results from the in-depth studies and extended additional data collection from five selected facilities.

## 4.2 Overview of Investigated Facilities

Figure 4-1 shows the location of the facilities included in the detailed survey and field testing. Grand Prairie and Nine Springs (shown in normal font) provided data for the detailed survey, but were not visited for field testing. Field testing was conducted at the remainder of the facilities (shown in bold italic font). Field testing targeted two regions of the United States (Pacific Northwest/western Montana and southeastern United States) to simplify logistics. Table 4-1 shows the surveys, tests, and reports that have been completed for various facilities.

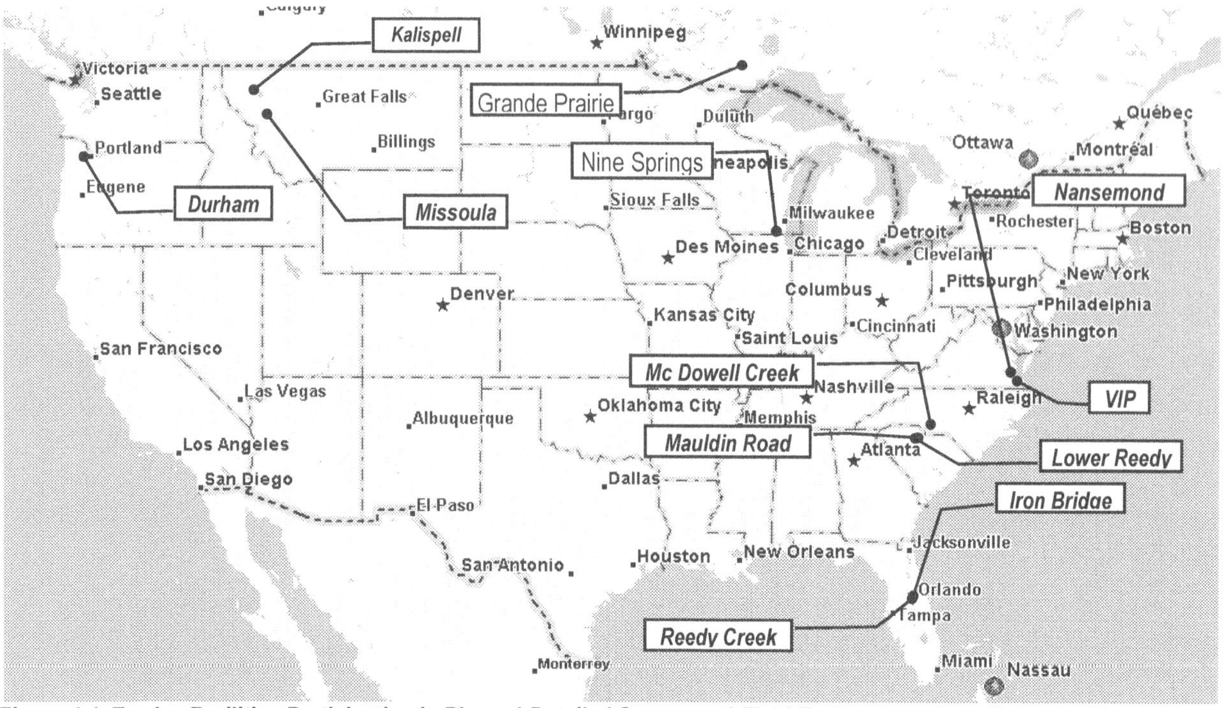


**Figure 4-1. Twelve Facilities Participating in Phase 1 Detailed Survey and *Field Testing* (map from DeLorme Street Atlas, USA, 2002).**

**Table 4-1. Surveys and Tests Conducted at Various Facilities.**

| # | Facilities | Detailed Survey Data | One Day Field Test | 4-Week Field Testing[1] |
|---|---|---|---|---|
| 1 | Durham | Yes | Yes | Yes |
| 2 | VIP | Yes | Yes | Yes |
| 3 | Nine Springs | Yes | No | No |
| 4 | Kalispell | Yes | Yes | No |
| 5 | Missoula | Yes | Yes | No |
| 6 | Reedy Creek | Yes | Yes | No |
| 7 | Lower Reedy | No | Yes | Yes |
| 8 | Mauldin Road | Yes | Yes | No |
| 9 | Iron Bridge | Yes | Yes | No |
| 10 | Grand Prairie | Yes | No | No |
| 11 | Mc Dowell Creek | Yes | Yes | Yes |
| 12 | Nansemond | Yes | Yes | Yes |

1. Results in Chapter 6

## 4.3 Detailed Survey Data Summary

Detailed process and operating data were collected from all facilities listed in Table 4-1, except Mauldin Road and Lower Reedy. The data management systems at these two plants were recently upgraded, and data could not be easily extracted. The purpose of the detailed survey was to gain additional information regarding the process design, operating stability, and process performance through review of plant data. Detailed survey reports were completed for three facilities.

The data available at facilities varied from plant to plant. Some plants had extensive electronic data, while others limit data collection and archiving. Table 4-2 shows a summary of the data received, the time period covered by the data, and the average sample population.

Table 4-3 summarizes summer plant influent conditions based on available data for the period April 2001 through October 2001. This time period was selected to compare conditions during a typical summer season. Plant flows ranged from 3.2 to 27.5 mgd, and average temperatures ranged from 14°C to 29°C.

**Table 4-2. Plant Data Summary.**

| # | Facilities | Location | Number of Measurements | Time Period | Average Sample Population |
|---|---|---|---|---|---|
| 1 | Durham | Tigard, OR | 780 | Jan 1998 – Jul 2002 | 345 |
| 2 | VIP | Norfolk, VA | 234 | Jan 2000 – Jun 2002 | 632 |
| 3 | Nine Springs | Madison, WI | 80 | Jan 1977 – Jul 2002 | 154 |
| 4 | Kalispell[1] | Kalispell, MT | 23 | Jan 1993 - May 2002 | 864 |
| 5 | Missoula[2] | Missoula, MT | 84 | Jan 1996 – Jul 2002 | 404 |
| 6 | Reedy Creek | Lake Buena Vista, FL | 32 | Sept 2001 – Aug 2002 | 352 |
| 7 | Lower Reedy[3] | Greenville, SC | No data | | |
| 8 | Mauldin Road[3] | Greenville, SC | 65 | Jan 2002 – Aug 2002 | 102 |
| 9 | Iron Bridge | Oviedo, FL | 193 | Aug 1998 – Jul 2002 | 831 |
| 10 | Grand Prairie | Grand Prairie, Alberta | 49 | Jan 2000 – May 2002 | 702 |
| 11 | Mc Dowell Creek | Huntersville, NC | 35 | Jan 1999 – May 2002 | 226 |
| 12 | Nansemond | Suffolk, VA | 222 | Jan 2000 – Jul 2002 | 604 |

[1] No phosphorus data available

[2] No process data available (i.e., MLSS, SRT)

[3] Limited or no data available due to data management system changes.

**Table 4-3. Average Influent and Effluent Characteristics Summary.**

| # | Facilities | State | Average Values between April and October 2001 | | | | | | | | |
|---|---|---|---|---|---|---|---|---|---|---|---|
| | | | Influent | | | | | | Effluent | | |
| | | | Flow, MGD | T, °C | BOD, mg/L | TP, mg/L | TKN, mg/L | SRT*, days | $PO_4$, mg/L | $NH_4$-N, mg/L | $NO_3$, mg/L |
| 1 | Durham | OR | 20.2 | 19.2 | 221 | 7.8 | 34.6 | 10.5 | 0.16 | 1.42 | 8.05 |
| 2 | VIP | VA | 27.5 | 26.5 | 217 | 6.6 | 28.4 | 8.9 | 0.86 | 2.9 | 6.6 |
| 3 | Nine Springs | WI | 8.3 | 19.7 | 205 | 5.7 | 30.1 | 10.3 | 0.38 | 0.05 | 12.6 |
| 4 | Kalispell | MT | 2.54 | 17.9 | 267 | - | - | - | - | 1.05 | 8.2 |
| 5 | Missoula | MT | 8.8 | 18.8 | 207 | 4.6 | 32.6 | - | 1.4 | 15.7 | 4.2 |
| 6 | Reedy Creek | FL | 10.2 | 29.3 | 233 | 7.2 | 40.2 | 12.2 | 0.39 | 0.03 | 1.3 |
| 7 | Lower Reedy | SC | - | - | - | - | - | - | - | - | - |
| 8 | Mauldin Road | SC | - | - | - | - | - | - | 0.52 | 0.27 | 3.5 |
| 9 | Iron Bridge | FL | 23.3 | 29.1 | 261 | 6.4 | 37.7 | - | 0.1 | 0.4 | 0.6 |
| 10 | Grand Prairie | Alberta | 3.2 | 14.2 | 351 | 6.6 | 42.3 | - | 0.5 | 0.6 | - |
| 11 | Mc Dowell Creek | NC | - | - | - | - | - | - | - | - | - |
| 12 | Nansemond | VA | 16.4 | 23.6 | 251 | 9.9 | 42.5 | 8.4 | 1.5 | 9.1 | 7.2 |

* Total SRT including entire basin volume (anaerobic, anoxic, and aerobic)

## 4.4 Process Descriptions

### 4.4.1 Liquid Stream

All facilities except Iron Bridge have primary clarification followed by conventional activated sludge. Iron Bridge uses an oxidation ditch with no primary clarification. About half of the facilities use filtration for tertiary treatment, but only the Durham facility uses chemical addition and tertiary clarification before filtration. Secondary and tertiary treatment processes for the plants participating in the detailed survey and/or field testing are summarized below.

**Table 4-4. Overview of Liquid Treatment Processes.**

| # | Facilities | State | Primary Treatment | Secondary Treatment | | Tertiary Treatment | |
|---|---|---|---|---|---|---|---|
| | | | | Aeration Basin | Oxidation Ditch | Tertiary Clarifier | Filter |
| 1 | Durham | OR | X | X | | X | X |
| 2 | VIP | VA | X | X | | | |
| 3 | Nine Springs | WI | X | X | | | |
| 4 | Kalispell | MT | X | X | | | |
| 5 | Missoula | MT | X | X | | | |
| 6 | Reedy Creek | FL | X | X | | | X |
| 7 | Lower Reedy | SC | X | X | | | |
| 8 | Mauldin Road | SC | X | X | | | X |
| 9 | Iron Bridge | FL | | | X | | X |
| 10 | Grand Prairie | Alberta | X | X | | | |
| 11 | Mc Dowell Creek | NC | X | X | | | X |
| 12 | Nansemond | VA | X | X | | | |

### 4.4.2 EBPR Process Mode

The individual EBPR process designs vary substantially. Table 4-5 lists the facilities and their EBPR process design.

**Table 4-5: Overview EBPR Processes.**

| # | Facilities | EBPR Process in Operation |
|---|---|---|
| 1 | Durham | $A^2/O$ |
| 2 | VIP | VIP |
| 4 | Nine Springs | $A^2/O$, Anoxic to Anaerobic Recycle |
| 5 | Kalispell | Modified UCT |
| 6 | Missoula | A/O |
| 7 | Reedy Creek | 5 Stage Bardenpho |
| 8 | Lower Reedy | VIP |
| 9 | Mauldin Road | $A^2/O$ |
| 10 | Iron Bridge | 5 Stage Bardenpho |
| 11 | Grand Prairie | A/O, Step Feed |
| 12 | McDowell Creek | Modified UCT |
| 13 | Nansemond | VIP |

### 4.4.3 Solids Stream

Table 4-6 summarizes the solids-processing facilities and solids-recycle management at each of the 13 plants.

Table 4-6. Overview of Solids Processing and Recycle Streams.

| # | Facilities | Solids Processing |
|---|---|---|
| 1 | Durham | ♦ WAS thickening with centrifuges<br>♦ Primary sludge pumped to static fermenter, where it is fermented and thickened sludge is remixed with fermenter overflow and resettled in downstream gravity thickener<br>♦ Thickened WAS and primary sludge pumped from combined feed tank to anaerobic digestion<br>♦ Digested sludge pumped to dewatering centrifuges from storage tanks<br>♦ Dewatering centrate stored for load equalization<br>♦ Centrate and fermenter overflow returned to primary effluent pumping station<br>♦ Thickening centrate returned to influent via plant drain |
| 2 | VIP | ♦ WAS directly sent to dewatering<br>♦ Primary sludge thickened in primary clarifier<br>♦ Dewatered primary sludge and WAS sent to incinerator |
| 4 | Nine Springs | ♦ WAS thickened in dissolved air flotation thickener<br>♦ WAS and primary sludge sent to anaerobic digestion<br>♦ Solids recycle returned to primary influent |
| 5 | Kalispell | ♦ WAS sent to dewatering<br>♦ Primary sludge sent to fermenter and fermented primary sludge sent to anaerobic digester<br>♦ Digested sludge blended with WAS before dewatering |
| 6 | Missoula | ♦ WAS and primary sludge sent to anaerobic digestion<br>♦ Dewatered sludge processed in composting facility |
| 7 | Reedy Creek | ♦ WAS thickened with gravity belt thickener<br>♦ Belt filter press dewatering<br>♦ Aerated static pile composting<br>♦ Solids recycle returned to primary influent |
| 8 | Lower Reedy | ♦ WAS thickening with gravity belt thickener and stored in mixed basin<br>♦ Primary sludge thickened in clarifier sent to anaerobic digester<br>♦ WAS sent to anaerobic digester<br>♦ Digested sludge dewatered with belt filter press |
| 9 | Mauldin Road | ♦ Primary sludge thickened in gravity thickener<br>♦ Return overflow to return to secondary treatment<br>♦ WAS thickened with gravity belt thickener<br>♦ WAS and primary sludge digested together<br>♦ Digested sludge thickened and liquid hauled |
| 10 | Iron Bridge | ♦ No primary sludge<br>♦ WAS sent to lime stabilization |
| 11 | Grand Prairie | ♦ WAS and primary sludge digested in aerobic digesters<br>♦ Dewatering centrate returned to primary effluent |
| 12 | Mc Dowell Creek | ♦ WAS thickened with gravity belt thickener<br>♦ Primary sludge thickened in clarifier<br>♦ Dewatering centrate returned continuously |
| 13 | Nansemond | ♦ Primary sludge thickened in primaries and sent to digester<br>♦ WAS thickened with gravity belt thickener and sent to digester<br>♦ Dewatering centrate returned to influent |

### 4.4.4 EBPR Design Aspects

Each facility has unique design features. Most plants have the option of operating in several different modes. This section presents some of the specific EBPR design features and how they are designed to influence the net phosphorus removal.

**Completely Mixed Fermenter—Kalispell**

Due to the geographical location in northern Montana and a relatively small collection system with short retention times, influent VFAs at Kalispell are lower than required for EBPR. The Kalispell plant operates a completely mixed fermenter to supplement influent VFA. According to operators, the fermenter operates year-round at an SRT of approximately 10 d.

The fermented primary sludge is settled out in a downstream gravity thickener. Some sludge is wasted to the digester, and the rest is recycled back to the fermenter.

**Reduced In-Service Volume During Summer—Kalispell, VIP**

These facilities both have significant excess capacity. To prevent secondary phosphorus release during the low flow summer season, cells can be taken offline as required.

**Non-Anaerobic Treatment of WAS—Kalispell, Iron Bridge, Reedy Creek, VIP**

By not anaerobically digesting EBPR WAS, the recycle phosphorus load is greatly reduced. This increases the influent readily available COD to phosphorus ratio for EBPR and also reduces the MLSS phosphorus content, thus reducing the impact of effluent TSS on effluent total phosphorus.

**Dewatering Centrate Storage—Durham**

Dewatering centrate is stored to allow load equalization over a seven-day period. Dewatering centrifuges are operated five days a week during the day shift. The centrate equalization is primarily driven by the need to limit the aeration demand from nitrification; however, it also contributes to enhanced EBPR performance and stability. The equalization limits the shock phosphorus loading and attenuates the high nitrate that results from oxidizing the centrate return ammonia load.

**Primary Effluent Lime Addition—Durham, Lower Reedy**

To supplement alkalinity, lime is added to the primary effluent at these facilities. Average lime addition at Lower Reedy ranged from 35 mg/L to 80 mg/L between 2001 and 2003. At Durham, lime addition is typically controlled to maintain a target concentration of 40-50 mg/L. The lime addition at Durham appears to contribute to the phosphorus removal through calcium phosphate precipitation in the anaerobic zone and in the anaerobic digester (Benisch, et al., 2001).

### 4.4.5 Chemical Feed

Most of the facilities with effluent phosphorus permit requirements have the ability to add phosphorus-precipitating chemicals in the event of EBPR failure or to improve EBPR performance by metal salt addition or RBCOD addition. Chemicals added for other types of process control can also impact phosphorus removal. Table 4-7 lists chemicals added for phosphorus removal and for other purposes.

Table 4-7. Overview of Chemical Addition Capabilities.

| # | Facilities | Chemical Addition |
|---|---|---|
| 1 | Durham | ♦ Tertiary alum addition for effluent polishing and EBPR supplementation (0.07 mg/L effluent TP limit). On average 15–20 ppm are added.<br>♦ Primary effluent lime addition to supplement alkalinity<br>♦ Ferric chloride addition for struvite control to dewatering centrate |
| 2 | VIP | ♦ Polymer, ferric on as-needed basis to secondary clarifier influent |
| 3 | Nine Springs | ♦ Ferric chloride addition to dewatering centrate |
| 4 | Kalispell | ♦ VFA from PSL Fermenter |
| 5 | Missoula | ♦ No chemical addition. |
| 6 | Reedy Creek | ♦ Ferric chloride addition to aeration basin and secondary clarifier influent |
| 7 | Lower Reedy | ♦ Lime for alkalinity |
| 8 | Mauldin Road | ♦ No chemical addition |
| 9 | Iron Bridge | ♦ Lime stabilization of WAS |
| 10 | Grand Prairie | ♦ No chemical addition |
| 11 | Mc Dowell Creek | ♦ Sugar waste for BOD supplement to activated sludge influent |
| 12 | Nansemond | ♦ Polymer, ferric on as-needed basis to secondary clarifier influent |

## 4.5 Field Investigation Results

### 4.5.1 Phosphate Profiles

No correlation between the level of anaerobic phosphorus release and effluent phosphate concentration was immediately obvious. Figures 4-2 and 4-3 show the flow-corrected phosphate profiles (i.e., concentrations are calculated based on influent flow). The Kalispell and Nansemond facilities stand out with respect to total release. Kalispell operated a high-yield completely mixed fermenter year-round, which likely contributes to high release due to the high VFA supply from the fermenter. At Nansemond, other than the high primary effluent phosphorus load, there were no obvious reasons for the unusually high P release. More detailed analysis of the Nansemond facility can be found in Chapter 6.0 and Appendix E.

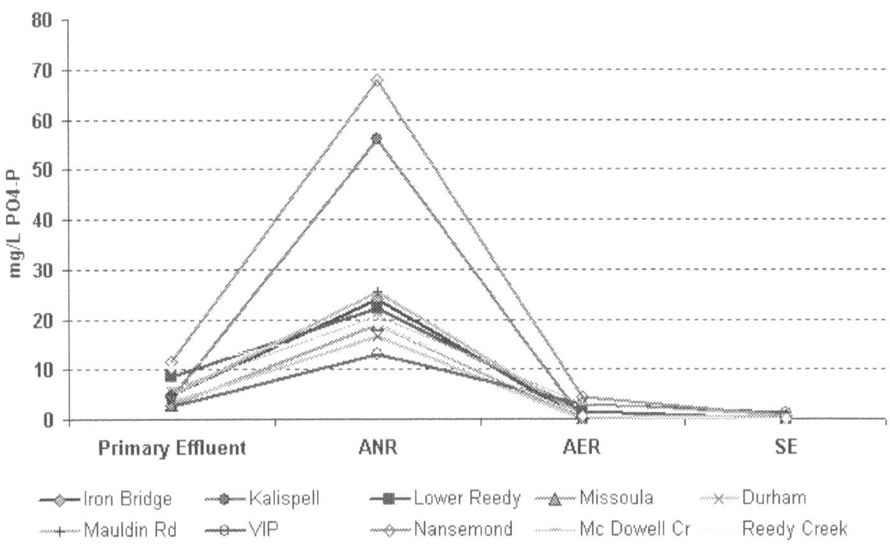


Figure 4-2. Flow-Corrected Phosphate Profiles.

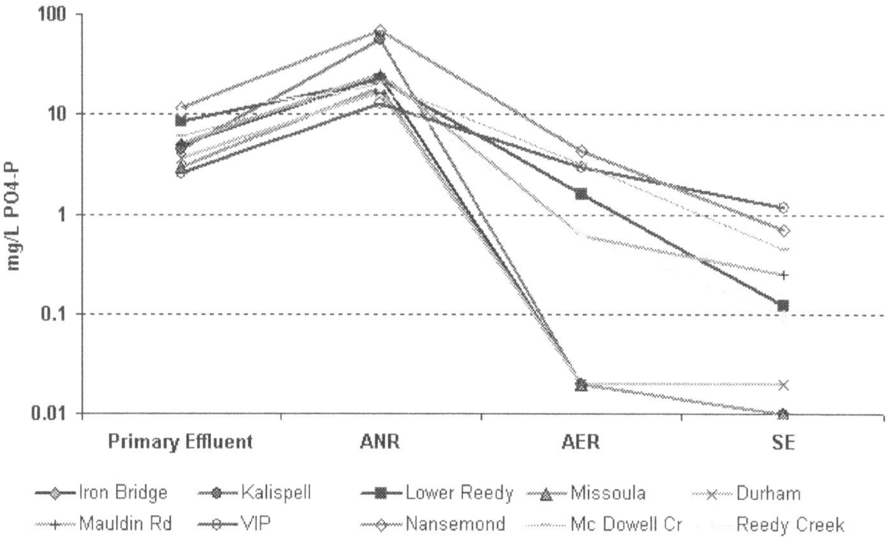


**Figure 4-3. Flow-Corrected Phosphate Profiles – Log Scale.**

### 4.5.2 Modeling Results

A steady-state simulation was performed on each of the models, and the output of the model was compared to the measured concentrations. Appendices K through N give detailed summaries of the modeling results. The model correctly predicted the general trends in the ammonia and phosphorus profiles at all of the facilities, even when the default kinetic parameters were used. Modifying the ratio of phosphorus released to COD consumed, based on the GAO testing, resulted in a modeled profile that better matched the measured results. Some influent characteristics required for model calibration were also estimated based on available data.

Generally the model closely predicts facility performance when influent fractions are well-known and the model is calibrated to yield, SRT, MLSS, and VSS/TSS. Very low effluent concentration or dynamic (diurnal) simulation is much more difficult to calibrate and requires far more data than was gathered during the Phase 1 research. Table 4-8 compares simulated phosphate and ammonia concentrations from the field testing and model simulations.

Table 4-8. Modeling and Field Testing Comparison.

| Sample Location | Phosphate, mg/L Measured | Phosphate, mg/L Modeled | Ammonia, mg/L Measured | Ammonia, mg/L Modeled |
|---|---|---|---|---|
| **Nansemond** | | | | |
| Primary Effluent | 15.2 | 14.7 | 47.4 | 46.5 |
| Anaerobic Zone | 40.4 | 44.8 | 29.3 | 32.0 |
| Anoxic Zone | 20.6 | 26.6 | 14.9 | 15.7 |
| Aerobic Zone | 0.1 | 0.0 | 0.3 | 0.1 |
| Secondary Effluent | 0.1 | 0.0 | 0.4 | 0.1 |
| **VIP** | | | | |
| Primary Effluent | 4.0 | 4.4 | 26.5 | 28.4 |
| Anaerobic Zone | 9.5 | 8.3 | 22.0 | 27.5 |
| Anoxic Zone | 11.6 | 12.6 | 19.8 | 23.0 |
| Aerobic Zone | 1.4 | 1.6 | 11.8 | 13.6 |
| Secondary Effluent | 1.7 | 1.6 | 11.3 | 13.6 |
| **Lower Reedy** | | | | |
| Primary Effluent | 7.6 | 8.0 | 29.8 | 30.0 |
| Anaerobic Zone | 18.5 | 17.6 | 17.4 | 18.8 |
| Anoxic Zone | 7.1 | 7.9 | 2.9 | 7.7 |
| Aerobic Zone | 0.3 | 0.4 | 0.6 | 0.2 |
| Secondary Effluent | 0.1 | 0.4 | 0.5 | 0.2 |
| **McDowell Creek** | | | | |
| Primary Effluent | 8.3 | 8.0 | 34.2 | 33.8 |
| Anaerobic Zone | 24.8 | 23.8 | 19.1 | 20.1 |
| Anoxic Zone | 15.3 | 13.8 | 6.8 | 8.2 |
| Aerobic Zone | 0.9 | 0.1 | 0.3 | 0.1 |
| Secondary Effluent | 0.4 | 0.1 | 0.3 | 0.1 |

CHAPTER 5.0

# GAO LITERATURE REVIEW

## 5.1 Introduction

EBPR has been applied in full-scale, activated-sludge facilities around the world for over 25 years. The process requires an anaerobic zone in which the mixed liquor is contacted with influent wastewater or other carbon source, followed by an aerobic zone. Characteristics of a successful EBPR process are acetate uptake and soluble COD removal, phosphorus release, an increase in cellular carbon storage material as PHAs, and a decrease in cellular polyphosphate and glycogen content in the anaerobic phase. In the aerobic phase phosphate is assimilated to form intracellular inorganic polyphosphate storage products, and the PHA content is decreased for metabolic intermediates, some conversion to glycogen, and oxidation. PHAs (commonly present as PHB) can be produced from acetate or propionate in the anaerobic zone but require energy for substrate uptake and conversion to acetyl-CoA, as well as reducing power for PHA formation. Comeau (1989) proposed that the reducing power is supplied through acetate assimilation and an incomplete tricarboxylic acid (TCA) cycle, but the more commonly accepted mechanism is that cellular degradation of glycogen through glycolysis (Mino, et al., 1994 and 1998) provides the necessary reducing power. Energy is obtained in the anaerobic phase by the PAOs through the release of phosphorus in the stored polyphosphate.

Other biological processes exist where organic substrates can be removed in the anaerobic zone without phosphorus release. Anaerobic/aerobic processes have been operated in which acetate and other soluble substrates have been removed in the anaerobic phase without phosphorus release but with PHA formation and without polyphosphate storage and enhanced phosphorus removal in the aerobic phase. This type of metabolism was first reported by Cech and Hartman (1990, 1993) as a result of laboratory studies on EBPR, and they used the term *G bacteria* to identify bacteria involved in anaerobic substrate uptake different from PAOs. In their study G bacteria outcompeted PAOs in an SBR, and PHA was formed in the anaerobic phase without the use of polyphosphates for energy. They noted that these bacteria could use glycogen instead of polyphosphate as the energy source in the anaerobic phase. They also noted a unique morphology for the G bacteria visible under a phase contrast light microscope that consisted of large cocci (2–3 μm) arranged in clusters of four (tetrads). Additionally, they were found to be gram-negative bacteria with Neisser-positive-stained cell walls, but not Neisser-positive cell inclusions as for PAOs. Further work by Satoh, et al. (1992) and Liu, et al. (1996) confirmed that non-phosphorus storing bacteria could outcompete PAOs in anaerobic/aerobic systems and referred to them as glycogen accumulating bacteria (GAB). Since then, these bacteria have more commonly been termed glycogen accumulating organisms (GAOs).

In bench-scale studies with SBRs, Carucci, et al. (1999) found a different mode of substrate uptake and storage by non-PAOs in an anaerobic/aerobic process. The SBRs were maintained at 20°C and the SRT was maintained at 8–10 d. The anaerobic and aerobic periods per cycle were 1.25 and 4.0 hours, respectively. A series of test runs was made with different influent carbon sources consisting of glucose and peptone, glucose only, and glucose and acetate.

Enhanced phosphorus removal was found with the glucose/peptone, glucose/acetate, and in one of the runs with glucose only. Glycogen but not PHA storage increased in the anaerobic phase. The unique aspect of this work was demonstrating glycogen storage with substrate uptake in the anaerobic phase, but no explanation was given for why enhanced phosphorus removal and PHA storage occurred with glucose feed in another run. This work may have little relevance to wastewater treatment because glucose is unlikely to be present in wastewater while acetate is produced by fermentation from readily biodegradable COD.

Following the discovery of GAOs in laboratory studies on EBPR processes, subsequent research was pursued by others with the objectives of determining factors that favored GAOs over PAOs in anaerobic/aerobic systems, finding methods to identify GAOs in activated-sludge mixed liquor, and making modifications to the International Water Association (IWA) activated-sludge model to include the GAO population, metabolism, and kinetics. Findings on these areas are summarized in this literature review.

## 5.2 Competition Between GAOs and PAOs

### 5.2.1 Effect of Substrate

Studies on factors that affect competition between GAOs and PAOs have been done only in laboratory bench-scale reactors with synthetic feed, and most of these have used SBRs instead of continuous flow reactors. Factors found to affect GAO versus PAO dominance are substrate type, pH, temperature, and SRT.

The main apparent experimental variable in Cech and Hartman's (1990) work leading to the discovery of the *G* bacteria was the feed composition. Two SBRs were operated in parallel at 20°C and an SRT of 18.5 d. The one that showed EBPR was fed acetic acid, and the other that had no EBPR but developed *G* bacteria was fed acetate and glucose at a COD ratio of about 1:1 in mg acetate/mg glucose as COD. Only two feed cycles were used per day with the times for the anaerobic and aerobic periods at 2 and 9 hours, respectively. Other differences observed between the reactors from the data presented showed a high sludge volume index (SVI) (310 mL/g) for the SBR with PAOs versus a low SVI (90 mL/g) for the SBR with GAOs and a lower pH range for the GAO SBR (6.8 versus 7.2).

Four laboratory SBRs were operated with acetate or glucose feed and with or without phosphorus limitation by Sudiana, et al. (1998). The SRT was set at 10 days, and the anaerobic and aerobic periods were 65 and 120 minutes per cycle, respectively. The temperature for the study was not given. With sufficient phosphorus (COD:P = 27), PAOs dominated, whether the feed was glucose or acetate. For the phosphorus limited SBRs (COD:P = 133) the GAOs dominated and were characterized by tetrad-shaped cells, as reported by Cech and Hartman (1990, 1993). Wang, et al. (2000) also showed that PAOs could be maintained in SBRs operated at 20°C with a 20-d SRT when fed either glucose or acetate. No data on reactor pH values were given. They also developed GAOs by reducing the feed phosphorus concentration to create phosphorus-limited conditions.

Other researchers have tried to sustain a PAO population by adding other complex substrates in addition to acetate in laboratory studies. Bond, et al. (1999) found that biological phosphorus-removal efficiency increased significantly when peptone was added to their acetate feed.

Liu, et al. (1997) varied the P/C ratio for SBRs fed acetate and peptone and operated at 20°C with a 7.5–8.0 d SRT. The anaerobic and aerobic times were 50 and 80 minutes per cycle, respectively, and the pH was controlled between seven and eight. Under phosphorus limiting conditions (P/C = 0.02) GAOs dominated, but with excess phosphorus (P/C = 0.2), PAOs dominated. At an intermediate level (P/C = 0.1), both types were claimed to be present, but it is not certain that the reactor was operated long enough for the population to stabilize after the feed operating change. They also noted that as the proportion of PAOs increased, the molar ratio of phosphorus released to acetate consumed increased from 0.7–1.3 (0.4–0.7 mg P/mg acetate). A review of literature values showed many reported molar P:acetate ratios in the range of 0.8–0.9 (0.46–0.51 mg P:mg acetate). They also suggested that low cycle frequency in SBR operations favored GAOs, as the PAOs likely lacked the capacity for uptake of all the acetate fed per cycle, leaving acetate available to develop the GAO population. They also speculated that lower SRTs favored PAO dominance.

### 5.2.2 Effect of pH

Filipe, et al. (2001a) showed the importance of pH in maintaining PAOs in the EBPR processes. The effect of pH was studied with laboratory scale SBRs operated at 22°C and maintained at a 7.0-d SRT. The anaerobic and aerobic periods per cycle were both at 2.25 hours. The feed consisted of sodium acetate (400 mg/L as COD), yeast extract (5 mg/L), and peptone (5 mg/L). Nitrification was inhibited by allythiourea. The work showed that GAOs consumed acetate faster than PAOs in the anaerobic zone when the pH is less than 7.25. PAOs removed acetate faster at pH values above 7.5. A stable system dominated by PAOs occurred when the operating pH was maintained above 7.25 at all times, according to the authors. Filipe, et al. (2001*b*) also showed in another similar study that if the aerobic pH was below 7.0, PAOs were less competitive against GAOs.

Other work has supported the advantage of high pH favoring PAOs over GAOs in SBR processes for EBPR. Bond, et al. (1999) reported that in a laboratory SBR study, the dominance of GAOs changed to PAOs when the pH in the anaerobic zone was increased from 7.0 to 8.5. The SBRs were operated at 22°C with a 7–8 d SRT. Serafim, et al. (2002) found much greater phosphorus-removal performance in a laboratory SBR system at a pH of 7.8 to 8.5 versus 7.0. The SBRs were fed a mixture of acetate, propionate, and butyrate and were operated at 22°C with a 10-d SRT.

### 5.2.3 Effect of Temperature and Solids Retention Time

Whang and Park (2001 and 2002) studied the effect of temperature at 20°C and 30°C on the dominance of PAOs or GAOs in laboratory SBRs. The operating SRTs ranged from 3–10 d in these experiments, and the anaerobic and aerobic period times were 1.5 and 3.7 hours, respectively. The synthetic feed organic source consisted of sodium acetate (380 mg/L), peptone (35 mg/L), and yeast extract (6 mg/L). The pH was controlled at 7.0, and nitrification was inhibited by the addition of allythiourea. Competition between PAOs and GAOs was judged based on the phosphorus-removal efficiency, phosphorus release in the anaerobic phase, and microscopic observations of tetrad-arranged coccoid bacteria cells. At 20°C and a 10-d SRT PAOs dominated, and the COD:P removed ratio was 10 mg/mg. The dominance was explained by a higher specific anaerobic acetate uptake and higher aerobic biomass yield for the PAOs. At 30°C and a 10-d SRT the GAOs dominated and the authors claimed that this was due to a higher specific anaerobic acetate uptake at this temperature and SRT by the GAOs. Even when the SRT was decreased to 5 d and the temperature decreased to 20°C, the PAOs could not overtake the GAOs over a 20-d period for each change. With a reseeding and operation at 30°C and a 5-d SRT, the PAOs could not establish a stable operation, but at a 3-d SRT they were able to dominate. The authors claimed that the SRT greatly affected the specific acetate uptake rate in the anaerobic phase for PAOs with improving rates at lower SRTs and improved specific acetate uptake rates for GAOs at 30°C. Using the results on acetate uptake rate, biomass yields, and PHA and glycogen storage from these experiments, the IWA activated-sludge model was modified to include GAOs as well as PAOs (Whang, et al., 2002). However, the model was not verified by application and calibration to other anaerobic/aerobic activated-sludge operating results.

Brdjanovic, et al. (1998) also reported greater phosphorus-removal efficiency at lower temperatures but not dominance by GAOs at 30°C, as seen by Whang and Park (2001). However, for their 30°C experiment, the 20°C mixed liquor with a high phosphorus-removal capability was used, and there may not have been sufficient time for the GAOs to develop. Barnard, et al. (1985) had also noted greater phosphorus-removal efficiency at lower temperatures based on full-scale plant observations over a range of 5°C to 24°C. Erdal, et al. (2002) compared biological phosphorus-removal performance at 5°C and 20°C in a laboratory University of Cape Town (UCT) process with anaerobic, anoxic, and aerobic reactors operated at 10-d and 18-d SRTs, respectively. This is one of the few laboratory studies done without SBRs. Their organic feed content consisted of acetate (450 mg/L as COD) and yeast extract (50 mg/L as COD). The pH was maintained between 7.1 and 7.6. Much greater phosphorus-removal efficiency occurred at 5°C with the phosphorus content of the mixed liquor volatile suspended solids (MLVSS) increasing over time from 14% to 37% as the temperature was changed from 20°C to 5°C. Based on microscopic observations, they could not attribute the difference in performance to competition between GAOs and PAOs, as they found no evidence of GAOs at either temperature. They did observe less glycogen storage and greater PHA storage at the lower temperature and suggested that the lower temperature caused either a change in the metabolic activity or selected for a different phosphorus-accumulating bacteria.

The effect of a lower SRT may explain the lack of GAOs and little uptake of RBCOD in the anaerobic zone of an anaerobic/aerobic full-scale plant recently evaluated by this author (H. David Stensel). An anaerobic/aerobic process is employed at the Jones Island Milwaukee, WI, wastewater treatment facility. Long narrow channels with detention times ranging from 20–90 minutes contact influent wastewater and return activated sludge under anaerobic conditions and distribute the feed to a number of long narrow aeration tanks. The primary effluent has a surprisingly high BOD (about 240 mg/L) and low phosphorus content (about 2.5 mg/L), and after cell synthesis there is little phosphorus left to support the development of PAOs. The RBCOD concentration in the influent wastewater has been measured in the range of 150–200 mg/L on three different plant intensive sampling studies between September and November 2002. Only about 10–20% of this has been removed in the anaerobic feed channel, and phosphorus release is minimal (2–4 mg/L) suggesting little PAO activity. However, the lack of RBCOD removal also suggests very little GAO activity. The system SRT is in the range of 4–5 d at mixed-liquor temperatures that range from 12–20°C. The GAOs have been noted to be slow-growing bacteria and unable to compete well with PAOs at low SRTs, so one explanation for these observations at the Jones Island facility is that the GAOs are washed out at the relatively low operating SRT.

## 5.3 Monitoring for GAOs in Anaerobic/Aerobic Activated-Sludge Processes

The ability to identify specific bacteria in an activated-sludge mixed liquor is challenging and limited mainly to the use of modern molecular tools, such as fluorescence in-situ hybridization (FISH) techniques. This method relies on the fact that each bacterium has a specific DNA sequence that can be identified through RNA extraction and analysis. Based on progress in this area, bacteria are now named and categorized according to their phylogenetic characteristics. Identifying a particular population based on morphology and staining methods had been done, especially for the identification of filamentous bacteria (Eikelboom, 2000), but requires the bacteria of interest to have unique physical characteristics. For most bacteria this is impossible due to the limited size and shapes of bacteria.

Fortuitously, the *G* bacteria noted in Cech and Hartman's work (1993) were associated with a unique morphology in the mixed liquor; the tetrad-arranged, large-size coccoid bacteria. An example of this unique morphology is shown in Figure 5-1, and numerous investigators for poor performing EBPR processes have reported this type of observation where dominance by GAOs has been suspected. Maszenan, et al. (1998) reported on isolation of the tetrad-arranged GAOs and recognized a new genus, *Amaricoccus*, with four species named after their place of isolation: *A. Kaplicensis*, *A. veronensis*, *A. macaueansis*, and *A. tamworcoccus*. A molecular probe specific to *Amaricoccus* has been developed and used to identify it as a commonly occurring bacteria in both conventional and EBRP activated-sludge systems (Seviour, et al. 2000).

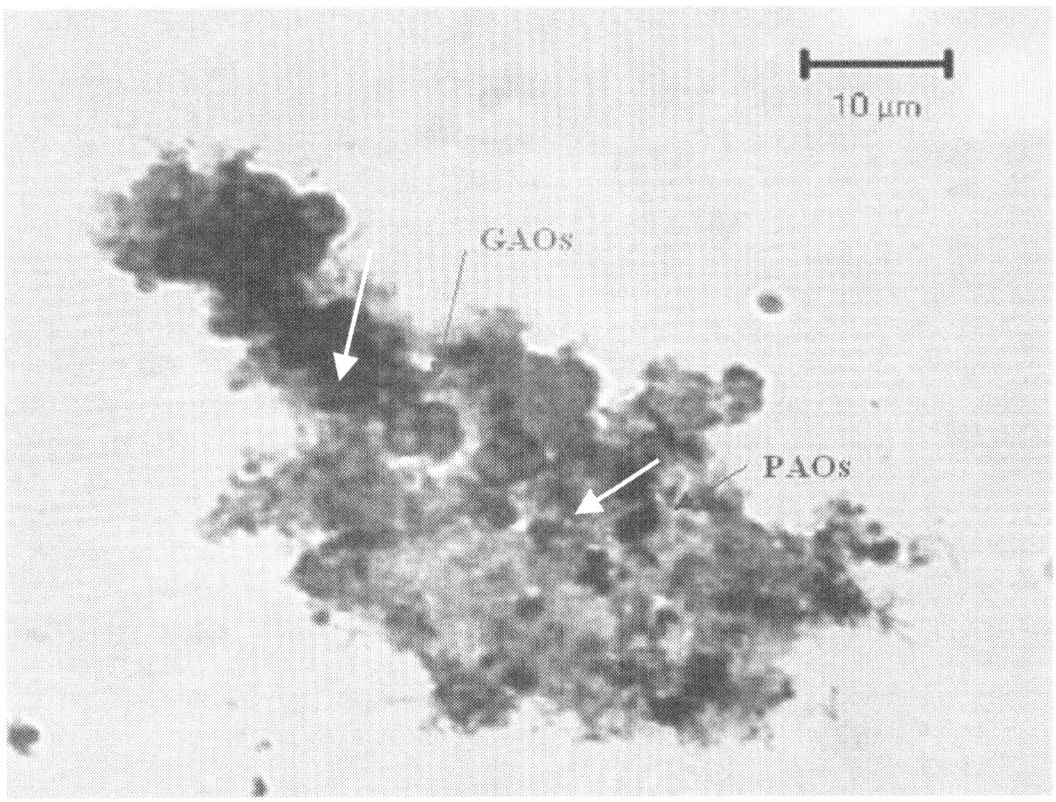


**Figure 5-1. Tetrad Arranged Coccoid Cells Associated with G Bacteria. (Durham WWTP, Train 2, August 26, 2003).**

A review by Seviour, et al. (2000) shows that a number of other bacteria can be classified as *G* bacteria based on their ability to take up substrate under anaerobic conditions and having similar morphology and similar Niesser-stain characteristics. These bacteria also had a common trait of being slow-growing microorganisms. Many of the bacteria found were not phylogenetically close to *Amaricoccus* and in general a wide diversity based on phylogenetic characteristics was found.

Bond, et al. (1999) report on efforts to use molecular probes to identify, in general, PAO and GAO bacteria under high and low biological phosphorus removal operations. With high biological phosphorus removal in a laboratory SBR, molecular probes revealed that the dominant populations were bacteria of the β-2 subclass of the phylogenetic classes *Proteobacteria* and *Actinobacteria* (55% and 35%, respectively). With poor biological phosphorus removal but uptake of acetate under anaerobic conditions, the dominant bacteria (55%) determined by FISH probing were of the β class of the class *Proteobacteria* other than β-1 and β-2. A unique finding of this work is that they did not find any tetrad-arranged coccoid bacteria in the poor phosphorus-removal system as noted in previous work. A probe for the tetrad-cluster coccoid bacteria contained in the class α-*Proteobacteria* was used, and negative results were found. Thus, these results show that GAOs may involve other bacteria than the tretrad arranged coccoid cells and that molecular probing provides a broader means of monitoring their existence. In some circles, the term "G" bacteria has now been applied to signify coccoid bacteria arranged in the tetrad formation capable of PHA formation under anaerobic conditions, while a broader range of bacteria capable of PHA formation in the anaerobic phase, without the ability for enhanced phosphorus removal, and identified phylogenetically, are referred to as GAOs.

## 5.4 Presence Of GAOs In Full-Scale Anaerobic/Aerobic Activated-sludge Plants

Though some recent work is expected to be reported soon in the referenced literature, there are presently minimal reports on efforts to determine if GAOs are present in full-scale facilities at a level having a significant affect on phosphorus-removal performance. Wanner, et al. (2000) have noted that competition between PAOs and GAOs is less common than would be expected from previous laboratory studies. Griffiths, et al. (2002) report on an attempt to evaluate the presence of GAOs at five different anaerobic/aerobic wastewater treatment plants in Australia designed to accomplish enhanced biological phosphorus removal. The basis for determining the presence of GAOs was poor phosphorus removal, results of staining for polyphosphate deposits in aerobic cultures, and morphological observations. The authors also claimed to judge the relative abundance of PAOs by staining for the presence of PHAs in the anaerobic phase cultures, but PHAs can be found for both GAOs and PAOs. The evaluation was not done in a systematic fashion, and much information that would be needed for a more convincing assessment, such as a sufficient level of wastewater characterization, pH, temperature, phosphorus release concentrations in the anaerobic zone, and percent phosphorus in the waste sludge, was lacking. Nevertheless the authors claimed to observe *G* bacteria in certain plants and put forth two concepts that favored the growth of GAOs. The first related to operating DO concentrations. They claimed that for initial operation of an EBPR plant, the aerobic DO concentration should be near 3.0 mg/L, based on observing improved biological phosphorus removal and a shift away from GAOs when the DO concentration was increased in two of the plants studied. They also suggested, based on plant performance results, that operating at an excessive DO concentration favored the GAOs (DO concentrations in the range of 4-5 mg/L). They also noted that for an underloaded plant, the higher DO concentration was especially conducive to the development of *G* bacteria. For two of the facilities it was necessary to add acetate to provide additional electron donor for EBPR. In one case the amount of acetate added and removed was at least three times the stoichiometric amount for biological phosphorus removal, and they concluded that the excess acetate encouraged the growth of *G* bacteria. Though these results did not provide a convincing reason for the presence of GAOs, they did show that GAOs do exist in full-scale EBPR systems, and when they are present, reduced biological phosphorus removal occurs.

## 5.5 Summary and Discussion

Results from laboratory studies and limited full-scale plant reports clearly establish the fact that operation of an anaerobic/aerobic activated-sludge process does not guarantee that the system will only select for PAOs for removal of soluble substrates in the anaerobic zone. A wide range of diverse bacteria can compete for substrate in the anaerobic zone and may outcompete the PAOs to seriously impair EBPR. Many of these mimic the PAOs in that they can store PHAs from substrate uptake in the anaerobic zone and produce glycogen in the aerobic zone. In contrast to the PAOs, they don't depend on polyphosphates for energy storage and thus do not release phosphorus in the anaerobic zone. Many of these are relatively slow growing bacteria, exhibit a tretrad arrangement of coccoid bacteria, and have been referred to as *G* bacteria, or in some cases GAO, which signifies their greater glycogen storage ability compared to PAOs. Other bacteria without the unique tetrad formation can also exist and compete for substrate with the PAOs. These have been identified by molecular probing methods using FISH techniques and belong to different phylogenetic grouping from the classic *G* bacteria that were first identified as affecting EBPR performance. These are referred to as GAOs but not *G* bacteria and have been monitored in activated sludge by molecular probes for the β class of the class *Proteobacteria* other than β-1 and β-2.

The presence of GAOs, whether the tetrad-arranged classic *G* bacteria or other of the *Proteobacteria*, can be ascertained by plant performance and confirmed by microscopic and/or molecular analyses. The presence of GAOs is suggested by the following:

1. SCOD or acetate removal in an anaerobic contact zone with little or no phosphorus release. For PAO-dominated populations about 0.5 mg of phosphorus release per mg of acetate uptake can be expected. Lower values suggest the presence of GAOs.
2. Poor phosphorus-removal performance in terms of higher effluent P concentration, in spite of having a favorable influent BOD:P ratio.
3. A high ratio of influent RBCOD concentration to P removal concentration (greater than 20:1).
4. Identification of a significant quantity of tetrad-arranged coccoid cells or by a significant response using molecular probes for *G* bacteria or for GAOs other than *G* bacteria. The tetrad-arranged cells are also gram negative and show Neisser staining of the cell walls.

Studies to determine factors that select for GAOs over PAOs have been limited to mainly bench-scale work with SBRs and simple substrates, in many cases only acetate or glucose. It should be noted that the initial soluble substrate or acetate concentrations are much higher in SBRs than can be expected for more conventional continuous-flow processes commonly found in the field. In addition, more complex compounds and other bacteria consortia are present in real world applications of EBPR processes. It should also be noted that in some studies designed to evaluate a single parameter, such as SRT or pH, other factors affecting PAO selection may also have affected the results, such as substrate complexity or temperature. However, some general guidelines concerning factors that favor GAOs over PAOs can be suggested from the research findings for SBR systems. These are as follows and should be evaluated for non-SBR systems in which the substrate gradient in the anaerobic zone is different.

1. Lower SRTs, certainly at 3–5 d, and possibly below 10 d, favor PAOs with proper pH and feed conditions.

2. Higher temperatures (greater than 20°C) and longer SRTs favor GAO dominance.
3. Higher pH (above 7.2 in both the anaerobic and aerobic zones) favor PAOs.
4. Lower temperatures, especially at 10°C and below, favor PAOs and improve phosphorus-removal performance.
5. More complex feed substrates, for example tests with yeast extract and peptone in addition to acetate, favor PAOs.
6. Once PAOs or GAOs are established in a given reactor, long times are needed to observe population changes, so that transient tests are inconclusive.
7. Lower DO concentrations (below 3 mg/L) may have an effect on the competitiveness of PAOs in full-scale facilities.
8. High COD:P ratios may favor GAOs.

The research reviewed did not evaluate transient loading conditions to any degree, such as periodic high or low acetate availability or long periods without feed, but these factors are also known to affect PAO performance and possibly competitiveness with GAOs.

While much can still be learned by additional laboratory studies on PAO versus GAO kinetics and metabolism, more useful information would likely be gained by evaluating full-scale facilities in terms of wastewater characteristics including design configuration; operating conditions such as pH, temperature, DO concentration, and SRT; and the populations developed that are capable of substrate removal in the anaerobic zone. At issue is how to most effectively use the influent degradable COD for biological phosphorus removal.

# CHAPTER 6.0

# YEAR 2: DETAILED TESTING

## 6.1 Introduction

The investigations during the first year of the study showed that plants with a history of EBPR operation exhibit a wide range of performance and effluent phosphorus quality. Average effluent total phosphorus concentrations were relatively consistent (between 0.3 and 0.8 mg/L) for most facilities. However, while some plants maintained effluent total phosphorus concentrations of under 0.5 mg/L more than 95% of the time, others experienced periodic decreases in performance resulting in concentrations as high as almost 4 mg/L for 5% of the time. Therefore, the initial analysis showed that more detailed study was required to evaluate biological phosphorus removal efficiency, reliability, and capacity. The objective during Year 2 of the investigation was to identify a limited number of facilities and update the dataset collected during Year 1 to further characterize EBPR performance.

Additional field visits and field sampling were conducted to assess the plant performance and process behavior. Six facilities (Durham, Lower Reedy, Mauldin Road, McDowell Creek, Nansemond, and VIP) were selected for detailed analysis based on the analysis and field studies conducted during the first year of the study, the range of operating conditions represented, availability of routine process and water quality data, and other unique characteristics. Intense sampling was conducted for a two-week period at each facility. (Though the Mauldin Road facility was initially included in field testing, it lacked adequate historical data to complete a full analysis.) Where possible, this testing was conducted on two separate occasions. The sampling results were augmented with additional bench-scale testing and microscopic analysis. McDowell Creek was forced to stop the intensive sampling prior to completion due to local budget constraints.

The biological phosphorus removal process was viewed in terms of our fundamental understanding of the process. General process behavior and performance were evaluated using the daily routine operating data. Specific relationships to fundamental parameters were evaluated based on the intense sampling period and special tests.

## 6.2 Goals and Objectives

The goals and objectives of Year 2 are described below.

**Determine process reliability**. EBPR process performance is a function of process design, influent wastewater characteristics, operation, and other factors. This evaluation looked at the overall performance of the process in terms of its ability to consistently produce effluent phosphorus concentrations at or below a given treatment goal. Reliability in the context of this study is considered the ability of the process to meet the treatment target on a day-to-day basis.

**Determine the biological phosphorus removal efficiency**, both in terms of the percent removal and performance under different conditions and in terms of the removal efficiency as a function of variables such as influent BOD:P or VFA:P ratios. Removal efficiency could also be affected by process configuration, SRT, temperature, and chemical addition.

**Determine the biological phosphorus removal capacity**. How is the phosphorus-removal efficiency related to fundamental mechanisms understood for the EBPR process? What effluent concentration is achieved?

**Determine how chemical addition affects EBPR**. Most plants have chemical addition facilities to assure compliance with effluent phosphorus permit limits. How does chemical addition affect the biological phosphorus-removal efficiency and overall performance?

The results presented in Section 6.5 and the discussion in Section 6.6 describe which of these objectives were achieved during the study. The data to evaluate phosphorus removal was obtained at full-scale facilities under actual operating conditions with variable wastewater flows and concentrations. It was not therefore possible to control the fundamental parameters that affect EBPR performance or to collect all the needed samples to meet all of the above objectives.

## 6.3 Overview of Investigated Facilities

Key features of the five treatment plants that participated in the detailed investigation are summarized in Table 6-1. Table 6-2 summarizes the basic process parameters of the five facilities.

The anaerobic-anoxic-oxic (A2O), Virginia Initiative Plant (VIP), and modified University of Cape Town (UCT) processes all have a basic liquid treatment reactor sequence of anaerobic, anoxic, and aerobic treatment but differ in how mixed liquor is returned to the anaerobic zone. In the A2O process, return activated sludge from the secondary clarifier underflow is directed to the anaerobic contact zone, where it is mixed with the influent wastewater. Mixed liquor from the aerobic zone is recycled to the anoxic zone for nitrate removal. With the A2O design, some nitrate remains in the return activated sludge flow, thus requiring some VFA in the anaerobic contact zone for denitrification and leaving less available for the PAOs.

The VIP, UCT, and modified UCT (not included in the five plants) processes remove nitrate from the mixed liquor before it is returned to the anaerobic contact zone. In both the VIP and UCT processes the return activated sludge, along with the internal mixed-liquor recycle from the anaerobic zone is directed to the first anoxic zone. Subsequently mixed liquor with a nitrate concentration near zero recycled to the anaerobic contact zone. The UCT process may employ one or two anoxic zones in series with recycle to the anaerobic zone from the last anoxic zone. The VIP process differs from the UCT process in that multiple stages are used for both the anoxic and aerobic zones, and in addition the original intent of the VIP process, was to operate the system as a high rate system, preferably at minimal SRTs needed for nitrification and denitrification. In the modified UCT process the first anoxic zone receives only return activated sludge. Mixed liquor, presumably near zero nitrate concentration, is recycled from that zone to the anaerobic contact zone. Some modified UCT process designs may have additional anoxic zones after the zone from which mixed liquor is returned to the anaerobic contact zone. These added anoxic zones receive internal mixed-liquor recycle from the anaerobic zone for additional nitrate removal.

Table 6-1. Summary of Wastewater Treatment Plants Participating in Detailed Study.

| | Durham | VIP | Nansemond | McDowell Creek | Lower Reedy |
|---|---|---|---|---|---|
| State | Oregon | Virginia | Virginia | North Carolina | Florida |
| Process | A2O | VIP | VIP | Modified UCT | Modified UCT |
| Chemical addition | Fermenter, Lime (Aeration Basin Influent)<br>Alum (Primary Influent, Aeration Basin Effluent, Secondary Effluent) | Polymer to SCL influent (when settling problems exist) | Ferric (Aeration Basin Effluent) | Alum (Primary Influent) | Lime (Aeration Basin Influent)<br>Alum (Aeration Basin Effluent) |
| VFA Addition | PSL fermenter overflow | None | None | Sugar waste from cola manufacturer | None |
| Notable feature | Stable process with low effluent phosphorus | Stable process, with apparent annual upset in August | Struggled to achieve EBPR until 2003 | Add sugar waste. | Consistently meets permit limit. Long HRT system |
| Filter? | Yes | No | No | Yes | No |
| Permit Limit (mg/L P) | 0.07 (seasonal) | 2.0 | 2.0 | 1.0 | 2.0 |

Table 6-2. Process Overview (Average Conditions 2000–2003).

| Parameter | Unit | Durham | VIP | Nansemond | McDowell Creek | Lower Reedy |
|---|---|---|---|---|---|---|
| Design Capacity | mgd | 25 | 40 | 30 | 6 | 9 |
| Current Flow | mgd | 22 | 31 | 20 | 5 | 4 |
| Temperature | °C | 19 | 21 | 21 | 20 | 21 |
| HRT | Hours | 7.8 | 7.2 | 13 | 13 | 7.7 |
| SRT (total) | days | 10.9 | 9.5 | 8.5 | 6.7 | 9.4 |
| Anaerobic Volume | % | 9 | 12 | 15 | 7 | 13 |
| Anoxic Volume | % | 23 | 25 | 30 | 20 | 20 |
| Aerobic Volume | % | 68 | 63 | 55 | 73 | 67 |
| RAS | % | 61 | 31 | 84 | 87 | 52 |
| MLSS | mg/L | 3,530 | 3,100 | 3,520 | 3,400 | 3,200 |

## 6.4 Facility Process Design

### 6.4.1 Durham Advanced Wastewater Treatment Plant

#### 6.4.1.1 Liquid Stream

The BNR process follows screening, grit removal, and primary clarification. The current BNR process is shown in Figure 6-1. PE is mixed with the fermenter overflow in the PE pump station wetwell. Lime is also added to the PE wetwell as necessary for alkalinity control. PE is distributed to each of the four secondary treatment trains as shown in Figure 6-2. In the first anaerobic zone, PE and return activated sludge are mixed. The mixed-liquor recycle is introduced into the first anoxic zone. Table 6-3 contains a summary of design secondary treatment parameters and their current average values.

Table 6-3. Key Process Parameters.

| Component | Unit | Design[1] | Average |
|---|---|---|---|
| Anaerobic HRT | hrs | 0.6 | 0.7 |
| Anoxic HRT | hrs | 1.0 | 1.8 |
| Aerobic HRT | hrs | 4.4 | 5.3 |
| Total HRT | hrs | 6.0 | 7.8 |
| RAS Rate | % | 90% | 61% |
| ML Recycle Rate | % | 220% | 300% |
| SRT | days | 10 | 10.9 |
| MLSS | mg/L | 4500 | 3525 |
| Solids Recycle Return | | To Primary Effluent Pump Station | |
| Flow | MGD | 34 MGD | 23.6 |

Based on dry weather maximum month design capacity

The four trains were designed as independent trains with their own clarifier and sludge return system and thus can be operated with separate SRTs. Operators have the option to seed one train from another. SE is mixed and sent to the tertiary treatment process where chemical phosphorus polishing is accomplished by alum addition. Polymer can also be added to the chemical clarifiers for improved solids capture in the clarifiers and downstream dual media filters. The final effluent is discharged into the Tualatin River.

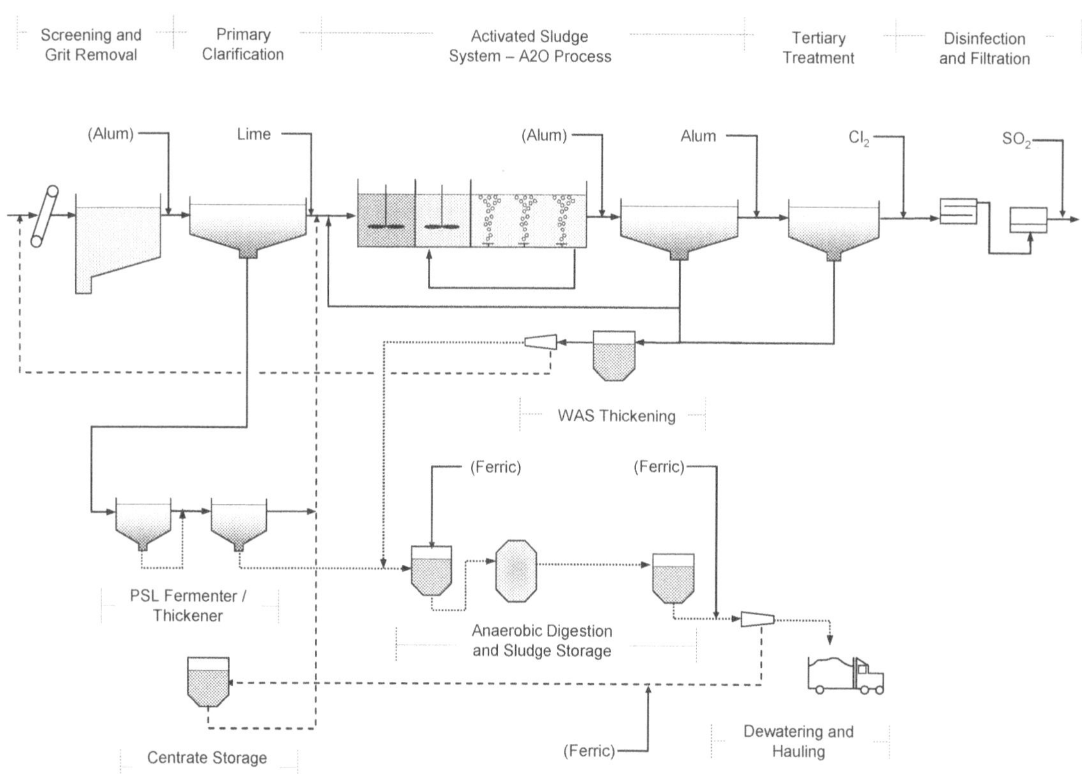


Figure 6-1. Process Schematic for Durham AWWTP Liquid and Solids Treatment.

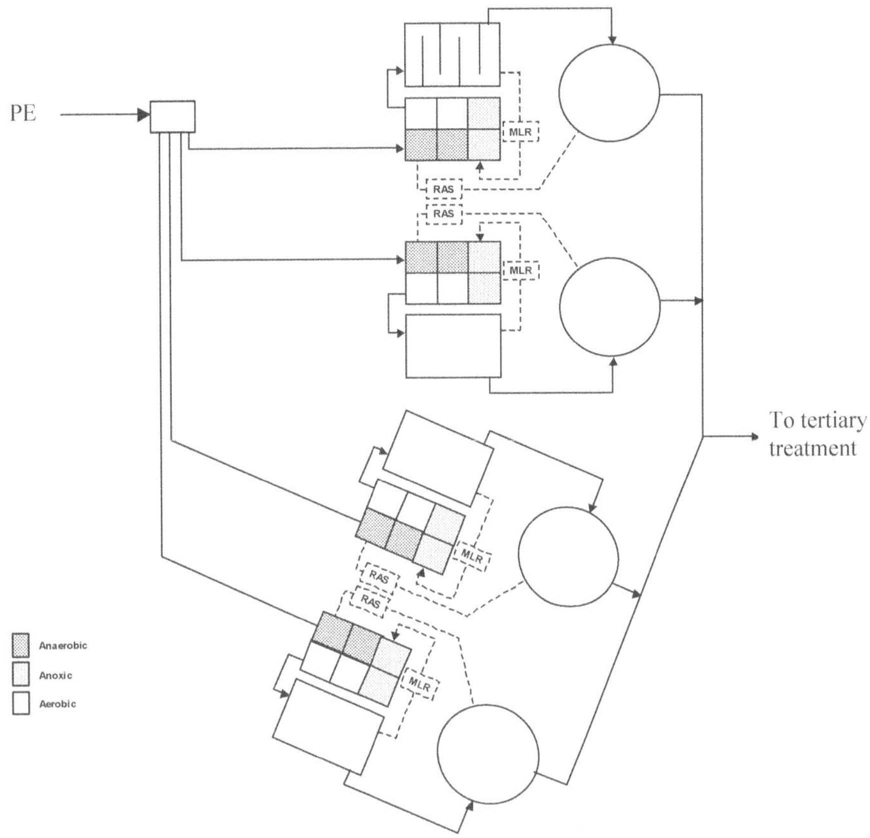


**Figure 6-2. Secondary Treatment Layout at Durham AWWTP.**

### 6.4.1.2 Solids Stream

All biosolids at Durham are stabilized by means of anaerobic digestion. Prior to digestion, PSL is retained and thickened in a two-stage fermenter/thickener. PSL is fermented for one to two days to generate additional VFAs. Underflow and overflow of the fermenter are remixed to elutriate the generated VFA and strip out trapped gas bubbles; the solids are removed and thickened in the downstream gravity thickener to approximately seven percent solids. WAS and chemical sludge (CSL) are mixed and sent to an aerated storage tank from which the thickening centrifuges are fed. Both thickened primary and WAS/CSL are mixed in a digester feed tank before being fed to the digester. Currently Durham operates four digesters in parallel, with two downstream digested sludge storage tanks.

Digested sludge is dewatered with a high solids centrifuge. Operators have the option to add ferric upstream or downstream of the centrifuge for struvite control, as well as to the digester feed tank for odor control. The dewatered cake is land applied. Table 6-4 summarizes the solids treatment at Durham.

Table 6-4. Solids Processing Summary.

| Component | Technology | Comments |
|---|---|---|
| PSL Thickening | Gravity Thickener | 7% TSS, if not monitored closely TSS can exceed 10% and generate conveyance problems |
| WAS Thickening | Thickening centrifuge | WAS and chemical sludge mixed prior to thickening |
| PSL and WAS Stabilization | Anaerobic Digestion | TS: 3.1%<br>SRT: 20-30 days<br>VSS destruction: 48% |
| Dewatering | Centrifuge | Operates 8 hours/day, 5 days/week |
| Centrate Storage | Yes | 7 day recycle load equalization capacity |

### 6.4.1.3 Chemical Addition

Operators at Durham can feed a variety of chemicals to the liquid and solids stream and do so as required. Chemical addition for the year 2002 is summarized in Table 6-5.

- Ferric can be added to the digester feed tanks for odor and struvite control. It can also be added upstream or downstream of the dewatering centrifuge for struvite control and upstream of the primary clarifier to aid flocculation if necessary.
- Alum can be added to the liquid stream upstream of primary and secondary clarifiers, as well as at the head of the tertiary treatment process. Alum is routinely added upstream of the tertiary clarifiers for final polishing and upstream of the secondary clarifiers as needed to augment EBPR. Alum was not added to primary clarifiers during intense sampling.
- Lime is added to the PE pump station wetwell to supplement alkalinity in the secondary treatment process.
- Sodium hypochlorite is used for disinfection and occasionally for filament control in the RAS. Sodium bisulfate is used for dechlorination.

Table 6-5. Chemical Addition Summary.

| Chemical | Unit | 2001 | 2002 | 2003 |
|---|---|---|---|---|
| Lime | lb/MGD | 560 | 201 | 280 |
| Alum | lb/MGD | 428 | 535 | 321 |
| Ferric | lb/MGD | 58.2 | 16.8 | 39.6 |

### 6.4.1.4 Process Control

During the dry weather season, operators aim to maintain the solids retention time (SRT) between 9 and 11 days. DO in the aerated cells is controlled and maintained at values between 2 and 3 mg/L. Online ammonia and alkalinity analyzers provide real-time performance feedback to the operators and process control specialists. The return activated sludge (RAS) rate is typically maintained around 50% and is flow paced. The mixed-liquor recycle rate varies with the influent flow because the pumps operate at a constant speed—on average about 300%. Efforts are currently underway to reduce the speed of the mixed-liquor recycle pumps to provide an average mixed-liquor recycle of approximately 100%. Recent process optimization investigation has shown that because of the short anoxic retention time, no benefit is gained from a higher recycle rate.

Operators aim to achieve 0.5 mg/L phosphate in the SE. Depending on EBPR performance, a tertiary alum dose of 10–20 mg/L[1] is used for effluent polishing. A secondary alum dose of 20–30 mg/L is added as needed to the secondary clarifier to meet the 0.50 mg/L secondary clarifier effluent phosphate concentration goal.

The PSL fermenter is operated at a constant SRT between one and two days. Operators monitor the underflow pH and sludge-blanket depth for SRT control. Fermenter overflow is returned to the PE pump station at a rate that is dependent on the PSL pumping. The majority of the time the fermenter overflow return occurs 24 hours per day, seven days per week.

### 6.4.2 VIP

#### 6.4.2.1 Liquid Stream

The BNR process at VIP follows screening, grit removal, and primary clarification. SE is disinfected and discharged to the Elizabeth River. The BNR process configuration is that of the VIP design, named after this facility where it was developed. Figure 6-3 shows the overall plant process schematic, and Figure 6-4 shows the VIP process layout.

The key feature of the VIP process is that RAS enters the process train at the head of the anoxic zone, combined with the mixed-liquor recycle from the aerobic zone (the nitrified recycle or NRCY stream). A second mixed-liquor recycle (the anoxic recycle or ARCY stream) returns biomass from the anoxic zone to the anaerobic zone to achieve EBPR. This configuration limits the transfer of nitrate into the anaerobic zone to maximize VFA availability for phosphorus release. PE and ARCY mix at the head of the anaerobic zone. In the first anoxic cell, the combined RAS/NRCY mixes with the anaerobic zone effluent. Operators also have the option to send the RAS/NRCY stream to one of the two downstream cells if longer anaerobic retention is desired.

VIP has four parallel trains, each with two anaerobic zones, four anoxic zones, and four aerobic zones (Figure 6-4). Effluent from the unaerated zones is split between the available aerobic basins. This configuration allows the operator to place any number of unaerated and aerated zones in operation as desired.

---

[1] Al/P dose in mole/mole is 4.3 to chemically remove phosphorus from 0.5 mg/L to 0.07 mg/L.

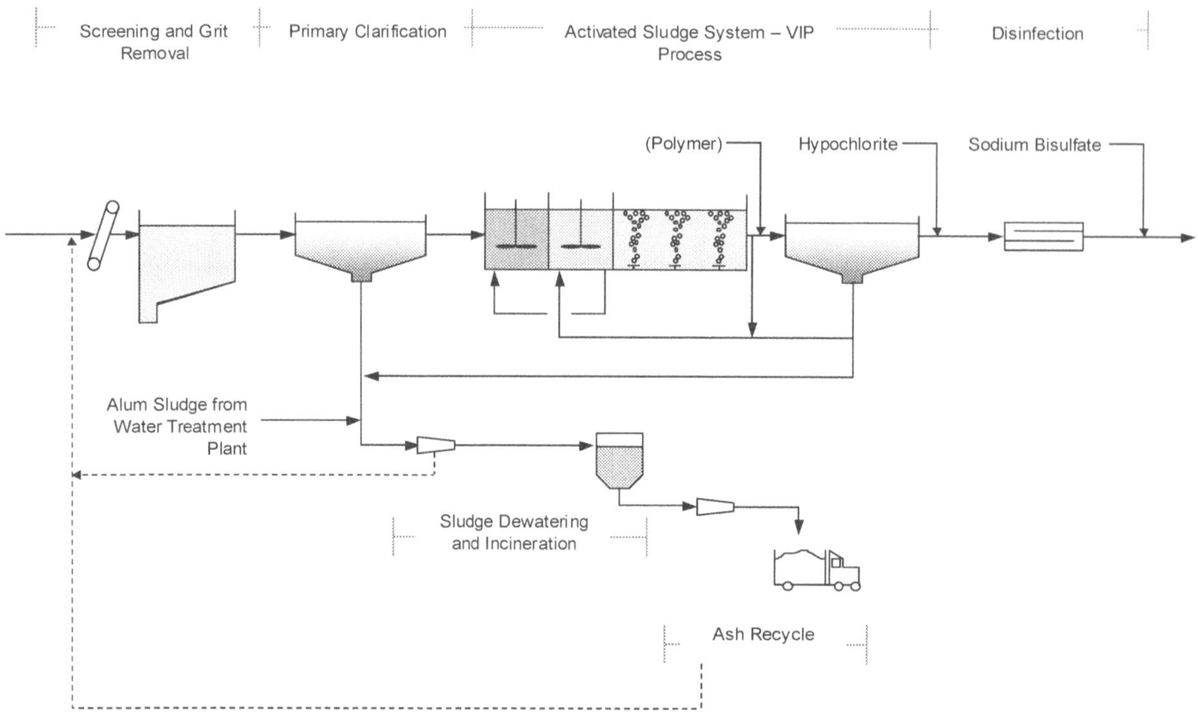


**Figure 6-3. Process Schematic for VIP Liquid and Solids Treatment.**

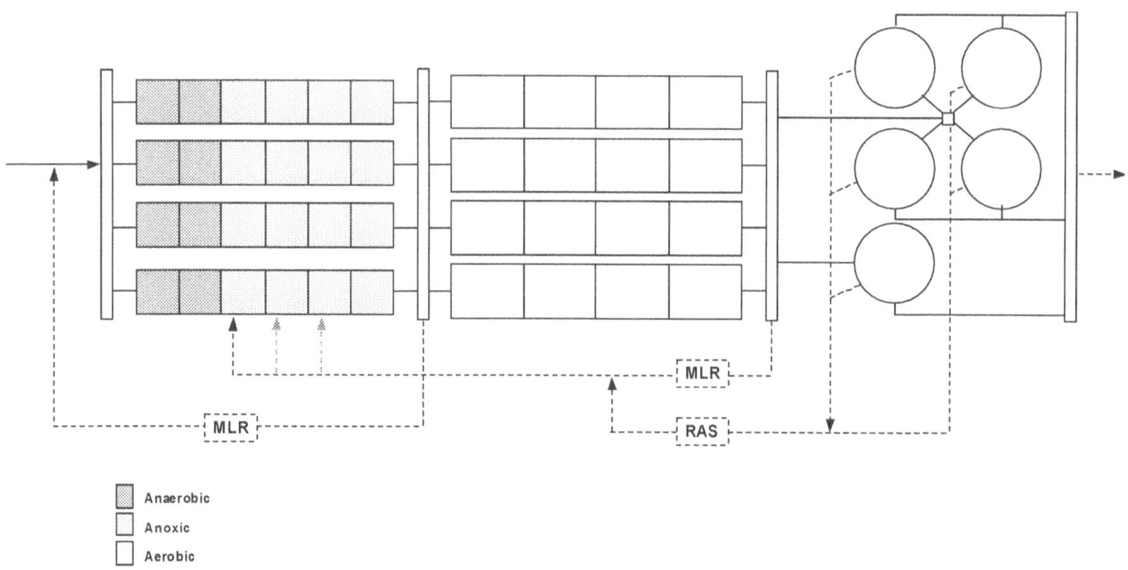


**Figure 6-4. Secondary Treatment Layout at VIP.**

Table 6-6. Key Process Parameters.

| Component | Unit | Design | Average |
|---|---|---|---|
| Anaerobic HRT[1] | hrs | 0.78 | 0.87 |
| Anoxic HRT | hrs | 1.56 | 1.76 |
| Aerobic HRT | hrs | 4.1 | 4.58 |
| Total HRT | hrs | 6.44 | 7.21 |
| RAS Rate | % | | 31 |
| Anaerobic ML Recycle Rate | % | | 73 |
| Anoxic ML Recycle Rate | % | | 77 |
| SRT | days | | 9.52 |
| MLSS | mg/L | | 3090 |
| Solids Recycle Return | | Upstream of primary clarifier (no anaerobic digestion) | |
| Flow | MGD | 40 | 28.45 |

1. Retention time is calculated based on aeration basin influent flow and does not include the recycle volume.

#### 6.4.2.2 Solids Stream

PSL is thickened in primary clarifiers to about 4 % solids. Primary and waste activated sludge are mixed with alum solids from a water treatment plant and sent directly to dewatering centrifuges. They are then dewatered and incinerated. Centrate is recycled to the plant headworks. Centrifuges operate continuously (24 hours per day, seven days per week). Scrubber water from the incinerator, which contains cyanide from the incineration process, is treated in a separate nitrification process before being recycled to the plant headworks. Table 6-7 summarizes the solids processing at VIP.

Table 6-7. Solids Processing Summary.

| Component | Technology | Comments |
|---|---|---|
| PSL Thickening | Gravity | Thickened in primary clarifier to 3%–5% |
| WAS Thickening | None | WAS sent directly to dewatering |
| Dewatering | Centrifuge | Operates 24/7 with no storage |
| Sludge | Incineration after dewatering | PSL and WAS mixed in dewatering feed tank prior to dewatering |

#### 6.4.2.3 Chemical Addition

The plant has the ability to feed alum to the secondary clarifiers in the event of lower than desired EBPR performance but rarely uses this capability. Polymer is occasionally added to the secondary clarifiers to aid TSS capture if necessary. Alum has been added occasionally to the primary clarifier to improve BOD removal, but this practice was not used during the intensive sampling periods of this study. Experience at the plant indicates that the plant operates best if the PE BOD loading is below 40,000 lb/d. No chemicals are added to supplement alkalinity or control pH.

### 6.4.2.4 Process Control

The operational strategy is to maintain the SRT just high enough to achieve nitrification. The operator reduces the SRT as low as possible while maintaining nitrification. The SRT averages approximately 9.5 days, but is operated as low as four days during the summer.

DO is controlled to maintain about 2.5 mg/L in the aerated zone. Operators monitor DO, MLSS, SVI, and effluent parameters. No particular response strategy is in place to deal with process upsets. Appropriate actions are determined based on operator experience.

## 6.4.3 Nansemond

### 6.4.3.1 Liquid Stream

The BNR process at Nansemond follows screening, grit removal, and primary clarification in the plant system. The configuration matches the VIP design developed by the Hampton Roads Sanitation District. Figure 6-5 shows the Nansemond WWTP process design. Table 6-8 summarizes key design parameters compared to 2002 average operating values.

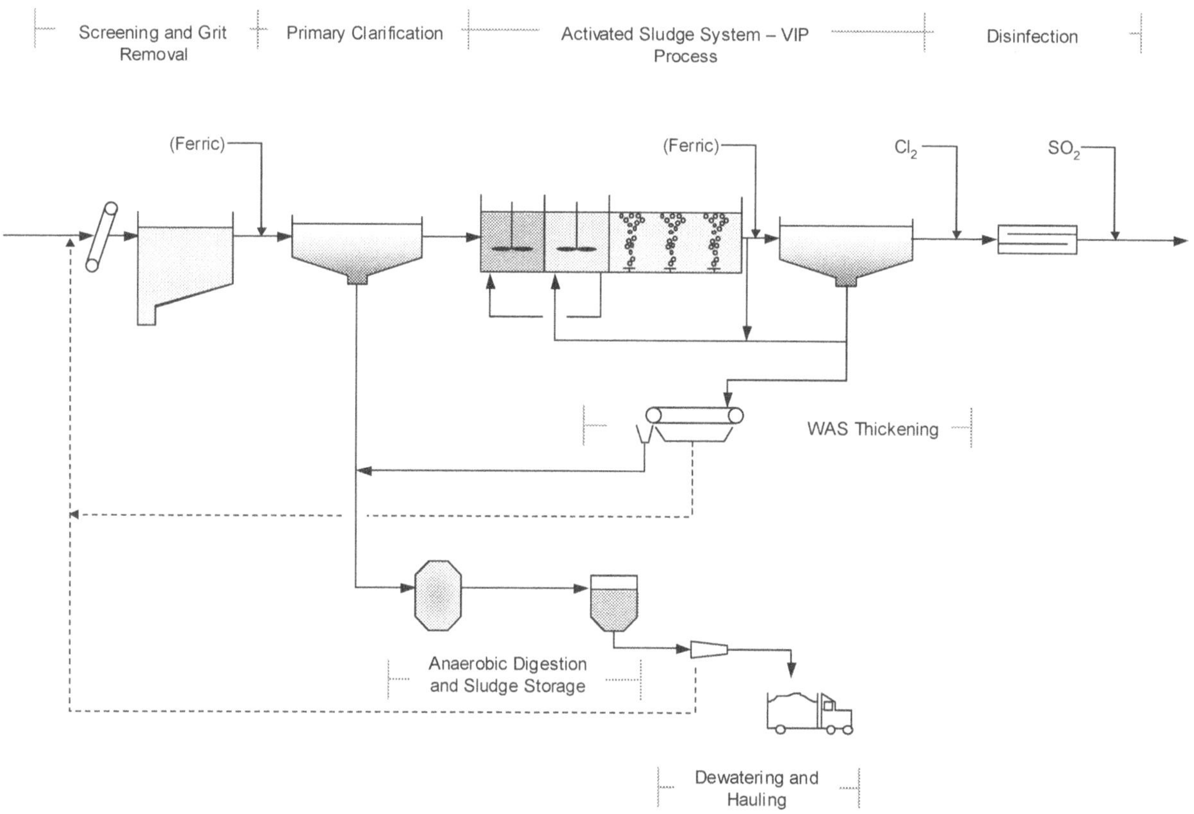


**Figure 6-5. Process Schematic for Nansemond WWTP Liquid and Solids Treatment.**

The key feature of the VIP process is that RAS is combined with mixed liquor recycled from the aerobic zone. RAS enters the process train at the head of the anoxic zone, and mixed liquor is recycled from the anoxic zone to the anaerobic zone to achieve EBPR. This configuration limits the transfer of nitrates into the anaerobic zone, optimizing VFA availability for phosphorus release. PE and the recycle stream from the anoxic zone mix at the head of the anaerobic zone. In the first anoxic cell, mixed liquor from the aeration basin effluent mixes with the anaerobic zone effluent. Operators also have the option to send the RAS/mixed-liquor recycle stream to one of the two downstream anoxic cells if longer anaerobic retention is desired.

At Nansemond there are seven unaerated trains, each with two anaerobic cells and four anoxic cells followed by five aerobic trains. Each aerobic train has four zones (Figure 6-6).

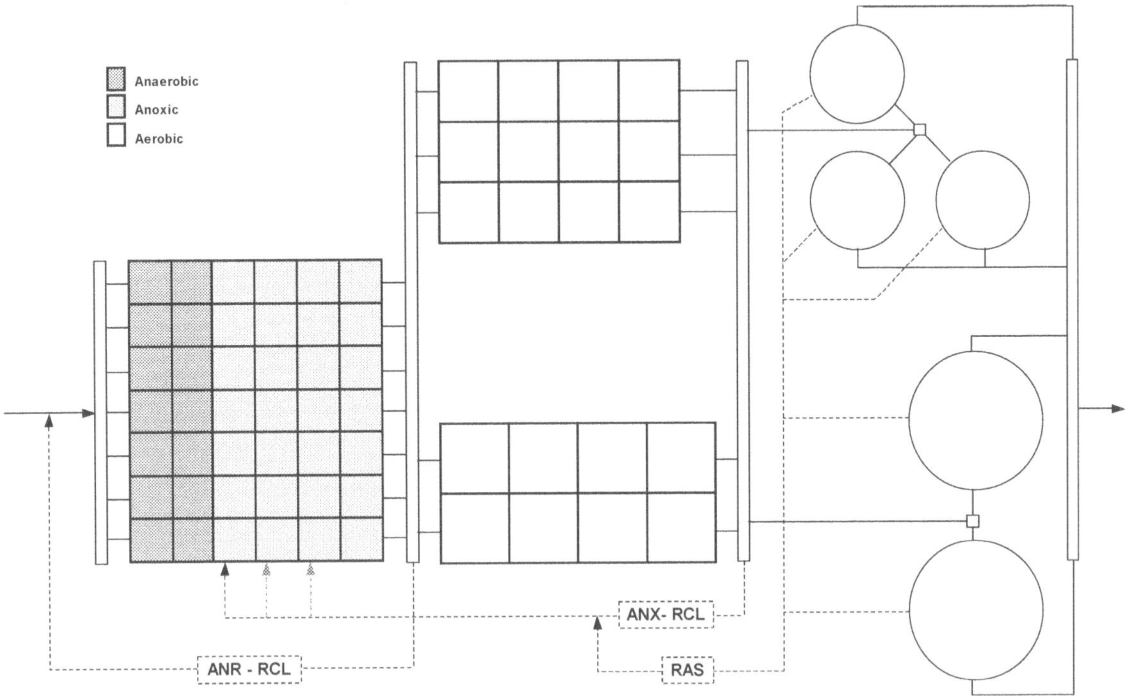


**Figure 6-6. Secondary Treatment Layout at Nansemond WWTP.**

**Table 6-8. Key Process Parameters.**

| Component | Unit | Design | Average[1] |
|---|---|---|---|
| Anaerobic HRT | hrs | 1.04 | 0.83 |
| Anoxic HRT | hrs | 4.15 | 2.91 |
| Aerobic HRT | hrs | 5.60 | 3.79 |
| Total HRT | hrs | 10.79 | 7.53 |
| RAS Rate | % | 75 | 84 |
| Anaerobic ML Recycle Rate | % | NA | 91 |
| Anoxic ML Recycle Rate | % | NA | 99 |
| SRT | days | NA | 8.53 |
| MLSS | mg/L | NA | 3520 |
| Solids Recycle Return | | to plant influent | |
| Flow | MGD | 30 | 18.1 |

[1] Based on 2002 data

### 6.4.3.2 Solids Stream

Solids treatment processes consist of thickening, anaerobic digestion, and digested biosolids dewatering. A gravity belt thickener (GBT) increases WAS TSS concentration to 5%–7% while PSL is thickened in the primary clarifier to about 3%–5% solids before conveyance to the digester. PSL and WAS are mixed prior to digester feed. The two digesters are operated as one and are fed intermittently, with the feed alternated between the two digesters every two hours.

The plant operates two mixed mesophilic digesters at a target mean cell retention time of 17–20 days to accomplish 38% volatile solids destruction on average. Digester operation has been plagued by periodic foaming problems. Controlling activated-sludge foaming to minimize the influx of filamentous organism has not translated into a reduction of digester foaming. Improved control of digester feeding has proven to reduce the foaming problem but has not entirely eradicated it.

Return centrate is sent to the headworks. Dewatering centrifuges operate about 8–10 hours per day, seven days per week during graveyard shift. Centrifuge operation starts around 11 p.m. and continues until completed. Digested biosolids are incinerated off-site. The agency has permits for composting, land application, and landfill disposal, all of which are only used in emergencies.

**Table 6-9. Solids Processing Summary.**

| Component | Technology | Comments |
|---|---|---|
| PSL Thickening | Gravity | PSL is thickened in the primary clarifiers to 3– 5 percent |
| WAS Thickening | GBT | 5– 7 percent |
| PSL and WAS Stabilization | Anaerobic Digestion | SRT = 17–20 days<br>TSS = 2.5–3 percent<br>VSS Destruction = 38% |
| Dewatering | Centrifuge | Operated during night shift between 11 p.m. and 7 a.m. |
| Filtrate/Centrate Storage | none | |
| Biosolids Disposal | Incineration | Option to land apply, compost, and landfill disposal |

#### 6.4.3.3 Chemical Addition

The facility was designed with the ability to add ferric chloride to the primary clarifier influent and secondary clarifier influent with flow pacing. However, experience with ferric addition showed that it negatively impacted EBPR, so this mode has been abandoned. Because EBPR produces an acceptable SE phosphorus concentration, ferric is only added to the dewatering feed in order to reduce the recycle phosphorus load and aid the dewatering process. Ferric was not added to primary clarifiers during the intensive sampling periods.

Sodium hydroxide is used for pH control and is added to the primary clarifier effluent as needed. Sodium hydroxide was not added during the intensive sampling period.

Anionic polymer is at times used in combination with ferric chloride to address settling problems in the secondary clarifiers. Cationic polymer is used for dewatering and thickening.

#### 6.4.3.4 Process Control

The operational strategy is to maintain a SRT between six and 12 days. SRT is calculated based on total activated sludge basins, excluding clarifiers. DO is controlled by one probe in the aeration basin to maintain at least 2 mg/L DO in the last one-third of the aeration tank. The number of aerated and non-aerated trains in service is determined based on flow. The operational goal is to stay within minimum and maximum design hydraulic retention time (HRT). Operators also have the option of changing the number of anoxic and anaerobic cells per train through a swing zone that can be operated as either anoxic or anaerobic, depending on where RAS/mixed-liquor recycle is returned.

### 6.4.4 McDowell Creek

#### 6.4.4.1 Liquid Stream

Figure 6-7 shows the McDowell Creek process layout for the modified UCT mode. Raw influent is degritted and screened before entering the primary clarifier. Dewatering recycle is returned to the influent channel upstream of grit removal. The activated sludge BNR process follows primary clarification. SE is filtered and disinfected. The notable feature of the process design at McDowell Creek is the sugar waste addition to the PE in combination with a modified UCT configuration. Originally PE VFA was supplemented using acetic acid. In 2001, the supplemental acetic acid was replaced with sugar waste from a soda bottling plant. Characteristics of this product are described in section 6.4.4.3.

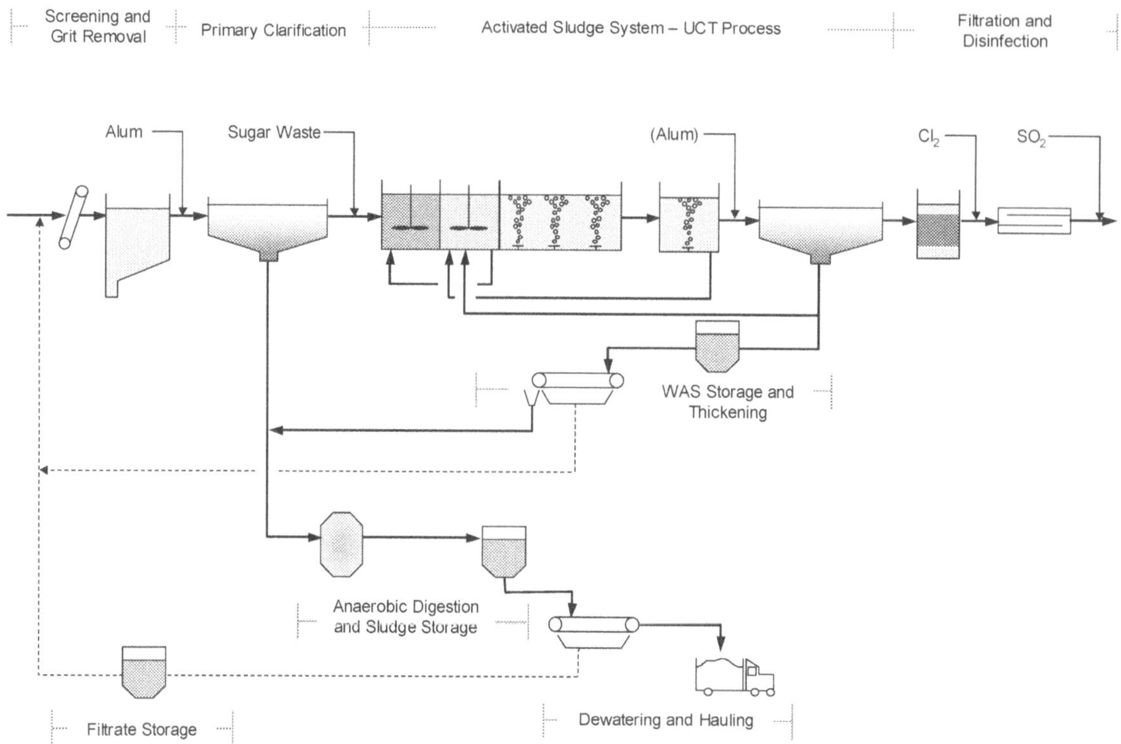


**Figure 6-7. Process Schematic for Mc Dowell Creek WWTP Liquid and Solids Treatment.**

The activated sludge system consists of a modified UCT process with an anaerobic zone that is equally divided into four cells in series. The anaerobic zone effluent is split equally to two trains, with each train consisting of two anoxic zones in series followed by single stage conventional plug flow aerobic zone. Each train has two anoxic cells and one long (plug flow) aerobic zone. Following each train is a post aeration basin from which the internal mixed-liquor recycle is pumped back to a junction box. It also receives RAS and the anaerobic zone effluent before splitting the flows to the two anoxic/aerobic trains.

The combined HRT at design flow is 11.5 hours, with an anaerobic retention time of 0.75 hours and anoxic retention time of 2.15 hours. Deep bed filters provide additional solids and phosphorus removal downstream of the secondary clarifier. Effluent BOD and TSS concentration of less than 2 mg/L are consistently accomplished. The filter also offers another alternative for post denitrification and phosphorus precipitation.

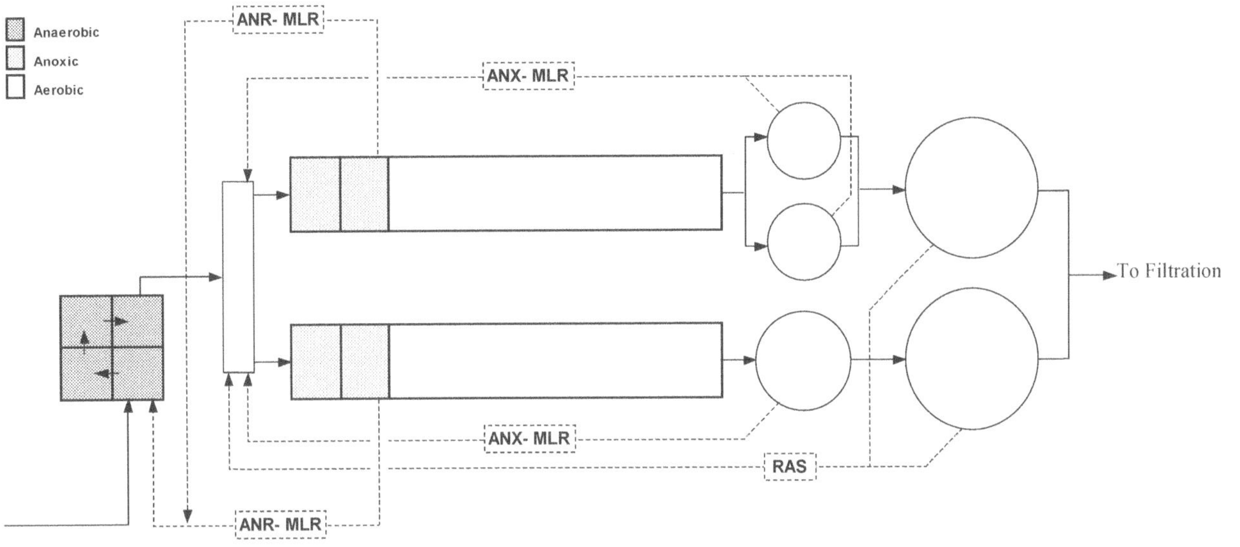


**Figure 6-8. Secondary Treatment Layout at Mc Dowell Creek WWTP (The plant has three trains but to date has never used more than two at a time.).**

**Table 6-10. Key Process Parameters (data from 2001– 2003).**

| Component | Unit | Design | Average |
|---|---|---|---|
| ANR HRT | hrs | 0.75 | 0.91 |
| ANX HRT | hrs | 2.15 | 2.62 |
| AER HRT | hrs | 8.5 | 10.4 |
| Total HRT | hrs | 11.5 | 13.4 |
| RAS Rate | % | 80% | 87% |
| ANR ML Recycle Rate | % | 65% | 70% |
| ANX ML Recycle Rate | % | 260% | 320% |
| SRT | days | 15 | 6.7 |
| MLSS | mg/L | 1800 | 3400 |
| Solids Recycle Return | | Returned to plant influent | |
| Flow | MGD | 6.0 | 5.04 |

### 6.4.4.2 Solids Stream

Solids treatment consists of thickening, anaerobic digestion, and digested sludge dewatering. PSL is thickened in the primary clarifier before being fed directly to the anaerobic digester. WAS is flow equalized in an aerated WAS holding tank and gravity belt thickened to 5%–7%. WAS thickening is operated continuously. Currently two anaerobic digesters are in operation at McDowell Creek, providing an average solids reduction of 55% at an SRT of 25 days.

Digested sludge is dewatered eight hours per day, five days per week. To minimize the adverse impact of the dewatering recycle load on BNR performance, filtrate is stored, and the return is distributed over 24 hours.

Table 6-11. Solids Processing Summary.

| Component | Technology | Comments |
|---|---|---|
| PSL Thickening | Gravity | Thickened in Primary Clarifier |
| Was Thickening | Gravity Belt Thickener | |
| PSL Stabilization | Anaerobic Digestion | SRT = 25<br>TSS = 3.0%<br>VSS Destruction = 55% |
| WAS Stabilization | | |
| Dewatering | Belt Filter Press | |
| Centrate Storage | Yes | |

#### 6.4.4.3 Chemical Addition

Since 2001, an industrial sugar waste has been used to supplement influent VFA and soluble BOD. Sugar waste addition rates range from 1,000 gallons per day (gpd) to 60,000 gpd. The utility is required by contract to accept all of the industrial plant's waste; therefore, the dosage rate is determined primarily by the delivery schedule.

The BOD concentration in this waste was measured in the past at between 40,000 mg/L and 130,000 mg/L, with an average concentration of approximately 80,000 mg/L (Goins, et al., 2003). Two samples of the sugar waste were taken during this project to evaluate the concentration of COD, BOD, and SCOD. The analytical results from these samples are shown in Table 6-12. The subsequent BOD dosage rate ranges from 10 to 30 mg/L based on plant flow.

Alum can be added to the primary clarifier and dewatering filtrate to reduce PE phosphate. It can also be added to the secondary clarifier whenever EBPR supplementation is required.

Table 6-12. Sugar Waste Analysis Results.

| Component | Unit | Sample |
|---|---|---|
| COD | mg/L | 93,500 |
| SCOD | mg/L | 83,300 |
| BOD | mg/L | 36,700 |

Table 6-13. Chemical Addition Summary.

| Chemical | Unit | 2001 | 2002 | 2003 |
|---|---|---|---|---|
| Lime | lb/MGD | 0 | 0 | 0 |
| Alum | lb/MGD | 342 | 244 | 203 |

#### 6.4.4.4 Process Control

The staff maintains an intensive operations effort to meet the low effluent phosphorus requirements. Effluent phosphate is measured three times per day with test kits. The key operation parameters are aeration basin DO, SRT, primary and final effluent phosphate, and PE alkalinity. The alum dosage to the PE is adjusted to maintain average PE phosphorus concentrations at 6.0 mg/L. Lime addition is controlled to maintain an aeration basin influent alkalinity of at least 150 mg/L. The internal recycle pumps are constant speed; therefore, the anaerobic and anoxic recycle rates cannot be controlled. The aeration basin DO is set to maintain a concentration of 2.5 mg/L. Dewatering filtrate storage and equalization is a critical element for process stability and for nitrogen and phosphorus-removal performance. Filtrate is stored and returned continuously.

## 6.4.5 Lower Reedy

### 6.4.5.1 Liquid Stream

The Lower Reedy facility has three identical activated sludge trains, each consisting of five equally-sized unaerated (anaerobic or anoxic) basins and one aerated aerobic basin. Only two of the three trains have been in operation at any time during the last three years. The system is currently operating in VIP mode, as shown in Figure 6-9.

Denitrified recycle from the last stage of the anoxic basins is combined with the PE upstream of the activated sludge basins and directed to the first of the two anaerobic stages in each train. One variable-speed denitrified recycle pump (also called anoxic recycle pump) is used per train. Each pump is rated at 1,750 gpm (2.5 mgd). At the time of testing, the anoxic pumps were operating at fixed rates of 1,680 gpm and 1,770 gpm for the two trains in operation.

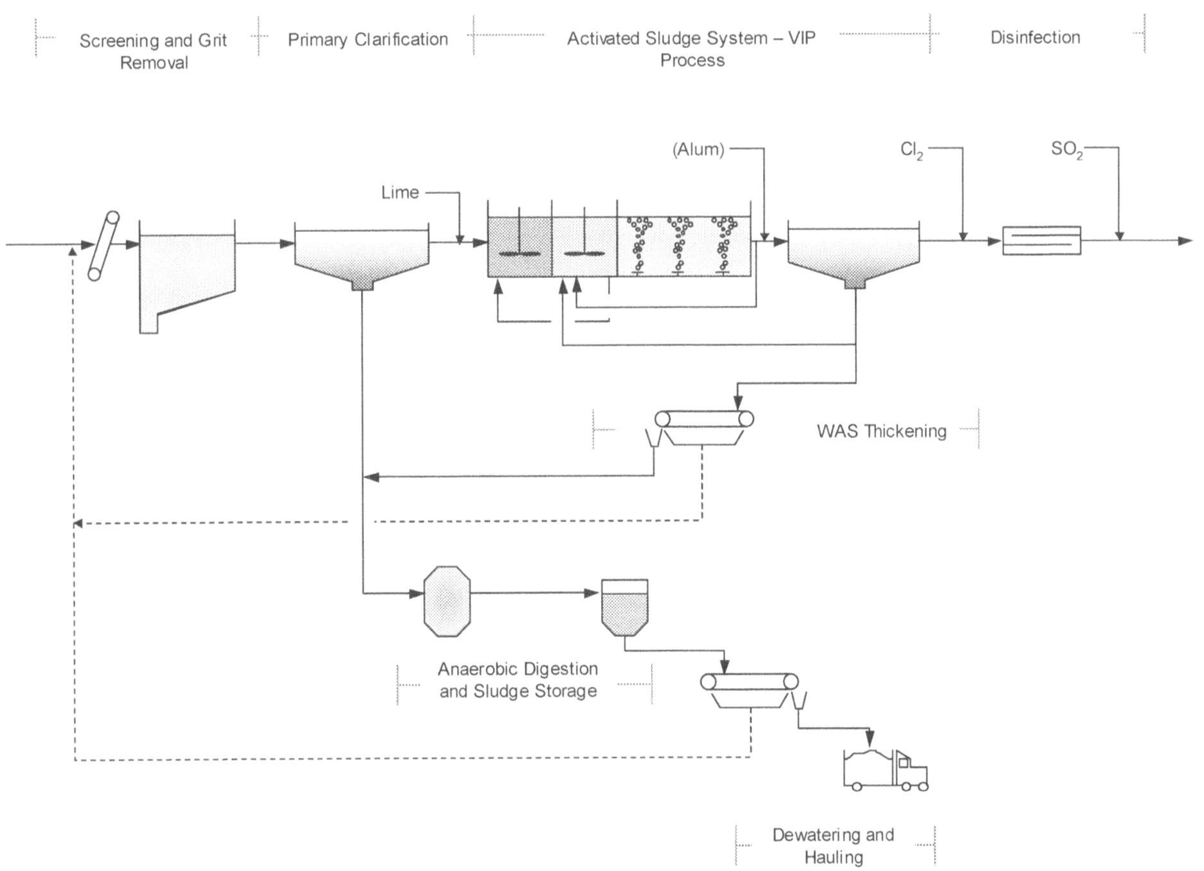


**Figure 6-9. Process Schematic for Lower Reedy WWTP Liquid and Solids Treatment.**

Aerobic recycle from each of the aeration basins is returned to the first stage of three anoxic stages in series. Five anoxic recycle variable-speed pumps (rated at 7,000 gpm each) are available to return mixed liquor from the end of the aerobic zone to the head of the anoxic zone. One pump is used with each train. The first anoxic basin of each activated sludge train also receives RAS, which is pumped in separately at 65% of the effluent flow.

One unique aspect of the facility is the long HRT. At the design flow and with all three trains in operation, the anaerobic zone provides 2.6 hours of HRT, the anoxic zone four hours and the aerobic zone 9.6 hours. Operators can use an upstream equalization basin to shave peak flows and help maintain relatively constant flow to the activated sludge basins. The plant currently only operates two of the three identical trains.

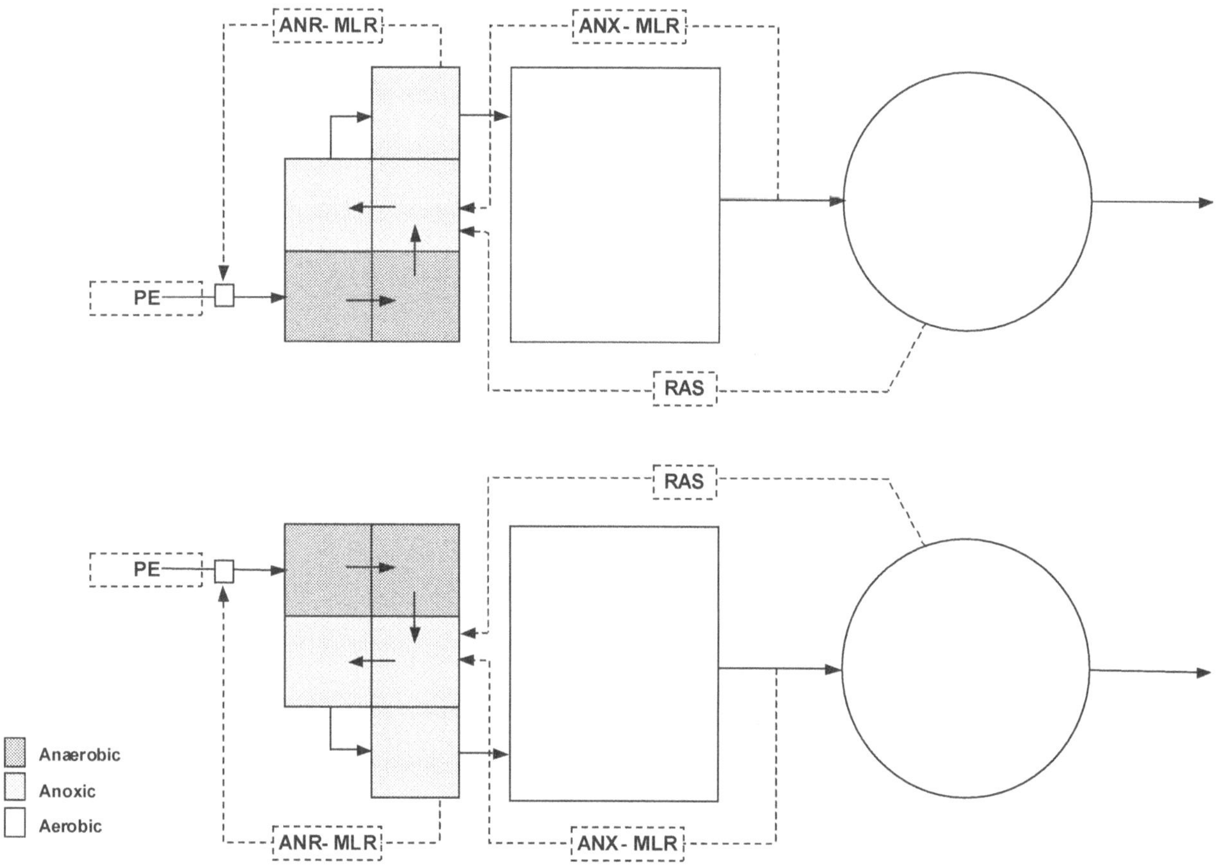


**Figure 6-10. Secondary Treatment Layout at Lower Reedy WWTP (shows the two operating activated sludge trains).**

Table 6-14. Key Process Parameters.

| Component | Unit | Design | Average (2000–2003) |
|---|---|---|---|
| Anaerobic HRT | hrs | 1.8–3 | 3.1 |
| Anoxic HRT | hrs | 2.7–4.5 | 4.5 |
| Aerobic HRT | hrs | 7–12 | 11.3 |
| Total HRT | hrs | 11.5–19.5 | 19.4 |
| RAS Rate | % | 50 % | 52% |
| Anaerobic Mixed-liquor Recycle Rate | % | 50 % | 55% |
| Anoxic Mixed-liquor Recycle Rate | % | 150 % | 116% |
| SRT | days | 7-12 | 9.4 |
| MLSS | mg/L | 2000–3500 | 3220 |
| Solids Recycle Return | | Upstream of Primary Clarifier | |
| Flow | MGD | 8.6 | 4.8 |

### 6.4.5.2 Solids Processing

PSL is thickened in the primary clarifier, and WAS is thickened using a GBT. WAS thickeners are typically fed 12 hours per day, starting at approximately 3 a.m. Thickened sludge is sent to a digester feed tank for flow equalization and mixing with the PSL and then fed continuously to the anaerobic digester. The dewatering filtrate is returned to the primary influent. Digested sludge is dewatered during weekdays between 7 a.m. and 3 p.m.

Table 6-15. Solids Processing Summary.

| Component | Technology | Comments |
|---|---|---|
| PSL Thickening | Gravity in Primary | Thickened in Primary Clarifier |
| WAS Thickening | Gravity Belt Thickener | Operated 8/7 |
| PSL Stabilization<br>WAS Stabilization | Anaerobic Digestion | TSS = 2% |
| Dewatering | Belt Filter Press | Liquid land application planned for near future |
| Centrate Storage | None | |

### 6.4.5.3 Chemical Addition

The Lower Reedy plant adds lime to the PE for alkalinity supplement. Since August 2002, alum has been available for addition to secondary influent for supplementary phosphorus removal. Table 6-16 summarized the chemical use from 2001 through 2003, which had an average lime addition of 66 mg/L for 2003 to accommodate the relatively low influent alkalinity of 122 mg/L. Aeration basin effluent alkalinity ranged from 100 mg/L to 300 mg/L. On demand acetic acid feed to the anaerobic zone is planned for the near future to supplement influent VFAs when necessary.

Table 6-16. Chemical Feed Summary.

| Chemical | Unit | 2001 | 2002 | 2003 |
|---|---|---|---|---|
| Lime | Tons/yr | 540 | 290 | 290 |
| Alum | Tons/yr | - | - | 57 |
| Lime | lb/MGD | 69 | 340 | 550 |
| Alum | lb/MGD | - | - | 110 |

#### 6.4.5.4 Process Control

The operational strategy is to maintain the sludge inventory at a constant value. Currently, the sludge inventory is maintained to produce a SRT of about 7–12 days depending on water temperature and influent load. Wasting is controlled on a daily basis by measuring the RAS/WAS concentration and then wasting or stopping the wastage to maintain the target mixed-liquor concentration.

RAS is controlled at a constant rate and adjusted to account for wasting; i.e., the underflow from the clarifiers is kept constant as the WAS flow varies during the day.

DO is controlled to maintain a target concentration of 2.5 mg/L in the aeration basin. Actual values are typically between 3 and 4 mg/L. The aeration basin is equipped with coarse bubble aeration.

## 6.5 Results

### 6.5.1 Approach

The process evaluation was based on routine processes and on water quality and special data collected during intensive testing periods.

#### 6.5.1.1 Routine Data Review

The detailed facility evaluation reviewed all available water quality and process monitoring information collected from January 1, 2000, through the fall of 2003. Data were analyzed to determine general process performance and reliability in meeting effluent quality goals, and to evaluate the interaction between process operating parameters and performance. These routine data are referred to as the *2000–2003 Data Set*. Unless otherwise noted, effluent values are SE.

#### 6.5.1.2 Special Data Collection

Special field and laboratory analysis tests were conducted at the five facilities. These included:

- Intensive sample collection and analysis over two 2-week periods. Samples were collected, filtered, and analyzed to characterize influent, PE, SE, final effluent, and recycle streams. Samples were analyzed for various measures of EBPR substrate — COD; SCOD; ffCOD, based on the methods of Mamais, et al., 1993; VFAs; and BOD — as well as various forms of suspended solids, nutrient species (nitrogen and phosphorus), and metals (calcium, potassium, magnesium).
- Activated sludge basin phosphorus concentration profiles were obtained several times over the 2-week intensive by collecting samples, filtering, and analyzing for phosphate concentrations in the anaerobic, anoxic and aerobic basins.
- Bench-scale phosphorus URTs. These tests were conducted at the University of Washington using mixed liquor collected during the intensive sampling periods. During

the URT, a mixed-liquor sample was spiked with a solution of sodium acetate to provide an 80 mg/L COD dose, and the amount of phosphorus and selected cations released (calcium, potassium, magnesium) were measured for the first 45 or 60 minutes. The sample was sparged with nitrogen gas prior to the test and during the anaerobic period to maintain anaerobic conditions. After the anaerobic period, the sample was aerated for two–three hours during which time the phosphorus and cation uptakes were measured.

- Microbial examination was done at the University of Washington to observe the presence of PAOs and GAOs using mixed liquor collected during the intensive sampling periods.
- Microbial examination to detect some of the known GAO bacteria using FISH tests was done by Dr. Linda Blackall's laboratory at the University of Queensland in Australia using mixed liquor collected during the intensive sampling periods.
- Facility phosphorus mass balances were done to determine the phosphorus recycle load and aeration basin phosphorus inventory.

The special data collection strategy was outlined in a Testing Plan for Year 2 (Appendix P). The testing plan identified sample locations, sampling frequency, and type of analysis required at each facility. The testing plan also outlined the methods of analysis for the phosphorus URTs, profile sampling, ffCOD analysis, and sample collection protocol for GAO testing.

Data collected at each of the facilities is summarized in individual plant reports. These reports are included in Appendices B through F. Key data are summarized in the following sections.

### 6.5.2 Plant Operating Data

#### 6.5.2.1 Influent Characterization—Routine Operating Data

Influent BOD concentrations ranged from approximately 150–200 mg/L, ammonia concentrations ranged from 20–25 mg/L, and total phosphorus ranged from 5–8 mg/L. The exception is the Nansemond facility, which had higher influent ammonia and phosphorus concentrations (35 mg/L and 14 mg/L, respectively). Table 6-17 summarizes average aeration basin influent characteristics at the facilities during the intensive sampling periods. The McDowell Creek plant also adds a high BOD waste from a cola manufacturing plant. The BOD concentration in this waste has been measured at between 40,000 mg/L and 130,000 mg/L, with an average concentration of approximately 80,000 mg/L (Goins, et al., 2003).

**Table 6-17. Representative Wastewater Characteristics from Facilities included in the Phase 2 Testing (2000–2003).**

| | Unit | Lower Reedy | VIP | Nansemond | McDowell | Mauldin Rd | Durham |
|---|---|---|---|---|---|---|---|
| Flow | MGD | 5.2 | 28.4 | 21.8 | 6.0 | 14.7 | 20.2 |
| TSS | mg/L | 96 | 55.9 | 210 | 83.3 | 86.4 | 125.7 |
| BOD | mg/L | 181.8 | 147.8 | 192.5 | 152.8 | 119.5 | 143.3 |
| SBOD | mg/L | 97 | 83.3 | 46.3 | 52 | - | 102 |
| COD | mg/L | 425.5 | 267.7 | 424 | 334.8 | 248 | 265 |
| TKN | mg/L | 32.6 | 23.2 | 51 | 31.7 | 27 | 34 |
| NH4-N | mg/L | 24.2 | 18.8 | 38.1 | 23.3 | 17.2 | 23.1 |
| TP | mg/L | 9.1 | 4.7 | 14.6 | 5.5 | 6.1 | 6.4 |
| PO-P | mg/L | 6.2 | 3.6 | 10.2 | 3.4 | 5.4 | 3.6 |
| Alkalinity | mg/L | 155 | 164.3 | 295.3 | - | 156 | 239.7 |
| pH | | 6.6 | 6.8 | 7 | - | 7 | 7.9 |
| Temperature | °C | 20.8 | 24.4 | 23.8 | 24.7 | 21 | 20.5 |
| VFA : P Ratio | mg COD / mg P | 8.2 | 12.1 | 1.5 | 8.8 | 0.4 | 6.2 |
| ffCOD : P Ratio | mg COD / mg P | 15.6 | 30.9 | 6.7 | 34.9 | 15 | 19.5 |
| COD : P Ratio | mg COD / mg P | 41.6 | 59.3 | 28.7 | 69.7 | 41 | 40.2 |
| BOD : P | mg BOD / mg P | 21.5 | 31.9 | 12.6 | 31.3 | 19.8 | 21.7 |
| ffCOD : COD Ratio | mg COD / mg COD | 0.4 | 0.5 | 0.2 | 0.5 | 0.4 | 0.5 |
| VFA : ffCOD Ratio | mg COD / mg COD | 0.3 | 0.4 | 0.2 | 0.3 | 0 | 0.4 |

Figure 6-11 below shows the aeration basin influent total phosphorus load (lb P/day) normalized to the aeration basin solids content (lb VSS). Relating the aeration basin influent phosphorus load to the volatile faction of the mixed liquor allows a better relative comparison of the different facilities. This figure illustrates Nansemond's high phosphorus load in relation to the other facilities.

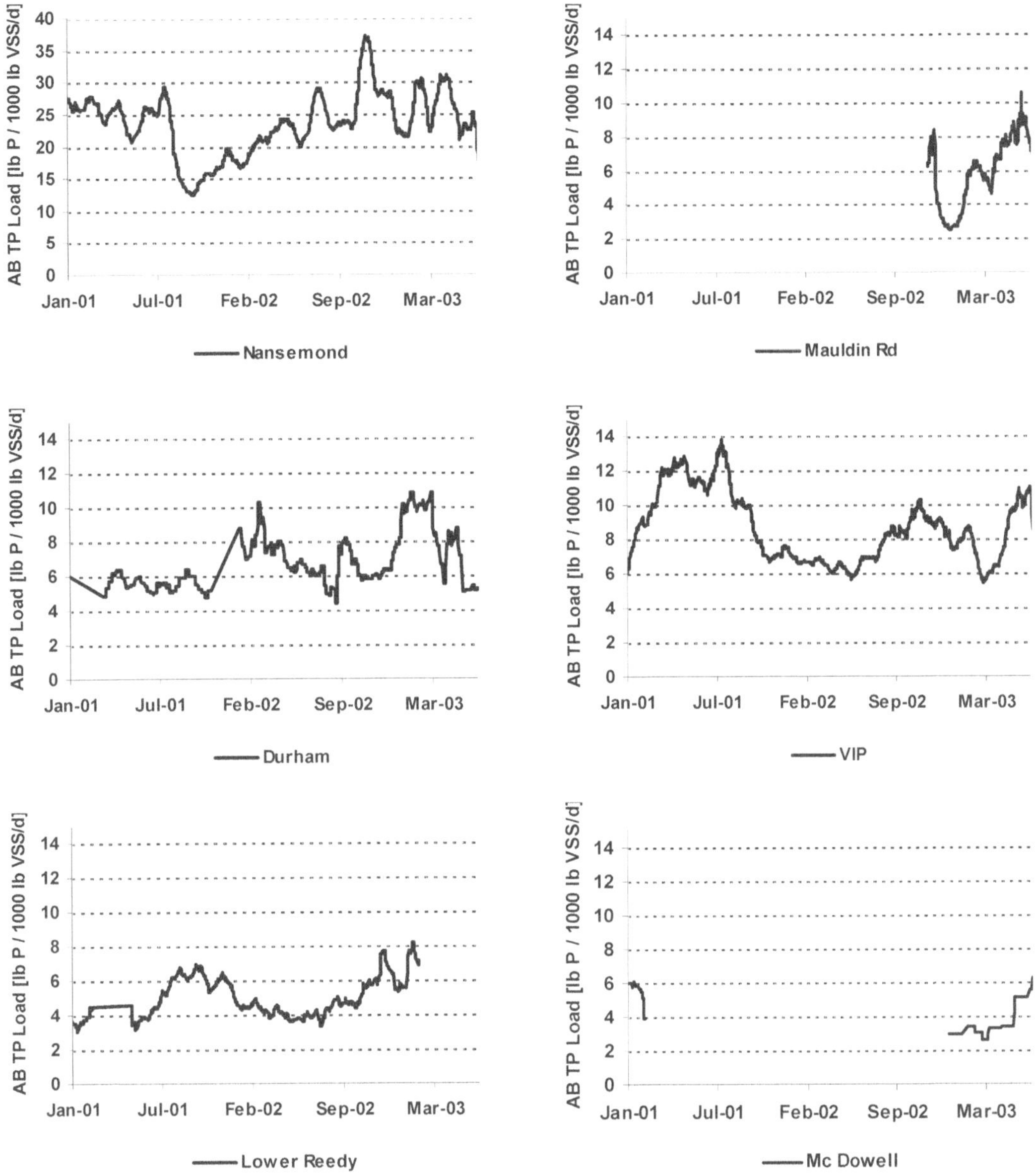


Figure 6-11. Comparison of Aeration Basin Phosphorus Loads.

### 6.5.1.3 Influent Characterization — Intensive Sampling Data

Table 6-18 shows the influent characteristics at each facility during the time of the two-week testing. EBPR was functioning at all facilities during the intensive sampling periods, as evidenced by the low SE phosphate concentrations and phosphorus release in the anaerobic zone.

Table 6-18. Summary Comparison of Aeration Basin Influent Parameters from Year 2 Intensive Testing.

| Aeration Basin Influent Parameter | | Durham | | Lower Reedy | | McDowell Creek | | VIP | | Nansemond | |
|---|---|---|---|---|---|---|---|---|---|---|---|
| | | Phase 1A | Phase 2B | Phase 2A | Phase 2B | Phase 2A | Phase 2B | Phase 2A | Phase 2B | Phase 2A | Phase 2B |
| Flow | MGD | 19.2 | 19.4 | 5.68 | 5.02 | 8.9 | 5.5 | 32 | 29 | 23.0 | 23.7 |
| TSS | mg/L | 130 | 121.1 | 44.0 | 168 | 92 | 87 | 59 | 50 | 79 | 136 |
| BOD | mg/L | 111 | 184 | 162 | 178 | 149 | 153 | 140 | 140 | 81 | 147 |
| SBOD | mg/L | - | 102 | - | 97 | - | | 97 | 86 | 27 | 31 |
| COD | mg/L | - | 265 | - | 421 | 320 | 341 | 280 | 270 | 288 | 414 |
| FFCOD | mg/L | - | 129 | - | 197 | - | - | 140 | 140 | 76 | 124 |
| VFA | mg/L | 16.4 | 53.9 | 90 | 48 | - | 47.2 | 26 | - | 14 | 48 |
| TKN | mg/L | - | 35.0 | 24.7 | 37.4 | 26.6 | 34.5 | 20 | 22 | 42 | 49 |
| $NH_4$-N | mg/L | 24.3 | 19.9 | 20.9 | 24.8 | 20 | 23.3 | 18 | 17 | 38 | 39 |
| TP | mg/L | 6.5 | 6.6 | 6.6 | 12.3 | 3.8 | 4.9 | 4.3 | 4.8 | 11.8 | 11.8 |
| $PO_4$-P | mg/L | 3.6 | 3.8 | 4.7 | 7.7 | - | - | 2.9 | 3.8 | 8.4 | 9.0 |
| Alkalinity | mg/L | 246 | 235 | 126 | 167 | - | - | 140 | 160 | 268 | 305 |
| pH | | 7.9 | 7.9 | 6.1 | 6.3 | - | - | 6.9 | 6.8 | 7.0 | 7.1 |
| Temperature | °C | 21.2 | 21.5 | 20.3 | 20.0 | - | 28 | 24 | 25 | 22 | 26 |
| VFA : P Ratio | mg COD mg P | 2.5 | 8.1 | 13.6 | 3.9 | - | 8.5 | 6.0 | - | 1.18 | 2.60 |
| FFCOD : P Ratio | mg COD/ mg P | - | 19.5 | - | 15.2 | - | - | 31 | 29 | 6.4 | 10.4 |
| COD : P Ratio | mg COD/ mg P | - | 40.2 | - | 34.2 | 84.2 | 69.6 | 64 | 56 | 24 | 35 |
| BOD : P | mg POD/ mg P | 17 | 27.8 | 24.5 | 14.5 | 39.2 | 31.2 | 33 | 29 | 6.8 | 12.5 |
| FFCOD : COD Ratio | mg COD/ mg COD | - | 0.48 | - | 0.46 | - | - | 0.49 | 0.52 | 0.26 | 0.30 |
| VFA : FFCOD Ratio | mg COD/ mg COD | - | 0.41 | - | 0.29 | - | - | 0.19 | - | 0.18 | 0.25 |
| SE $PO_4$-P | mg/L | 1.5 | 0.1 | 0.77 | 1.38 | 0.15 | 0.2 | < 0.2 | 0.46 | 0.2 | 0.2 |

#### 6.5.1.4 Effluent Characterization

All of the plants achieved high levels of EBPR on an average basis. Table 6-19 below compares the average SE water quality at the facilities based on operating data collected between January 2001 and Fall 2003. As the table indicates, not all facilities collect SE phosphate data as part of their routine sampling. EBPR performance will be evaluated in more detail in the Discussion section.

**Table 6-19. Summary Comparison of Average Secondary Effluent (SE) Parameters between 2000 and 2003.**

| Parameter | Unit | Durham | VIP | Nansemond | McDowell Creek | Lower Reedy |
|---|---|---|---|---|---|---|
| SE $NH_4$-N | mg/L | 0.65 | 2.4 | 12 | 0.25 | 0.62 |
| SE $NO_3$-N | mg/L | 8.4 | 6.5 | 6.4 | 4.1 | 4.7 |
| SE TN | mg/L | 10 | 9.4 | 19 | NA | NA |
| SE TP | mg/L | 0.15 | 0.5[1] | 1.2 | 0.15 | 0.67 |
| SE $PO_4$-P | mg/L | 0.1 | 0.6[1] | NA | NA | 0.5 |
| SE BOD | mg/L | 1.6 | 6.2 | 6.1 | 3.4 | 4.1 |
| SE TSS | mg/L | 1.0 | 3.9 | 7.7 | 3.1 | 4.7 |
| SE Alkalinity | mg/L | 113 | 76 | 169 | 86 | 111 |

1. Different sets of data were used for TP and $PO_4$-P, resulting in higher average $PO_4$-P

## 6.6 Discussion

### 6.6.1 Impact of Influent Characteristics

With the exception of the Nansemond facility, all plants show influent compositions within the typical range for municipal treatment. Based on the nutrient ratios one would expect very good conditions for biological phosphorus removal at McDowell and Lower Reedy, both of which have a BOD:P ratio of more than 30 lb BOD/lb P.

The Nansemond facility differentiates itself from the other plants by its high industrial wastewater fraction due to a number of industrial food processing facilities with in the service area. The result is a very high influent phosphorus concentration and very low BOD:P ratio.

Throughout the data evaluation it became apparent that the strongest affiliation between influent composition and effluent phosphorus concentration as well as biological phosphorus-removal reliability was found in the BOD:P ratio. Similar affiliations are likely to be found between phosphorus removal and ffCOD:P or VFA:P ratio. However, the sample populations for both ffCOD and VFA were too small to allow conclusions regarding their relationship to phosphorus removal.

Figure 6-12 illustrates that the reliability of EBPR increases with BOD:P ratios at and above approximately 25 lb BOD:P. The charts in Figure 6-12 show that facilities with BOD:P ratios over 30 lb BOD/lb P operate more stably with a smaller range of effluent phosphorus concentrations. Furthermore, those facilities with BOD:P ratio of 20 lb BOD/lb P or less show a strong cause and effect relationship between reduction of BOD:P ratio and the loss of EBPR performance. This can be seen most strongly at the Nansemond facility where effluent EBPR performance deteriorates when the BOD:P ratio decreases and vice versa. This cause-and-effect relationship was not observed when BOD:P ratios are above 30 lb BOD/lb P.

Figure 6-12 shows the aeration basin influent BOD:P ratio and effluent phosphorus for the facilities. Aside from the cause-and-effect relationships that can be observed for BOD:P ratios of less then 20 lb BOD/lb P the charts also illustrate the variability of the influent composition. The latter variability appears to cause variability in conditions for EBPR.

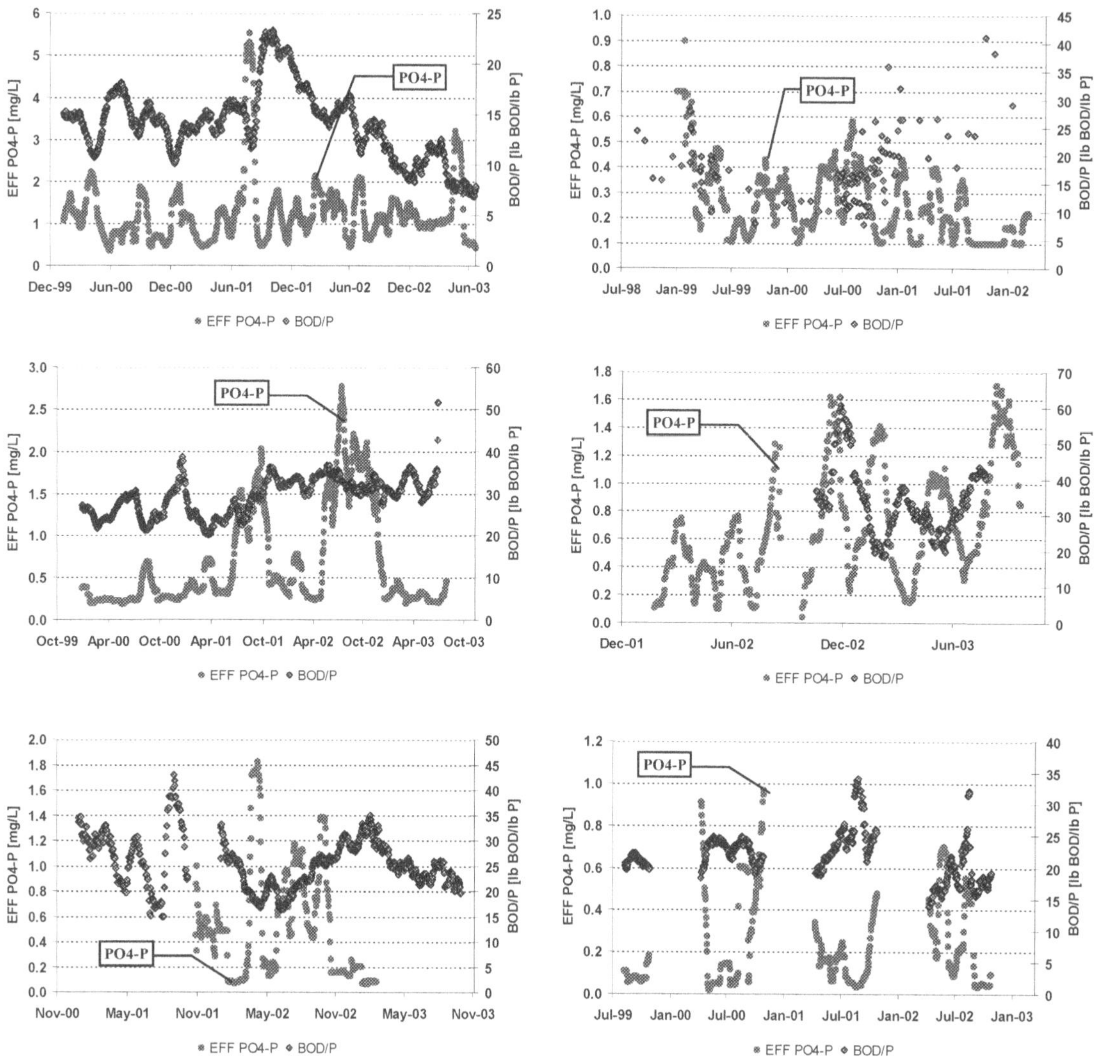


Top Left: Nansemond
Middle Left: VIP
Bottom Left: Lower Reedy

Top Right: McDowell (BOD:P does not included added sugar waste)
Middle Right: Mauldin Rd
Bottom Right: Durham (BOD:P includes Fermenter VFA)

**Figure 6-12. Relationship Between Aeration Basin Influent Bod:P Ratio And Secondary Effluent Phosphate (30-day averages).**

## 6.6.2 Reliability

Each of the facilities has different permit limits and operational goals, making it difficult to compare process reliability. However, it is clear that for plants with effluent permit limits of 1 mg/L or less, the EBPR process alone is not stable enough to provide reliable permit compliance. Figure 6-13 shows the 30-day running average effluent phosphate concentrations and operational targets of each facility based on routine operating data collected between January

2001 and Fall 2003. All charts are shown at the same scale to allow visual comparison. Full-size charts for each plant can be found in the individual facility reports. The performance shown in this figure underscores the observation that while average EBPR concentrations may be similar, overall plant performance varies significantly.

Figure 6-13 shows that full scale EBPR plants can achieve very low effluent phosphoate concentrations. McDowell maintains low phosphate concentrations very reliably, in part due to the ability to add sugar waste to maintain a healthy PAO population. Most other facilities show a random variation in performance. Durham is a good example, where SE phosphate concentrations are typically below 0.5 mg/L, but some excursions occur even in the summer permit season. VIP is a special case, showing a clear seasonal trend in the effluent phosphate concentrations. In summer, the phosphorus increases. Rabinowitz, et al. (2004) showed that this summer excursion could be associated with high temperatures (above 30°C).

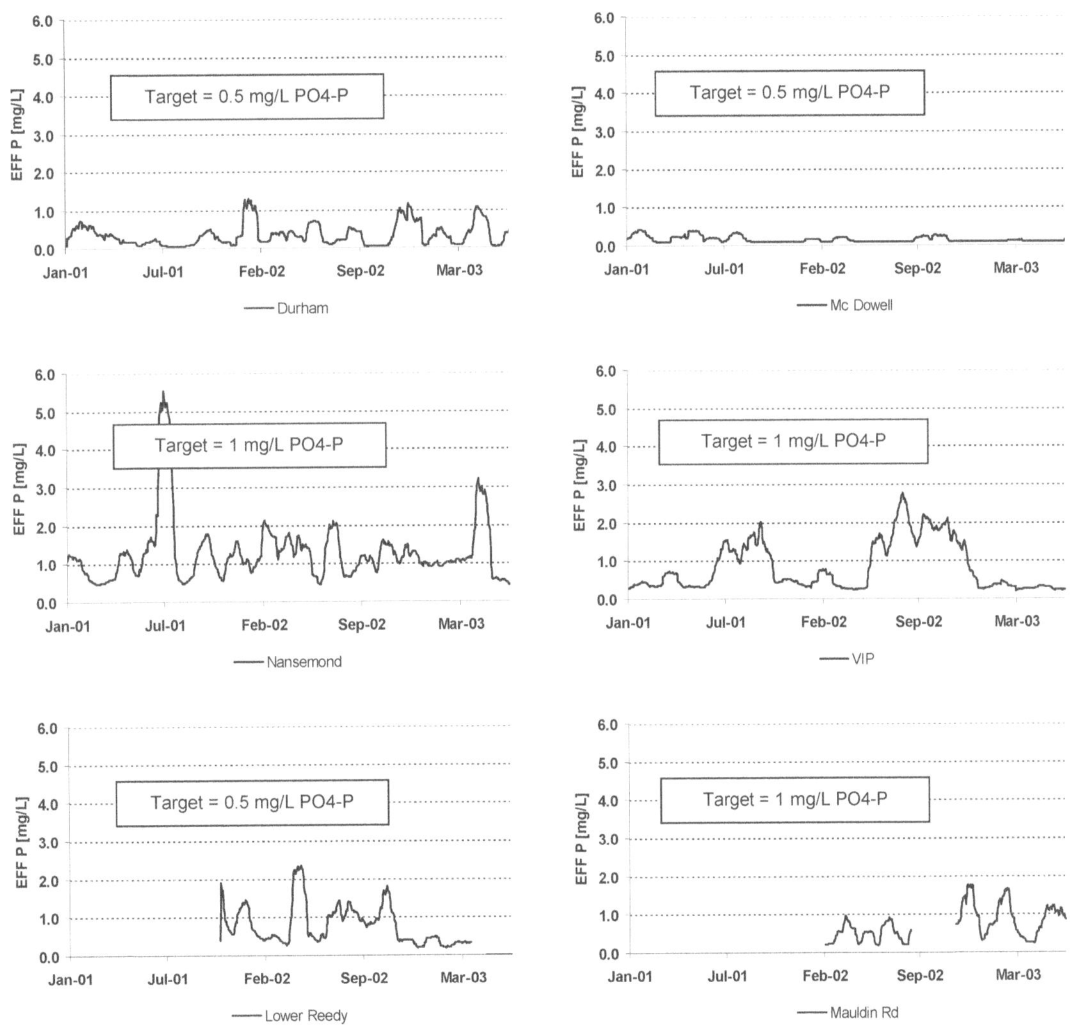


**Figure 6-13. Secondary Effluent Phosphate Concentrations.**

Table 6-20 shows the frequency with which the plants achieve effluent phosphorus concentrations of 0.5 mg/L, 1.0 mg/L, and 2.0 mg/L based on routine operating data over a three-year period. The table also compares the concentrations achieved in 50% and 90% of the samples, respectively. Smaller differences in these values represent greater reliability in treatment. The only facility that reliably removes phosphorus below 0.5 mg/L is McDowell Creek, where the addition of sugar waste greatly increases the BOD concentration in the secondary treatment process. However, many permits do not require that the permit limit be met on a daily basis, but rather on a monthly average, median, or annual basis. Therefore even a reliability of 70% can be sufficient to satisfy phosphorus discharge permits. In addition, chemical phosphorus-removal backup is a standard feature for EBPR plants with permit limits of less then 2 mg/L total phosphorus.

**Table 6-20. EBPR Reliability for Different Treatment Goals.**

| | Lower Reedy | VIP | Nansemond | McDowell | Durham | Average |
|---|---|---|---|---|---|---|
| OP < 0.5 mg/L | 76% | 70% | 24% | 95% | 50% | 68% |
| OP < 1 mg/L | 81% | 82% | 64% | 99% | 80% | 82% |
| OP < 2 mg/L | 91% | 92% | 85% | 100% | 98% | 93% |
| | | | | | | |
| 50% (Geometric Mean) | 0.09 mg/L | 0.29 mg/L | 0.76 mg/L | 0.10 mg/L | 0.05 mg/L | 0.26 mg/L |
| 90% | 1.9 mg/L | 2.0 mg/L | 2.5 mg/L | 0.20 mg/L | 1.2 mg/L | 1.6 mg/L |

The reliability of EBPR decreases with lower effluent phosphorus concentration. Figure 6-14 shows the overall reliability of all five Phase 2 facilities. When comparing the EBPR performance from the five Phase 2 facilities one finds that the reliability of EBPR to meet a 2 mg/L SE phosphate goal is 93%, compared to 68% when the goal was 0.5 mg/L. While these results include various process designs and chemical addition, the data do provide a good representation of the observed EBPR reliability for full-scale operating facilities.

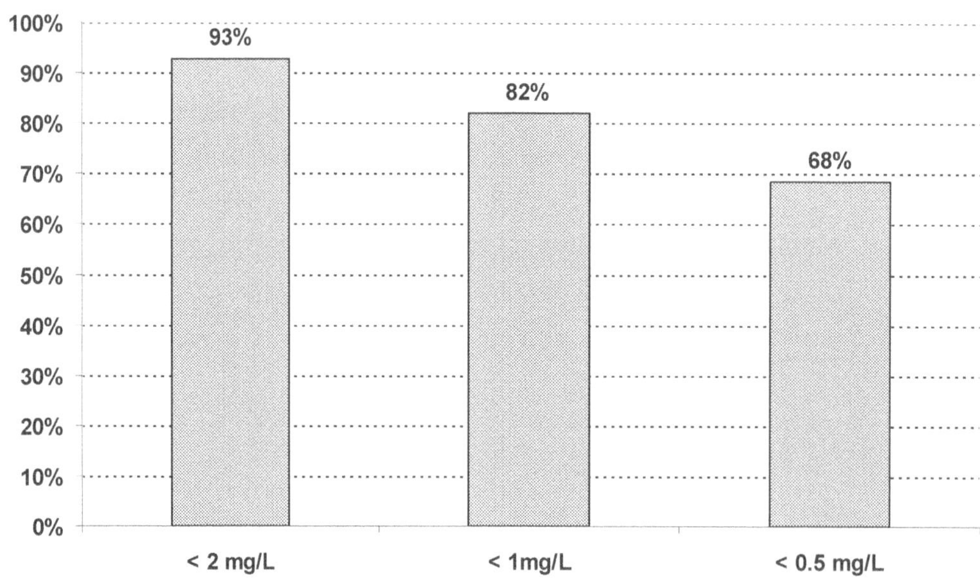


**Figure 6-14. EBPR Reliability to Meet Secondary Effluent Phosphate Concentrations of Less Than 2 Mg/L, 1 Mg/L and 0.5 Mg/L for All Five Phase 2 Facilities Combined (Based on Plant Data from 2000–2003).**

Overall, the variability in EBPR performance can be related to the variability of the operating conditions for EBPR. Typical facilities experience fluctuating influent loads, flows, and additional internal recycle loads. These results in a significant variability in:

- Anaerobic and aerobic retention time
- Aeration basin influent VFA
- Aeration basin influent ffCOD
- Anaerobic zone effluent phosphorus release

In addition, facility operational variations in chemical feed, recycled nitrate, or mixed-liquor wasting can contribute to the biological process variation.

### 6.6.3 Aeration Basin Phosphate Profiles

Figure 6-15 shows flow normalized secondary treatment phosphate profiles from six facilities investigated during Phase 2.

Contrary to expectation, the level of phosphorus release in the anaerobic zone did not correlate with the effluent phosphate concentration or the net phosphorus removal. It is mainly dependent on the available VFA (influent, added or generated) in the anaerobic zone.

The level of phosphorus release varies dramatically, even at the same facility. At Lower Reedy, where the most phosphate profile data were available, the flow-corrected phosphate at the end of the anaerobic zone ranged from 22 mg/L to 65 mg/L with effluent phosphate concentrations ranging between 0.1 mg/L and 1.8 mg/L. Figures 6-16 through 6-18 show Lower Reedy profiles at different periods. The profiles show that the phosphorus release varied significantly while maintaining similar effluent phosphate concentrations. This means that a wide range of available RBCOD or VFA resulted in similar net phosphate removal and almost identical effluent phosphate. Lower Reedy sometimes shows phosphorus release in the anoxic zone, something that has been observed at other facilities with VIP- or UCT-type process design. This is likely due to depletion of nitrate and development of anaerobic conditions in the anoxic zone.

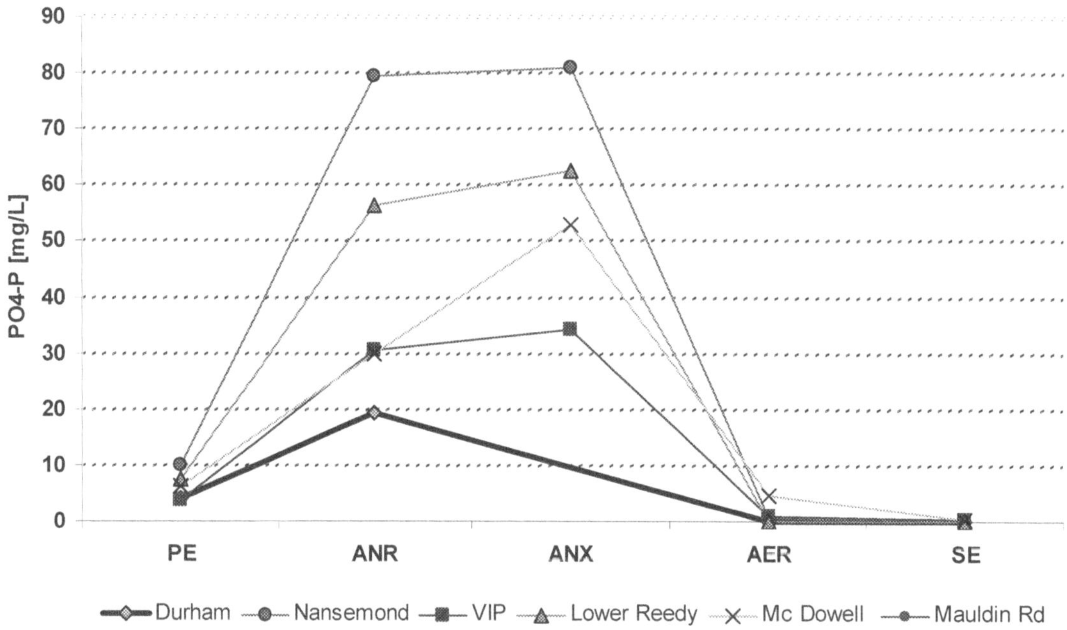


**Figure 6-15. Flow-Corrected Phosphorus Profiles.**

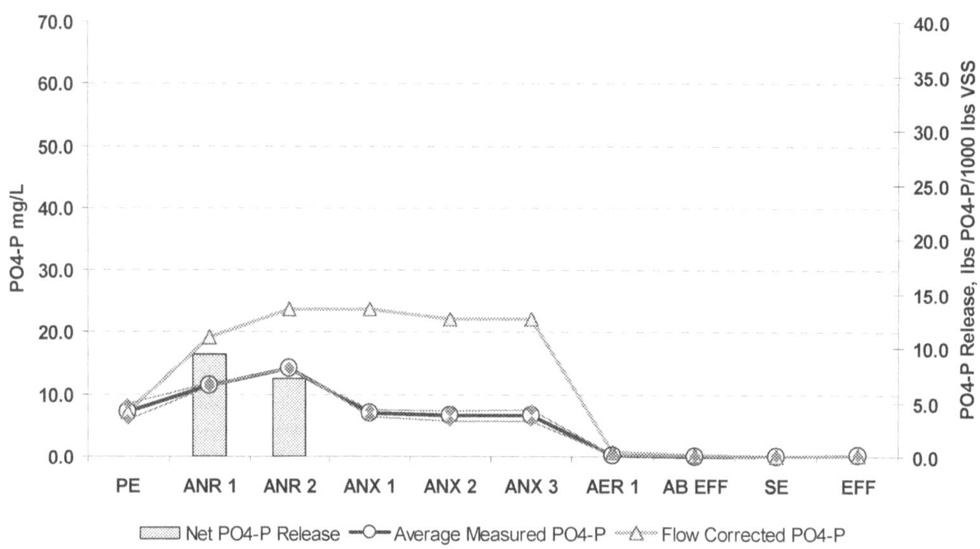


**Figure 6-16. Phosphate Profiles Measured During August 2001 Field Testing Showing Low Release with Low Effluent Phosphorus (shaded lines represent individual measured profiles).**

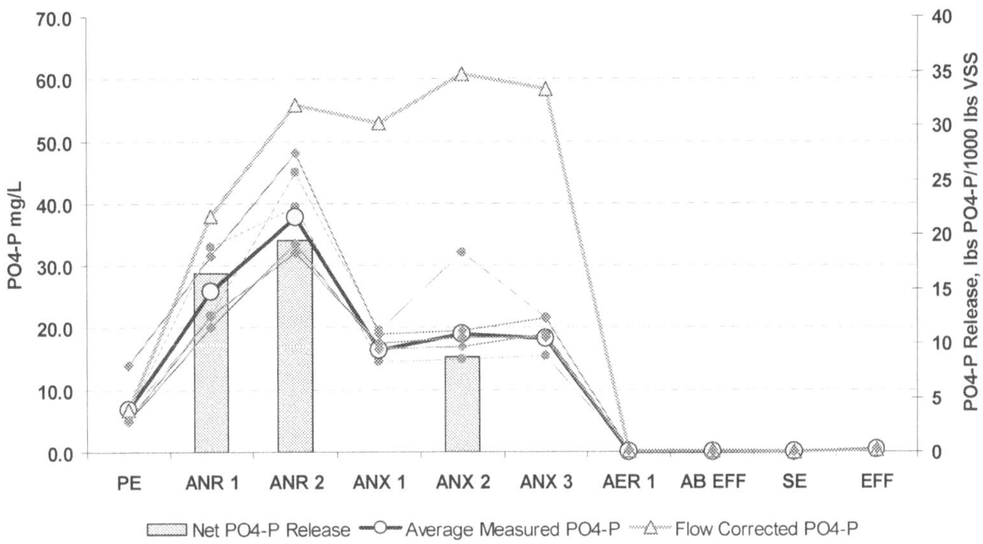


**Figure 6-17. Phosphate Profiles from February–March 2002 Showing High Release with Low Effluent Phosphorus (shaded lines represent individual measured profiles).**

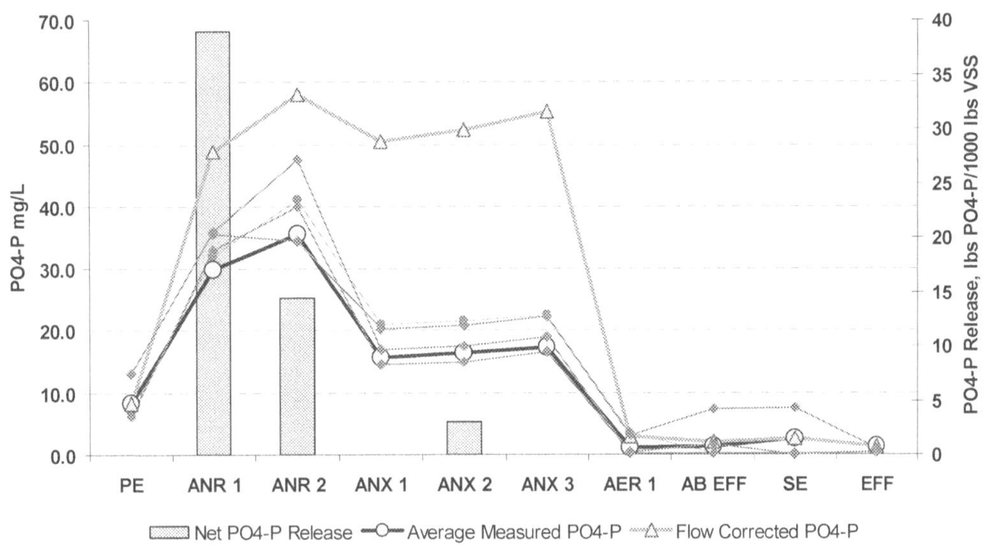


**Figure 6-18. Phosphate Profiles from April–May 2002 Showing Very High Release with Low Effluent Phosphorus But a Slight Increase in the Secondary Clarifier (shaded lines represent individual measured profiles).**

### 6.6.4 Uptake and Release Tests

The URT results (Table 6-21) completed in the second year were conducted either in the field or at the University of Washington using plant mixed-liquor samples. These URTs showed that the phosphorus release rate ranged between 4.0 mg P/g VSS/h and 30.6 mg P/g VSS/h and the uptake ranged from 1.4 mg P/g VSS/h and 7.6 mg P/g VSS/h. The ratio between the phosphorus release and uptake rates averaged 3.3 (ranging from 1.4 to 5.0). The results were normalized to VSS to allow some comparison with respect to release and uptake rates. Care should be exercised when comparing the data from one plant to another because of differences in

VSS composition. The test procedure and a discussion about the DO sensitivity of the test can be found in Appendices J and Q.

Table 6-21. Summary of Uptake and Release Rates Calculated from URT Results of 2003 (Initial Acetate COD Concentration = 80 Mg/L).

| | Month of 2003 | P-Release mg P /g VSS /h | P-Uptake mg P /g VSS/ h | P-Release/P-uptake |
|---|---|---|---|---|
| Durham Train 3 | May | 8.0 | 2.3 | 3.5 |
| Durham Train 4 | May | 7.4 | 2.3 | 3.2 |
| Durham Train 1 | Aug | 13.4 | 3.3 | 4.1 |
| Durham Train 2 | May | 5.6 | 1.4 | 4.0 |
| Nansemond | June | 5 | 3.4 | 1.5 |
| Nansemond | Sep | 7.7 | 2.3 | 2.3 |
| McDowell | June | 16.6 | 5.5 | 3.0 |
| McDowell | Sep | 7.1 | 1.7 | 4.2 |
| VIP | June | 12.4 | 7.6 | 1.6 |
| VIP | September | 11.6 | 3.0 | 3.8 |
| Lower Reedy | June | 30.6 | 6.1 | 5.0 |

### 6.6.5 Recycle Phosphorus Load

Recycle loads cause dynamic shock loads, change the influent BOD:P ratio, and cause process instability. Table 6-22 shows the recycle loads from the Phase 2 facilities. With the exception of the VIP plant, all facilities had anaerobic digesters (for both PSL and WAS). McDowell and Durham both add substantial amounts of chemical to supplement EBPR because of their low permits. This explained the lower recycle concentration and very low BOD:P ratio. The Nansemond facility stands out from the other plants due to its high industrial wastewater fraction from industrial food processing facilities within the service area. The result is a very high influent phosphorus load.

Table 6-22. Average Recycle Phosphorus Loads (2000–2003 averages).

| | Lower Reedy | VIP | Nansemond | McDowell | Durham |
|---|---|---|---|---|---|
| Total Recycle P | 54% | 9% | 86% | 26% | 24% |
| Anaerobic Digestion Recycle P | 54% | 9% | 86% | 23% | 14% |

### 6.6.6 Performance Expectation and Comparison of Facilities

The following conditions are considered to contribute to the ability to achieve consistent low phosphorus concentrations:

- Influent composition — high influent BOD:TP
- Exclusion of oxidants (nitrate and DO) from the anaerobic zone
- Excluding or minimizing recycle flows and load fluctuations from solids processing
- Favorable operating conditions (low SRT, modest temperature, sufficient DO in aeration basin, balanced anaerobic and aerobic HRTs).

Using these four conditions, the expected success of the plant to achieve EBPR consistently can be rated. The expectation can then be compared to the actual performance of the plants.

#### 6.6.6.1 Influent Composition

As more substrate is available to the PAOs relative to the influent phosphorus concentration, a higher phosphorus-removal efficiency due to EBPR is expected, given other conditions are equal. One of the problems in evaluating an EBPR process is the ability to measure the substrate available to the PAOs in the anaerobic zone. The available substrate is derived from two sources: first the influent RBCOD that includes VFA and other components that can be quickly fermented to VFA in the anaerobic zone, and second the VFA produced in the anaerobic zone from the hydrolysis and fermentation of colloidal and particulate material. The amount of VFA produced from these particles depends on the anaerobic contact time, and the nature of the influent particulate material. No method was available in this study to determine this portion of the PAO substrate.

The influent BOD:P ratio can be calculated for all the facilities and used to rate the suitability of the influent for EBPR. Higher influent BOD:P ratios have been associated with greater phosphorus removal by the EBPR process, suggesting that higher RBCOD concentrations may occur at higher BOD concentrations. Thus the BOD:P ratio provides an approximate indicator of potential EBPR performance.

#### 6.6.6.2 Exclusion of Oxidants from Anaerobic Zone

Nitrate and DO are known to affect the function of the anaerobic zone. Processes or designs that preclude nitrate or DO return to the anaerobic zone would therefore have a better chance to achieve consistent EBPR. Process configurations like VIP or UCT (or facilities that do not nitrify) would have a better ability to exclude nitrate than an A2O process.

#### 6.6.6.3 Recycle Loads and Fluctuating Loads

Recycle loads have been shown to cause less stable performance. Plants with anaerobic digesters and dewatering operation can experience large and sometimes magnified return phosphorus loads to the EBPR process. These facilities would therefore have lower expectation to achieve good performance.

Periods of low substrate availability are also expected to be detrimental to phosphorus-removal efficiency as it upsets the PAOs internal carbon storage and the VFA uptake in the anaerobic zone (Stephens and Stensel, 1998).

#### 6.6.6.4 Favorable Operating Conditions

Operating conditions such as low SRT, modest temperatures, and balanced aerobic and anaerobic times contribute to good EBPR performance.

SRT impacts EBPR performance. Shorter SRTs tend to produce more consistent results with a more active biomass. For a given influent RBCOD concentration, system with longer SRTs are expected to achieve higher effluent phosphorus concentrations due to the lower population of PAOs in the waste sludge due to endogenous respiration. In addition, it is possible that under more endogenous conditions the PAOs are less efficient for phosphorus storage (Stephens 1998).

Temperature impacts EBPR performance. High temperatures appear detrimental to EBPR performance (as in VIP performance and noted by Rabinowitz, et al., 2004).

Field studies have demonstrated that the phosphorus release rate is much higher than the phosphorus uptake rate. The ratio between the uptake rate and the release rate in Tables 6-21 and

6-26 are remarkably consistent—typically between three and four when expressed as release rate:uptake rate. This implies that in full-scale treatment facilities, one may expect that the aerobic (and some anoxic) contact time should be three to four times as long as the release time (anaerobic plus excess-anoxic time). Secondary release has also been found in plants with excessive aerobic time and can be seen in the field results (see, for example, Figure 6-17). Maintaining a consistent ratio appears to be beneficial to EBPR performance.

#### 6.6.6.5 Rating Scale for Expectation

The expectation of a plant to achieve consistently low effluent phosphate concentrations can be rated using the four fundamental factors described above. The following scale is used:

- 10 = Meets all four factors favorably
- 8 = Meets three of four factors favorably
- 6 = Meets two factors favorably, and one partially
- 4 = Meets one factor favorably and some partially
- 2 = Meets some factors partially
- 0 = Meets none of these factors

Intermediate ratings are used for facilities that exceed one level but do not quite meet the next level.

#### 6.6.6.6 Rating Scale for Plant Performance

The effluent performance reported can be rated on a scale of 0 to 10 according to the scale in Table 6-23. Ratings are based on evaluating the performance results. If the performance is better than one rating scale, but not quite at the next level, an intermediate rating is used. For example, if performance exceeds a "6" but is not quite an "8," then the rating is shown as a "7."

**Table 6-23. Rating Scale for Performance of Facilities.**

| Rating | Concentration | Percentile |
|---|---|---|
| 10 | 0.2 | 80% |
| 8 | 0.5 | 80% |
| 6 | 1.0 | 80% |
| 4 | 1.0 | 50% |
| 2 | 2.0 | 50% |
| 0 | No EBPR | |

#### 6.6.6.7 Results

Table 6-24 summarizes process and influent wastewater characteristics, SE phosphorus performance, and expected and actual ratings based on the criteria described above.

Table 6-24. Surveys, Tests, and Reports Completed at Investigated Facilities.

| Parameter | Durham | VIP | Nansemond | McDowell | Lower Reedy |
|---|---|---|---|---|---|
| Process | A2O | VIP | VIP | UCT | UCT |
| Lime addition | Yes | No | No | Yes | Yes |
| VFA addition | Fermenter | - | - | Sugar waste | - |
| SRT, days | 10.9 | 9.5 | 8.5 | 6.7 | 9.4 |
| % Anaerobic volume | 9 | 12 | 15 | 7 | 13 |
| Ratio Uptake/Release HRT[4] | 3.9 | 3.1 | 2.3 | 4.9 | 3.3 |
| Infl BOD:P[1] | 17-28 | 29-33 | 7-13 | 31-35 | 15-25 |
| Infl BOD:P[2] | 21.7 | 31.9 | 12.6 | 31.3 | 21.5 |
| ffCOD:P[1] | 19.5 | 30.9 | 6.7 | 34.9 | 15.6 |
| Avg sec effl NO3-N[2], mg/L | 8.4 | 6.5 | 6.4 | 4.1 | 4.7 |
| RAS, % | 61 | 31 | 84 | 87 | 52 |
| % Internal recycle rate | 300 | 73 / 77 | 91 / 99 | 70/ 320 | 116 |
| Avg effl sec effl ortho P[2], mg/L | 0.1 | 0.6 | NA | NA | 0.5 |
| Percent of SE ortho P data <0.50 mg/L[2] | 50 | 70 | 24 | 95 | 76 |
| Percent of SE ortho P data <1.0 mg/L[2] | 80 | 82 | 64 | 99 | 81 |
| Percent of SE ortho P data <2.0 mg/L[2] | 98 | 92 | 85 | 100 | 91 |
| Phosphorus Recycle Load[3], % | 24 | 9 | 86 | 26 | 54 |
| Expected Success Rating | 7 | 10 | 4 | 9 | 7 |
| Observed Performance Rating | 7 | 6 | 4 | 9 | 5 |

1. From intense sample periods (Table 6-17) – exclude fermenter or sugar additive
2. From routine plant operating data (2000–2003)
3. From mass balance
4. Release HRT = Anaerobic plus ½ Anoxic HRT; Uptake HRT = Aerobic + ½ Anoxic HRT

The McDowell and VIP facilities have high expectation ratings because they have the most optimal conditions for efficient phosphorus removal compared to the other facilities. Both facilities have a high influent BOD:P ratio, a low SRT, and minimal recycle of nitrate to the anaerobic zone due to the use of these UCT process. The VIP plant also has no return loads from the solids incinerator but operates at a higher temperature in summer. The Durham facility has a good expectation rating because it has a relatively low BOD:P ratio and nitrate entering the anaerobic zone as a result of using the A2O process. The expected performance is increased due to the continual VFA addition from the PSL fermenter. Lower Reedy has a similar expectation rating with poorer influent BOD:P ratio and high recycle load, but has the positive attribute of using the UCT process, which reduces nitrate return to the anaerobic zone. Nansemond has the worst expectation rating due to the low average influent BOD:P ratio, even though it does use the VIP process, which minimizes nitrate feed to the anaerobic zone.

None of the facilities received a performance rating of 10 (below 0.2 mg/L for 80% of the time). The McDowell facility performance meets the best rating with as 95% of its effluent soluble phosphorus concentrations under 0.5mg/L—this figure is in line with the expected performance. Similarly, the Durham expectation and performance rating were both good.

The VIP plant performance is worse that the expected rating would indicate. This can be attributed primarily to the summertime upsets noted at this plant. Also, the effluent-rating criteria (0.5 mg/L and lower) are below the permit requirements for this plant, and the performance may be more indicative of the need to meet permit limits. The performance at Lower Reedy is somewhat lower than expected.

The Nansemond plant is an interesting case. At first look, the performance is in line with the expectation—both relatively low. However, this plant has an extremely high recycle load and high influent phosphorus load. The performance in terms of phosphorus *removal* (not effluent quality) is therefore actually very high at this facility, even though effluent concentrations are higher than at other plants.

### 6.6.7 GAO Observations

#### 6.6.7.1 Sample Collection and Evaluation

Mixed-liquor samples were collected from each plant and examined to observe the presence of polyphosphate granules (indicative of PAOs) and to look for the presence of GAOs. GAOs often consist of large cocci (2–3 μm) arranged in tetrad clusters within the floc but can also be present as non-tetrad clustered bacteria (Crocetti, et al., 2002).

Both microscopic study and molecular techniques were used to examine the presence of GAOs in plant mixed-liquor samples. Light microscopy observation was used to observe the tetrad-forming GAOs, and molecular probing using specific oligonucleotide probes targeting GAOs was completed to identify both tetrad-forming and non-tetrad-forming GAOs. In addition Neisser and DAPI staining were applied to observe the presence and relative abundance of PAOs.

Figure 6-19 and Figure 6-20 show representative observations of PAOs and GAOs from the VIP facility using a Neisser stain. Figure 6-21 shows the results of a DAPI stain. The GAOs have Neisser-positive-stained cell walls, but are not positive for cell inclusions as are PAOs. The Neisser-positive-stained PAOs are dark blue to purple in color and usually group in clusters within the flocs. The DAPI-stained PAOs are yellow in color, which is characteristic of polyphosphate granules observed under the fluorescence microscope. Other non-polyphosphate-accumulating bacteria including GAOs show up in blue with DAPI stain.

Examination of the mixed-liquor samples was also conducted using molecular methods (FISH) using 16S rDNA-based probes targeting *Accumulibacter* (representing PAOs) (Crocetti, et al. 2000) and some non-tetrad-forming GAOs—*Competibacter* (GAOQ989 and GB-G2, Crocetti, et al., 2000 and Kong, et al., 2002).

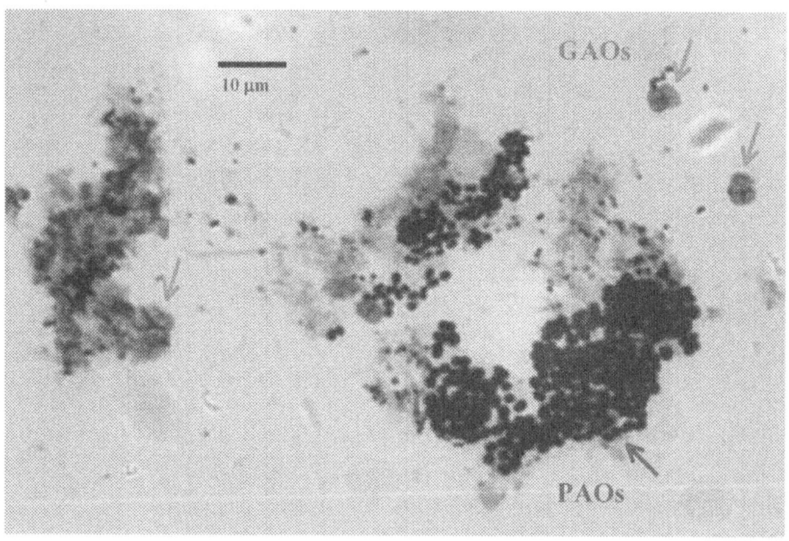


**Figure 6-19. Neisser Stain Observation of PAOS in a Mixed Liquor Sample from VIP (7/3/2003).**

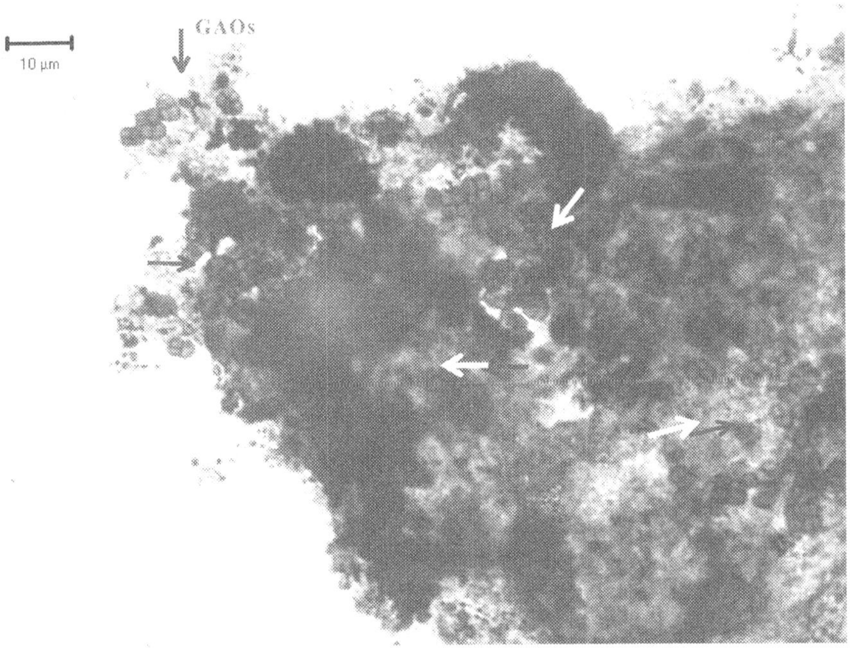


Figure 6-20. Neisser Stain Observation of GAOs in a Mixed Liquor Sample from VIP (9/12/2003).

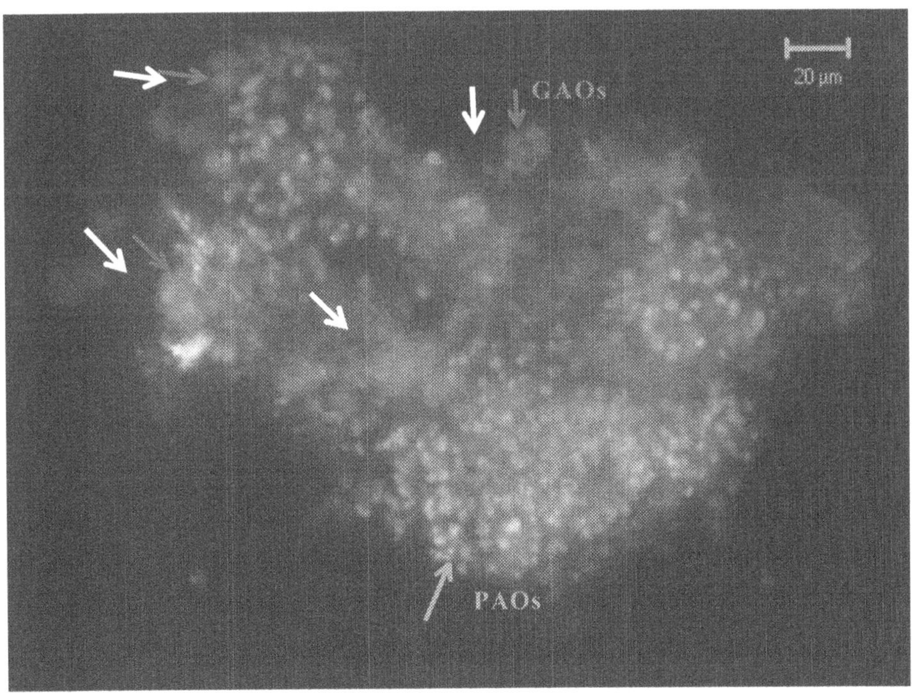


Figure 6-21. DAPI Stain Observation of PAOs and Tetrad-forming GAOs in a Mixed Liquor Sample from VIP (9/3/2003).

#### 6.6.7.2 Results of GAO Analysis

Table 6-25 summarizes the observation of PAOs and GAOs for mixed-liquor samples collected during Year 2 of the study using both the staining and molecular probing methods.

Table 6-25. Summary of PAOs and Tetrad-forming GAOs Observations.

| Sample | GAO observation (tetrad-forming) | GAO observation (*Competibacter*) | PAO observation | Comments |
|---|---|---|---|---|
| Durham, Side 1 | Some, 3 out of 10 fields observed GAOs, some grouped together | Not observed. | Abundant, big flocs of PAOs | PAOs dominant , some GAOs were easily observed |
| Durham, Side 2 | Easily found, 5 out of 10 fields observed GAOs, some grouped together | Not observed. | Abundant, big flocs of PAOs | PAOs dominant , some GAOs were easily observed |
| VIP, July 2003 | Some, 3 out of 10 fields observed GAOs, 2-27 tetrapods per field and spread, not in big flocs grouped together | 1-10% of total Eubacteria | Abundant, cocci shaped cells group together in big flocs | PAO dominant, both tetra pod-shaped and *Competibacter*-type GAOs observed |
| VIP, September 3, 2003 | Easily found, 5 out of 10 fields observed GAOs, some group together | 1-10% of total Eubacteria | Abundant, in big flocs | PAO dominant, both tetra pod-shaped and *Competibacter*-type GAOs easily observed |
| VIP, September 12, 2003 | Abundant, 8 out of 10 fields observed GAOs, in big flocs, some fields had almost all GAOs | 1-10% of total Eubacteria | Easily found, in flocs | Abundant GAOs, almost as abundant as PAOs. |
| Nansemond, July 2003 | Not observed | 10-20% of total Eubacteria | Abundant, cocci shaped cells group together in big flocs | PAO dominant, no tetra pod-shaped GAOs were observed, but *Competibacter*-type GAOs were observed. |
| Nansemond, September 2003 | Not observed | | Abundant, cocci shaped cells group together in big flocs | PAO dominant, no GAOs was observed |
| McDowell | Abundant, 8 out of 10 fields observed GAOs, grouped together in focs, some fields almost had all GAOs. | Not observed. | Not many PAOs flocs, small poly-P granules | GAOs seemed to be dominant, more abundant than PAOs. |
| Lower Reedy | A few,1 out of 10 fields observed GAOs, 3-5 tetrapods per field | Not observed. | Abundant, cocci shaped cells group together in big flocs | PAOs dominant in big flocs, only a few GAOs were found |

### 6.6.7.3 Discussion of GAO Evaluation

Table 6-26 below compares the PAO and GAO observations with phosphorus release and uptake rates (from URT tests) and effluent phosphate concentrations at the time samples were collected for URT and PAO/GAO analysis. The specific phosphorus release rate shows the relative PAO abundance in the sludge. The P-release: acetic acid uptake ratio indicates the relative PAO and GAO activities at a given pH. The ratio is also pH dependent, however, no significant pH impact is expected since the pH was between 7–8 in all tests. Higher P-release: acetic acid -uptake ratios indicate high relative PAO activity, and low ratios indicate low relative GAO activity. Table 6-22 shows that the P-release: acetic acid uptake ratio varies among plants ranging from 0.14 to 0.78 gP/g acetic acid. The values are consistent with those reported in the literature (Schuler and Jenkins, 2003; Maurer, et al., 1997). The different P/acetic acid ratios observed at two parallel trains at the Durham plant (0.14 to 0.40 mg P/mg acetic acid as COD) indicated that Side 1 has more PAO activity and Side 2 has less PAO activity. This agrees with the microscopic tetrad-forming GAO observation. The results of the August 2003 testing at Durham showed that at that time, Side 1 was performing quite well (average effluent phosphate of 0.05 mg/L during the month), and Side 2 was struggling (average effluent phosphate of 1.22 mg/L during the month). The more abundant tetrad-forming GAOs observed in Side 2 could contribute to the relative poor phosphorus-removal performance on Side 2.

**Table 6-26. Comparison of PAO/GAO Observations and Uptake and Release Test Results.**

| Sample | Summary of Microbial Investigation | Phosphorus Release Mg P/mg Acetic acid | Phosphorus Release Rate (mg P/g VSS/hr) | Phosphorus Uptake Rate (mg P/g VSS/hr) | RR/UR ratio[2] |
|---|---|---|---|---|---|
| Durham, Side 1 | PAOs dominant , some tetra pod-shaped GAOs[3] were easily observed | 0.4 | 10 | 3.6 | 2.8 |
| Durham, Side 2 | PAOs dominant , some tetra pod-shaped GAOs were easily observed | 0.14 | 4.2 | 1.5 | 2.8 |
| VIP July 2003 | PAO dominant, both tetra pod-shaped and *Competibacter*-type GAOs observed | 0.39[1] | 11.6[1] | 3.0[1] | 3.9 |
| VIP September 3, 2003 | PAO dominant, both tetra pod-shaped and *Competibacter*-type GAOs easily observed | 0.32 | 11.1 | 2.2 | 5.0 |
| VIP September 12, 2003 | Abundant tetra pod-shaped and *Competibacter*-type GAOs, almost as abundant as PAOs. | NA | NA | NA | NA |
| Nansemond July 2003 | PAO dominant, no tetra pod-shaped GAOs were observed, but *Competibacter*-type GAOs were observed. | NA | NA | NA | NA |
| Nansemond September 2003 | PAO dominant, no tetra pod-shaped GAOs were observed, but *Competibacter*-type GAOs were observed | 0.30 | 7.7 | 2.6 | 3.0 |
| McDowell | Tetra pod-shaped GAOs seemed to be dominant, more abundant than PAOs. | 0.28 | 5.5 | 1.4 | 3.9 |
| Lower Reedy | PAOs dominant in big flocs, only a few GAOs were found | 0.78 | 24.3 | 4.6 | 5.2 |

1. URT run in August 2003
2. RR/UR = Ratio of Release Rate to Uptake Rate
3. Tetra-pod shaped organism observed were presumed to be GAOs since they are the most commonly found morphology type of known GAOs. It was possible that some tetrad-shaped organisms may not be GAOs (Tsai and Liu, 2002).

Lower Reedy has a high phosphorus release rate of 0.78 mg P/mg acetic acid (COD) indicating a PAO-dominated sludge. This is confirmed by both the microscopic observation and the FISH study, which indicate that there were nearly no GAOs in the sludge and PAOs were dominant. The one plant with dominant GAOs (McDowell) has a relative low release ratio (0.28 mg P:mg acetic acid), which seems to agree with the suggested quantitative values for characterizing relative GAO and PAO activity by Schuler and Jenkins (2003) that GAO-dominant sludge has P/acetic acid ratio <0.27 mg P:mg acetic acid.

The specific uptake and release rates indicate the relative PAO activity level and shed light on the dynamics in the EBPR process. The average phosphorus release rates were found to be significantly higher than the average phosphorus uptake rates for all plants tested. Release rates varied from 4.2 mg P/g VSS/h in Train 2 of the Durham facility to 24.3 mg P/g VSS/h at Lower Reedy. The initial P uptake rates varied among plants, and the average P uptake rate was generally quite slow (as demonstrated in other URT tests) and indicates the importance of providing sufficient aerobic time to complete the phosphorus uptake in full-scale facilities.

The results of the GAO analysis indicate that the environmental conditions at all facilities allowed the growth of GAOs during the study period. Many factors are known to have an effect on the competition between the PAOs and GAOs including feed substrate, COD:P ratio, pH, SRT, temperature and DO (Stephens, et. al, 2003). More complex substrate favors the growth of GAOs, so changes in influent wastewater characteristics may have an influence on the stimulated growth of GAOs. Phosphorus limiting conditions (COD:P >40-50) and low pH (under 7.3) also favor the growth of GAOs. Higher temperature (>20$^{o}$C) and longer SRT (>7-10 days) are also known to favor the growth of GAOs. Other factors such as the availability of other electron acceptors in the anaerobic zone may also have an effect on the phosphorus-removal process. The amount of nitrate introduced to the anaerobic zone could affect the intracellular poly-P storage formed and the stability of the phosphorus removing population and performance.

Some of the observations made during the study can be related to the characteristics described above. At the Nansemond facility where the BOD:P ratio was typically lower than the other facilities, no tetrad-forming GAOs were observed, and a small number of GAOs were identified by molecular probing. By contrast, at the McDowell facility where the cola waste with a concentration of up to 130,000 mg/L BOD is added to the EBPR process, GAOs were more abundant than PAOs. This likely reflects the impact of BOD-limited conditions at the Nansemond facility and phosphorus-limited conditions at the McDowell facility. In other cases, the GAO observations could not be easily attributed to operating conditions. For example, mixed-liquor samples collected at the same time from two parallel treatment trains at the Durham facility showed different levels of GAOs. These results suggest that while the presence of GAOs does not necessarily preclude EBPR, an increase in the GAO population (as observed at the VIP plant) can indicate a decrease in EBPR performance.

## 6.7 General Recommendations to Improve EBPR Reliability

Based on the field investigations and corrective actions implemented by the various facilities, the following recommendations can be made to improve EPBR reliability. The improvements are presented in terms of both operational strategies and design features.

### 6.7.1 Operational Strategies

- Minimize the impact of phosphorus recycle loads from solids processing. This can be accomplished through a variety of approaches, including:

- Side stream management. Avoid or minimize the return of solids processing recycle loads during peak load hours. This can be accomplished through temporary storage, dewatering during off-peak hours, or dewatering continuously.
- Phosphorus precipitation from a side stream. Precipitating the recycle phosphorus improves EBPR performance and reliability by reducing the high variable load to the plant. Side stream chemical precipitation is also more effective then precipitation in secondary or tertiary treatment because the phosphorus concentrations are higher and fewer metal hydroxides are formed.
- Metal addition to solids processing feed. Adding metal ions such as Mg, K, Ca, Fe, or Al upstream of solids processing will allow phosphorus to be removed with the solids and will minimize the phosphorus load in the dewatering recycle.
- Minimizing phosphorus release from WAS prior to solids processing. Process operation should avoid extended unaerated storage time or mixing of WAS and PSL prior to thickening or dewatering as these conditions can result in rerelease of stored phosphorus.

♦ Maintain consistent operation conditions.

- Avoid frequent changes in operation condition. At typical EBPR SRTs, the activated sludge system may take several weeks to adjust to operational changes. Optimum performance is observed when EBPR systems operate at a steady state. Therefore, frequent changes can prevent the system from achieving stable performance.
- Adjust SRT and HRT to seasonal changes. Provide EBPR with ideal conditions by maintaining the proper SRT and HRT. Seasonal changes in wastewater composition and temperature may require different unaerated and aerated periods. For example, anoxic zone requirements may change as denitrification kinetics change.
- Adjust anaerobic and aerobic retention times to match release and uptake kinetics. The release rate is typically three times higher than the uptake rate in this study suggesting that aerobic retention times should be at least three times higher than anaerobic retention times.
- Avoid extended anaerobic conditions that can result in phosphorus release without VFA uptake (secondary release).
- Use flow-paced RAS and internal mixed-liquor recycle flow rates.
- Equalize influent phosphorus loads to avoid dynamic loading.

♦ Optimize sampling and process monitoring to enable real time process control and troubleshooting.

- Identify reliable online analytical equipment to provide real time data for process and water quality parameters, promoting early detection of process upsets.
- Monitor aeration basin influent composition to facilitate process control adjustments and support troubleshooting.
- Monitor water quality in recycle streams, especially those from solids processing.

- Minimize sample analysis turnaround time to promote early corrective actions. Where online monitoring is not feasible, use portable test kits to measure critical water quality parameters.

♦ Include troubleshooting procedures in the plant operation strategy.

  - Collect reference data to establish baselines for stable EBPR performance. Baseline testing should include aeration basin phosphate profiles, phosphate URTs, GAO tests, and RAS $PO_4$-P and $NO_3$-N measurements.

♦ Use chemical addition as required to enhance phosphorus removal. This can include:

  - Polymer addition to improve solids capture in dewatering,
  - Addition of iron- or aluminum-containing chemicals to enhance phosphate precipitation in secondary or tertiary treatment, or
  - Addition of VFAs from sludge fermenters, solids thickeners, or other sources to enhance EBPR. Because these chemicals may shift the population dynamics of the activated sludge culture, changes should be implemented gradually.

### 6.7.2 Design Features

♦ Provide means of recycle load management or removal. Side stream management is a powerful tool to improve EBPR performance. Provide storage tanks, chemical feed systems, or other provisions for side stream management.

♦ Design flexibility in anaerobic/anoxic/aerobic operation of various activated sludge basin compartments to provide operational flexibility to match release and uptake kinetics.

♦ Design aeration basin influent and recycle streams to minimize dissolved oxygen and nitrite/nitrate concentrations entering the anaerobic zone of the activated sludge basin.

♦ Design flexibility in aeration basin operation to allow basins to be placed online and offline to adjust HRT in response to seasonally-changing flows.

♦ Avoid mixing of primary and secondary sludge prior to thickening to avoid rerelease of stored phosphorus prior to digestion.

♦ Design chemical feed systems to be load-paced instead of flow-paced to better meet demand.

♦ Provide chemical phosphorus removal for polishing. Consider using online feed control.

♦ Minimize potential for hydraulic short circuiting, which changes the operating HRT and reduces the available treatment volume.

## 6.8 Research Needs

This investigation has identified some specific topics and questions that remain unanswered. These present some opportunities for future research, either by WERF or others.

♦ Determine the impact of GAOs on performance of full-scale EBPR plants. In this research GAOs and PAOs appeared to coexist in EBPR plants without a major impact on the EBPR performance. However, since GAO organisms remove substrate needed for PAO metabolism, their presence poses a threat to having sufficient substrate available to maintain efficient EBPR.

♦ Determine whether the species of PAOs vary in EBPR facilities, and if so, what causes the variation and what is the impact of PAO species in EBPR plants.

- Determine the impact of fluctuations in influent load (both organics and phosphorus) on EBPR variability. Changes in BOD:TP ratio or RBCOD:TP ratio appear to affect EBPR performance. Since these changes are uncontrolled events at full-scale facilities, the impact and potential control of load variability should be determined.
- Recycle loads have a significant impact on EBPR performance. In particular, the fate of phosphorus in digesters (aerobic and anaerobic) and other solids processing streams should be investigated in order to determine the role of chemical precipitation and phosphorus release and uptake.
- Determine the role of chemicals in EBPR processes. In particular, determine the fate of phosphorus with added chemicals. Determine the role of chemicals during solids processing—thickening, digestion, and dewatering. To what extent do the added chemicals continue to precipitate phosphorus during solids processing?
- Evaluate the performance of cold and warm temperature EBPR systems. Confirm and expand on recent reports and results from this study that suggest that EBPR deteriorates when temperatures exceed 30°C.
- Determine the significance of the relationship between phosphorus release (anaerobic time and available VFA) and uptake in batch testing. Does the batch test indicate both the quantity and some metabolic state of the PAOs? Can it be used to determine the EBPR capacity for a given system and changes in this capacity as a function of operating conditions?
- Develop protocols for calibrating the IWA-activated sludge models to full-scale or pilot plant facilities. What are key process parameters that should be measured and used to calibrate the activated sludge models for EBPR?

## 6.9 Conclusions

The results of this study lead to following conclusions.

- EBPR is capable of achieving very low effluent phosphate concentrations. When operating well, phosphate concentrations <0.1 mg/L can be achieved for extended periods (more than a month), 0.03 mg/L for a week, and even below 0.02 mg/L for several sequential days. Excursions above these levels are common.
- EBPR performance can be estimated by examining the plant influent BOD and TP loading, the level of oxidants in the anaerobic zone, the degree of recycle phosphorus control, and operating parameters (SRT, temperature, DO, anaerobic and aerobic HRT).
- To reliably achieve effluent phosphorus concentrations of 0.5 mg/L at 50$^{th}$ percentile, chemical backup or BOD augmentation (through fermentation or adding readily biodegradable substrate source) are necessary.
- The performance of phosphorus removal improves when the BOD:TP ratio is above 25:1. Absolute values of the BOD:TP ratio or PE BOD load did not correlate directly with phosphorus removal as long as the BOD:TP ratio was high.
- Changes in activated sludge influent load (as indicated by the BOD:TP ratio) correlated to changes in effluent phosphorus concentration. This variation appears to be caused by a change in the steady-state operation of the PAOs, resulting in excessive release of the large accumulated phosphorus mass in the biomass.

- Phosphate release varied widely (from under 20 mg/L to over 70 mg/L) between plants and even under different loading conditions at one plant. A high anaerobic phosphate release did not necessarily correlate with low effluent phosphate concentrations. Other factors such as chemical precipitation with calcium in the anaerobic zone or excess (secondary) phosphorus release could impact the measured phosphate concentration in the anaerobic zone.
- A range of phosphate release to VFA uptake ratios can be associated with functioning EBPR and low effluent phosphorus concentrations.
- Recycle streams play a significant role in adding variability to EBPR performance. Eliminating or attenuating the recycle load fluctuations greatly improves the process reliability.
- The presence of GAOs in full-scale facilities does not necessarily preclude the plant from achieving EBPR provided there is sufficient carbon in the influent.
- Chemical addition (such as ferric, alum, or lime) for phosphate precipitation or to meet other operational needs, such as alkalinity or odor control, can be recycled to the aeration basin and react with phosphate.
- Optimizing EBPR by adding a carbon source (VFA) is a viable method for reliably achieving very low (<0.02 mg/L) effluent phosphate concentrations.

# APPENDIX A

# SCREENING SURVEY
# WATER ENVIRONMENT RESEARCH FOUNDATION
# ENHANCED BIOLOGICAL PHOSPHORUS REMOVAL
# (EBPR) SURVEY

## Facility Name and Location

Facility Name

Facility Owner

City

State

Country

## Contact Information

First Name

Last Name

Phone (Include Country Code if not USA)

Fax (Include Country Code if not USA)

Email

## General Information

Treatment Design Capacity (Under EBPR Conditions). Enter MGD, m3/d, or population equivalents (PE).

Capacity: MGD

Capacity: m3/d

Capacity: PE

Average Plant Flow during Phosphorous Permit Season

Flow: MGD

Flow: m3/d

Secondary Treatment Goal (select all that apply)

- [ ] Phosphorous Removal
- [ ] Nitrification
- [ ] Nitrification/Denitrification

What type of phosphorus removal limit is in place?

- [ ] Seasonal
- [ ] Year-round
- [ ] No Permit

For how many years has the plant operated in EBPR mode?

- [ ] < 1
- [ ] 1 – 2
- [ ] 2 – 5
- [ ] 5 – 10
- [ ] > 10

What is the target secondary effluent SOLUBLE phosphorus concentration?

- [ ] < 0.1 mg P/L
- [ ] 0.1 – 0.5 mg P/L
- [ ] 0.5 – 2.0 mg P/L
- [ ] 2.0 – 5.0 mg P/L
- [ ] > 5.0 mg P/L

What is the target secondary effluent TOTAL phosphorus concentration?

- [ ] < 0.1 mg P/L
- [ ] 0.1 – 0.5 mg P/L
- [ ] 0.5 – 2.0 mg P/L
- [ ] 
- [ ] 

2.0 – 5.0 mg P/L

> 5.0 mg P/L

What is the target secondary effluent ammonia-nitrogen concentration?

- [ ] No target
- [ ] < 0.1 mg N/L
- [ ] 0.1 – 0.5 mg N/L
- [ ] 0.5 – 2.0 mg N/L
- [ ] 2.0 – 5.0 mg N/L
- [ ] 5.0 – 10.0 mg N/L
- [ ] > 10.0 mg N/L

What is the target secondary effluent total inorganic nitrogen (Ammonia + Nitrate + Nitrite) concentration?

- [ ] No target
- [ ] < 0.2 mg N/L
- [ ] 2.0 – 5.0 mg N/L
- [ ] 5.0 – 10.0 mg N/L
- [ ] > 10.0 mg N/L

What is the sewage temperature range during the phosphorus permit season?

Minimum: [ ] °C

Average: [ ] °C

Maximum: [ ] °C

What range of SRT is typically operated during the phosphorus permit season?

- [ ] Don't know
- [ ] < 3 Days
- [ ] 3 – 5 Days
- [ ]
- [ ]

5 – 10 Days

> 20 Days

## Influent Information

Enter Average Plant Influent Concentrations (leave empty if unknown)

| | | |
|---|---|---|
| COD | ☐ | mg/L |
| TOC | ☐ | mg/L |
| BOD | ☐ | mg/L |
| SBOD | ☐ | mg/L |
| VFA | ☐ | mg/L |
| TSS | ☐ | mg/L |
| TKN | ☐ | mg N/L |
| NH4-N | ☐ | mg N/L |
| NO3-N | ☐ | mg N/L |
| PO4-P | ☐ | mg P/L |
| TP | ☐ | mg P/L |
| Alkalinity | ☐ | mg/L as $CaCO_3$ |

Which metals are routinely being measured? (select all that apply)

- ☐ Calcium (Ca)
- ☐ Potassium (K)
- ☐ Magnesium (Mg)
- ☐ Iron (Fe)
- ☐ Aluminum (Al)

## Secondary Treatment Process Information

Which of the following treatment processes are in operation? (select all that apply)

- [ ] Primary Clarifier
- [ ] Trickling Filter
- [ ] Oxidation Ditch
- [ ] Aeration Basin
- [ ] Sequencing Batch Reactor (SBR)
- [ ] Rotating Biological Contact (RBC)
- [ ] Membrane Bioreactor
- [ ] Secondary Clarifier
- [ ] Effluent Filtration
- [ ] Disinfection (Chlorine)
- [ ] Disinfection (Non-Chlorine)
- [ ] Dechlorination
- [ ] Other (please specify) ____________

Select the components of your EBPR process in operation. (select all that apply)

| | Yes | No |
|---|---|---|
| Aeration | | |
| Anoxic Zone | | |
| Anaerobic Zone | | |
| Mixed Liquor Recycle from Aerobic to Anoxic Zone | | |
| Mixed Liquor Recycle from Anoxic to Anaerobic Zone | | |
| Mixed Liquor Recycle from Aerobic Zone to Anaerobic Zone | | |
| Phosphorus Stripper | | |
| Sequence Batch Reactor | | |
| Step Feed | | |
| Simultaneous Nitrification/Denitrification | | |

Characterize supplemental BOD (VFA, Methanol) addition, if applicable. (select all that apply)

| | Always | On Demand |
|---|---|---|
| VFA from Primary Sludge Fermenter | | |
| VFA from Other Source (purchased etc.) | | |
| Methanol | | |
| Other | | |

Additional comments:

Characterize the supplemental alkalinity addition.

| | Magnesium Hydroxide | Lime | Caustic | Other |
|---|---|---|---|---|
| No Addition | | | | |
| Sometimes | | | | |
| On Demand | | | | |
| Always (Continuously) | | | | |

Additional comments:

## Supplementary Chemical Addition for Phosphorus Removal

What type of chemical is used for phosphorus precipitation? (select one if applicable)

- ☐ No Chemical Added
- ☐ Ferric
- ☐ Ferrous
- ☐ Alum
- ☐ Lime
- ☐ Other

Indication supplemental chemical addition points for phosphorus removal. (select all that apply)

- ☐ Primary Clarifier Influent
- ☐ Aeration Basin
- ☐ Secondary Clarifier Influent
- ☐ Tertiary Clarifier Influent
- ☐ Effluent Filter Feed
- ☐ Dewatering Filtrate
- ☐ Anaerobic Digester
- ☐ Other Recycle Streams
- ☐ No Chemicals Added

Enter the average chemical phosphorus precipitation dosage rate. Use total dose in case of multiple addition points.

Feed Dosage: ☐ mg/L

Daily Dosage: ☐ lbs/d

Daily Dosage: ☐ kg/d

## Solids Handling Process Information

Select waste activated sludge thickening method

☐ Thickening Centrifuge ☐ Belt Thickener ☐ DAF ☐ Other

Characterize Biosolids Processing. (select all that apply) Note: Check the "Recycle" column if the solids process returns recycle flow to the main liquid stream process.

| | Primary Sludge | WAS | Recycle? |
|---|---|---|---|
| Anaerobic Digestion | | | |
| Aerobic Digestion | | | |
| Sludge Lagoon | | | |
| Sludge Incineration | | | |
| Lime Stabilization | | | |

Additional comments: ☐

Select biosolids dewatering method.

- [ ] Dewatering Centrifuge
- [ ] Belt Filter Press
- [ ] Recessed Chamber Filter Press
- [ ] Drying Bed
- [ ] Other

Return location for sludge dewatering recycle flows

- [ ] Primary Influent
- [ ] Primary Effluent
- [ ] Other (please specify)

If you selected other, please specify: ______________________

## Secondary and Plant Effluent

Enter average secondary and/or final effluent concentrations. (leave empty if unknown)

| | | |
|---|---|---|
| Secondary effluent COD | | mg/L |
| Secondary Effluent TOC | | mg/L |
| Secondary Effluent BOD | | mg/L |
| Secondary Effluent TSS | | mg/L |
| Secondary Effluent TKN | | mg N/L |
| Secondary Effluent NH4-N | | mg N/L |
| Secondary Effluent NO3-N | | mg N/L |
| Secondary Effluent PO4-P | | mg P/L |
| Secondary Effluent TP | | mg P/L |
| Final Effluent COD | | mg/L |
| Final Effluent TOC | | mg/L |
| Final Effluent BOD | | mg/L |
| Final Effluent TSS | | mg/L |
| Final Effluent TKN | | mg N/L |
| Final Effluent NH4-N | | mg N/L |
| Final Effluent NO3-N | | mg N/L |
| Final Effluent PO4-P | | mg P/L |
| Final Effluent TP | | mg P/L |

## Process Data Availability

Describe the available data. (select all that apply)

| | Paper Records | Electronic Records |
|---|---|---|
| Daily Values | | |
| Monthly Average | | |
| Special Study Data | | |

## Project Participation Information

Can data be provided to the WERF Team for statistical analysis?

☐ Yes ☐ No ☐ Not Sure

**Thank you for completing this survey. We will send you a copy of the compiled results.**

# APPENDIX B

# DURHAM FACILITY ENHANCED BIOLOGICAL PHOSPHORUS REMOVAL (EBPR) EVALUATION REPORT

## B.1 Summary of Findings

The Durham Facility located in Tigard, Oregon, is operated by Clean Water Services of Washington County (District). The plant was designed to operate in either UCT or $A^2O$ mode, and typically operates in $A^2O$ mode. The plant has a dry weather design capacity of 25 million gallons per day (MGD). Average dry weather flows have decreased recently due to a change in the District's service area, and are currently approximately 17 MGD. Alum can be added upstream of the primary, secondary, and tertiary treatment processes to meet the seasonal total phosphorus (TP) permit limit of 0.07 mg P/L. Primary sludge is fermented in a two-stage fermenter/thickener, and volatile fatty acids (VFAs) are elutriated and returned to the secondary treatment process. Primary, waste activated and chemical sludges are anaerobically digested and dewatered prior to land application in eastern Oregon. Dewatering centrate is returned to the primary effluent pump station upstream of the aeration basins.

The aeration basin influent composition at Durham is quite constant with favorable nutrient ratios for BNR and EBPR. The $BOD_5$/TP ratio averages 20:1. A primary sludge fermenter/thickener was recently brought online to provide additional VFAs to support the EBPR process. Centrate recycle from the anaerobic digestion and dewatering process contributes approximately 24% of the influent phosphorus load.

Durham has a seasonal (dry weather) effluent phosphorus limit. The limit is flow-tiered, and approaches 0.07 mg/L when influent and receiving water flows are low. Plant staff operates the EBPR process to achieve a target secondary effluent soluble phosphorus concentration of 0.5 mg/L This target is achieved with 50% reliability in all four secondary treatment trains; with some trains meeting this target over 70% of the time. Alum is added to augment EBPR and polish secondary effluent prior to discharge. An alum dose of 20-30 mg/L is typically added upstream of the tertiary treatment process for polishing, and alum is added upstream of the primary or secondary clarifiers as needed for additional phosphorus removal.

Despite high influent VFAs, soluble phosphorus release rates at Durham are low. Influent soluble phosphorus concentrations are relatively low, and the total aeration basin phosphorus inventory is also lower than other facilities. These results suggest that the facility has additional phosphorus removal capacity.

The following table summarizes the key findings from this study.

| | |
|---|---|
| Reliability | $50^{th}$ percentile secondary effluent soluble phosphorus concentrations range from 0.06 mg/L to 0.31 mg/L in four treatment trains. With chemical polishing, average final effluent concentrations are 0.10 mg/L phosphate and 0.15 mg/L total P. |
| Operational Criteria | Average solids retention time (SRT) of 10.8 days during the summer low P permit season. Average anaerobic, anoxic, and aerobic hydraulic retention times are 0.7 hours, 1.2 hours, and 5.3 hours, respectively. |
| Influent Organic Characterization | Aeration basin influent BOD:P ratio of 20:1 and VFA:P ratio of 4.3:1, (includes fermenter overflow). VFA concentrations based on less than one year of data. VFA production from fermenter increases as wastewater temperature increase. |
| P Removal Capacity | Aeration basin influent VFAs are higher than in most plants due to the contribution of the primary sludge fermenter, but soluble phosphorus release and total phosphorus inventory are low, indicating additional soluble phosphorus removal capacity. |
| Secondary Release | Secondary release was not observed at Durham. In phosphate profiles, all release was measured in the first anaerobic zone. |

## B.2 Introduction

The analysis of this facility was completed as part of a project sponsored by the Water Environment Research Foundation (WERF). The project objective is to develop information for evaluating the ability of municipal wastewater treatment plants to efficiently and cost-effectively remove phosphorus to low levels through processes that involve EBPR. This was completed in large part through detailed studies at several treatment facilities that employ EBPR. The goals of the study were to:

- Determine process reliability.
- Determine the biological phosphorus removal efficiency.
- Determine the biological phosphorus removal capacity.
- Determine how chemical addition affects EBPR.

Definitions of terms used in this analysis are described in Appendix P.

### B.2.1 Facility Introduction

The Durham advanced wastewater treatment plant (AWWTP) is located in Tigard, Oregon and is operated by Clean Water Services.

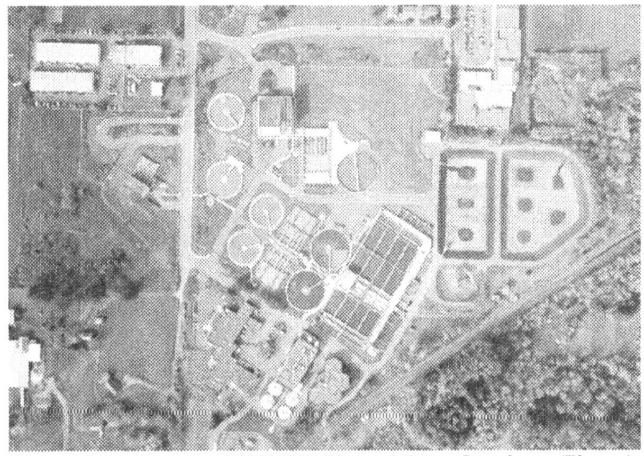

Figure B-1. Durham AWWTP, Clean Water Services, Tigard OR

Durham AWWTP has a dry weather capacity of 25 MGD design treatment capacity with a current annual average flow of approximately 22 MGD. Plant influent is composed of mostly municipal wastewater, and is heavily influenced by inflow and infiltration (I&I) during the wet winter season.

Durham has a seasonal permit and is only required to nitrify and remove phosphorus between the months of April and November. For this reason average data presented in this report uses data from the permit season only, ignoring data from all other months.

The original plant was modified for EBPR in the early 1990s. The facility design allows operation in either the UCT or $A^2O$ mode; however the plant is typically operated in $A^2O$ mode. Between 1998 and 2000, the District implemented EBPR optimization improvements, including addition of a primary sludge fermenter to provide additional volatile fatty acids (VFA). This addition has improved EBPR performance and reliability significantly, though the plant experience occasional upsets (often triggered by fermenter upsets).

Durham operates four identical trains with separated sludge systems (Figure B-1 shows the facility prior to the last expansion). During the summer (dry weather) season, the facility operates to achieve 0.07 mg/L final effluent total phosphorus and 1 mg/L ammonia on a monthly median basis, consistent with the most restrictive permit limits. While the maximum allowable discharge concentration actually changes based on influent flow and flow in the receiving stream, plant staff strives to meet the most restrictive standards regardless of influent and river flow during the summer permit season. Denitrification is practiced to recover alkalinity and oxygen, but there is no total nitrogen permit requirement.

Primary sludge, waste activated sludge, and sludge from the chemical clarifiers are stabilized by means of anaerobic digestion. Prior to digestion the primary sludge is thickened in a fermented/thickener, and WAS and chemical sludge are thickened using centrifuges. The centrifuge-dewatered sludge is land applied for agricultural soil augmentation in Eastern Oregon.

### B.2.2 Facility Process Design

#### B.2.2.1 Liquid Stream

The biological nutrient removal (BNR) process follows screening, grit removal, and primary clarification. The current BNR process is shown in Figure B-2. Primary effluent is mixed with the fermenter overflow in the primary effluent pump station wetwell. Lime is also added to the primary effluent pump station wetwell as necessary for alkalinity control. Primary effluent is distributed to each of the four secondary treatment trains as shown in Figure B-3. In the first anaerobic zone (ANR), primary effluent and return activated sludge are mixed. The mixed liquor recycle is introduced into the first anoxic zone (ANX). Table B-1 contains a summary of design secondary treatment parameters and their current average values.

**Table B-1. Key Process Parameters.**

| Component | Unit | Design[1] | Average |
|---|---|---|---|
| ANR HRT | hrs | 0.4 | 0.7 |
| ANX HRT | hrs | 0.7 | 1.8 |
| AER HRT | hrs | 3.3 | 5.3 |
| Total HRT | hrs | 4.4 | 7.8 |
| RAS Rate | % | 90 | 61 |
| ML Recycle Rate | % | 220% | 300% |
| SRT | days | | 11 |
| MLSS | mg/L | 4500 | 3500 |
| Solids Recycle Return | | Primary Effluent Pump Station | |
| Flow | MGD | 34 MGD | 24 |

1. Based on dry weather maximum month design capacity

The four trains can be operated with separated sludge ages. Operators have the option to seed one train from another. Secondary effluent is mixed and redistributed to the tertiary treatment process where chemical phosphorus polishing is accomplished by alum addition. Polymer can also be added to the chemical clarifiers for improved solids retention in the clarifiers and downstream sand filters. The final effluent is discharged into the Tualatin River.

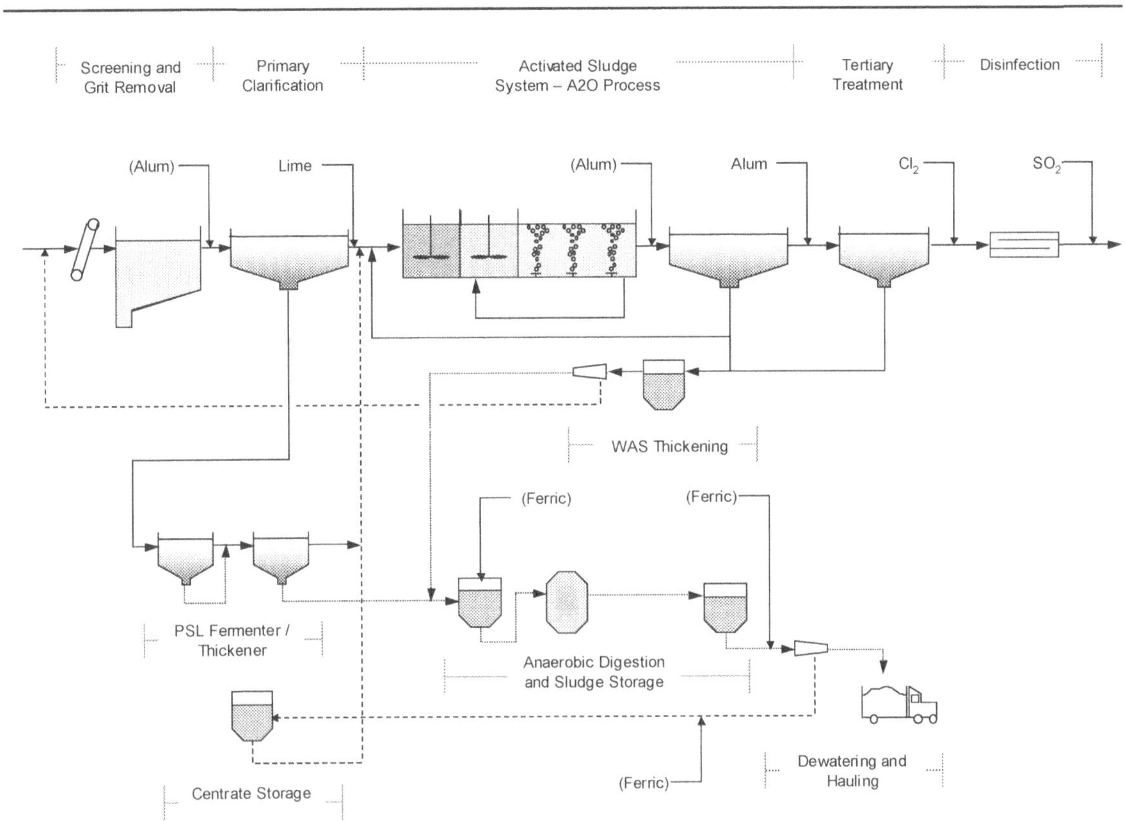


**Figure B-2. Process Schematic for Durham AWWTP Liquid and Solids Treatment.**

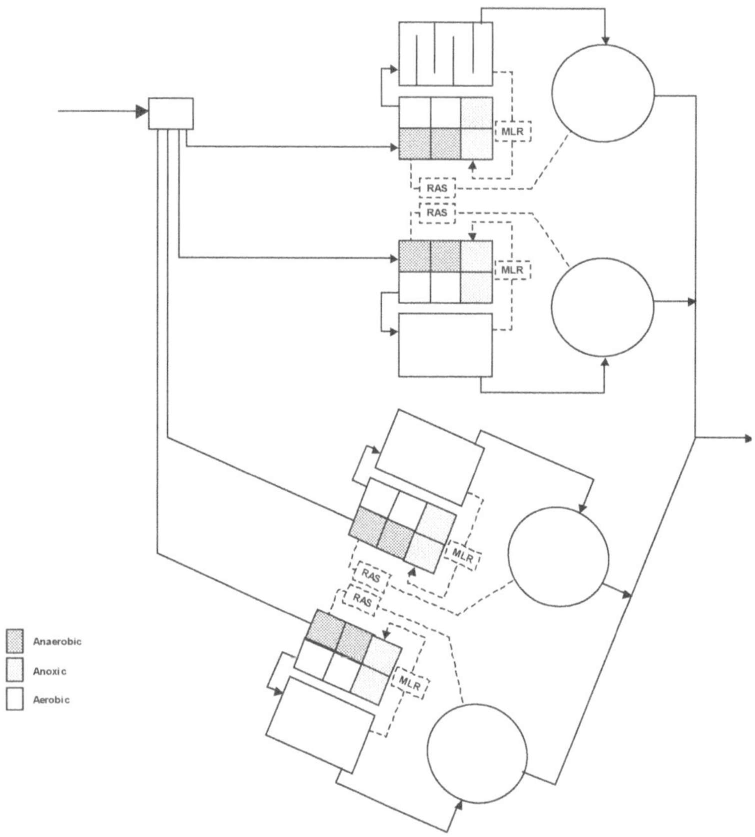


**Figure B-3. Secondary Treatment Layout at Durham AWWTP.**

### B.2.2.2 Solids Stream

All biosolids at Durham are stabilized by means of anaerobic digestion. Prior to digestion, primary sludge is retained and thickened in a two-stage fermenter/thickener. Primary sludge is fermented for one to two days to generate additional VFAs. After underflow and overflow of the fermenter are remixed to elutriate the generated VFA and strip out trapped gas bubbles, the solids are removed and thickened in the downstream gravity thickener to approximately 7% solids. Waste activated and chemical sludge (CSL) are mixed and flow equalized in an aerated storage tank from which the thickening centrifuges are fed. Both thickened primary and WAS/CSL are mixed in a digester feed tank before being fed to the digester. Currently Durham operates four digesters in series, with two downstream digested sludge storage tanks.

Digested sludge is dewatered with a high solids centrifuge. Operators have the option to add ferric upstream or downstream of the centrifuge for struvite control as well as to the digester feed tank for odor control. The dewatered cake is land applied. Table B-2 summarizes the solids treatment at Durham.

Table B-2. Solids Processing Summary.

| Component | Technology | Comments |
|---|---|---|
| Primary Sludge Thickening | Gravity Thickener | 7% TSS, if not monitored closely TSS can exceed 10% and generate conveyance problems |
| WAS Thickening | Thickening centrifuge | WAS and chemical sludge mixed prior to thickening |
| Primary Sludge Stabilization | 2 Stage Anaerobic Digestion | TS: 3100 mg/L<br>SRT: 20-30 days<br>VSS destruction: 48% |
| WAS Stabilization | | |
| Dewatering | Centrifuge | Operates 8 hours/day, 5 days/week |
| Centrate Storage | None | 7 day recycle load equalization capacity |

## B.2.3 Chemical Addition

Operators at Durham can feed a variety of chemicals to the liquid and solids stream. Chemical addition for the year 2002 is summarized in Table B-3.

- Ferric can be added to the digester feed tanks for odor and struvite control. It can also be added upstream or downstream of the dewatering centrifuge for struvite control and upstream of the primary clarifier to aid flocculation if necessary.
- Alum can be added to the liquid stream upstream of primary and secondary clarifiers, as well as at the head of the tertiary treatment process. The goal is to limit upstream alum dose to zero. Alum is routinely added upstream of the tertiary clarifiers for final polishing, and upstream of the secondary clarifiers as needed to augment EBPR.
- Lime is added to the primary effluent pump station wetwell to supplement alkalinity in the secondary treatment process.
- Sodium hypochlorite is used for disinfection, and occasionally for filament control in the RAS. Sodium bisulfate is used for dechlorination.

Table B-3. Chemical Addition Summary.

| Chemical | Unit | 2002 |
|---|---|---|
| Lime | Tons/year | 407 |
| Alum | Tons/year | 1080 |
| Ferric | Tons/year | 34 |
| | | |
| Lime | lb/MGD | 201 |
| Alum | lb/MGD | 535 |
| Ferric | lb/MGD | 16.8 |

### B.2.3.1 Process Control

During the dry weather season, operators aim to maintain the solids retention time (SRT) between 9 and 11 days. Dissolved oxygen (DO) in the aerated cells is controlled and maintained at values between 2 mg/L and 3 mg/L. Online ammonia and alkalinity analyzers provide real-time performance feedback to the operators and process control specialists. The RAS rate is typically maintained around 50% and is flow paced. The mixed liquor recycle rate varies with the influent flow because the pumps operate at a constant speed—average about 200%. Efforts are currently underway to reduce the speed of the mixed liquor recycle pumps, providing an average mixed liquor recycle of approximately 100%. Recent process optimization investigation has shown that because of the short anoxic retention time, no benefit is gained from a higher recycle rate.

Operators aim to achieve 0.5 mg/L soluble phosphorus in the secondary effluent. Depending on EBPR performance, a tertiary alum dose of 10–20 mg/L is used for effluent polishing. A secondary alum dose of 20–30 mg/L is also used when needed for additional phosphorus removal. Alum is also added to the secondary clarifier.

The primary sludge fermenter is operated at a constant SRT between two and three days. Operators monitor the underflow pH and sludge blanket depth for SRT control.

## B.3 Process Performance Evaluation

### B.3.1 Approach

The process evaluation entailed two data collection activities: a review of historical plant operating data, and the conduct of special sampling to supplement the routinely recorded water quality and process measurements. The following data sets were collected (see Table B-4):

**Routine Data Review.** The reviewed data represent all available water quality and process monitoring data collected between January 1, 2000 and July 31, 2003. Data were analyzed to determine general process performance, reliability of meeting effluent quality, and evaluation of the interaction between process parameters and performance. These routine data are referred to as the **2000-2003 data set**.

**Special Data Collection.** During both Phase 1 (2002) and Phase 2 (2003), project team members visited the Durham facility to conduct field tests and to interview plant staff. During Phase 1, the focus of the site visits was on supplementing the detailed survey information and data. Aeration basin phosphate and ammonia profiles were measured, and phosphorus uptake and release tests conducted. The results of the Phase 1 uptake and release tests are not presented in this report because the methodology was modified for Phase 2 to permit better interpretation of the results. Wastewater characteristics were collected for one day only during Phase 1. At Durham one intensive sampling period was also completed during Phase 1 as part of a separate process optimization project.

The focus of the field testing and intensive sampling during Phase 2 was to characterize the influent wastewater, determine the plant phosphorus mass balance; repeat the uptake and release test using the modified method; and collect samples for microscopic glycogen accumulating organism (GAO) analysis. During the intensive sampling period, plant staff collected samples including influent VFA, COD, $BOD_5$, readily biodegradable COD as measured by the flocculated filtered COD test (FFCOD), and soluble $BOD_5$ ($sBOD_5$) as well as phosphorus and nitrogen fractions. To capture the phosphorus and metal mass balance influent, effluent and digested sludge phosphorus and metals were also included in the sampling plan.

The results of special data collection efforts at Durham are included in the following data sets:

- **Phase 1F data set.** This data set was field data collected in Year 1 of the project during September 2002.

- **Phase 2F data set.** This data set was field data collected in Year 2 of the project during April and May 2003.
- **Phase 1A data set.** This intensive sample data set was collected in Year 1 of the project during the first intensive sampling period in September 2002.
- **Phase 2B data set.** This intensive sample data set was collected in Year 2 of the project during September 2003.

**Table B-4. Summary of Field Tests and Sampling.**

| Field Testing Segment | Month | Data Set ID | Purpose Conducted |
|---|---|---|---|
| Historical Data Review | January 2000 – July 2003 | 2000 – 2003 | Influent, Effluent Averages<br>Process Parameter Averages |
| Phase 1 – Field Tests | September 2002 | Phase 1F | Aeration Basin Profiles<br>Uptake and Release Test<br>Operator Interviews |
| Phase 2 – Field Tests | April – May 2003 | Phase 2F | Uptake and Release Test<br>Operator Interviews<br>Data Collection |
| Phase 1 – Intensive Sampling Period 1 | September 2002 | Phase 1A | Influent Characterization<br>Aeration Basin Profiles<br>Phosphorus Mass Balance<br>Metal Mass Balance |
| Phase 2 – Intensive Sampling Period 2 | September 2003 | Phase 2B | Influent Characterization<br>Uptake and Release Test<br>Phosphorus Mass Balance<br>Metal Mass Balance |

## B.3.2 Overall Process Characterization and Performance

This section of the report summarizes the aeration basin influent characteristics and secondary treatment process performance based on the data collection described in Section 3.1. It also discusses relationships between primary effluent loading and process performance.

### B.3.2.1 Aeration Basin Influent Characterization

The collection system within the Durham AWWTP service area has a separated sewer system with no direct stormwater inflow. However, as Figure B-5 illustrates, the plant does have significant I&I. Because the plant has a seasonal permit that only requires nutrient removal during the dry weather season from May 1$^{st}$ through October 31$^{st}$, all statistical numbers presented in this section refer to the dry weather months only and do not include values from the winter months.

From 2000 to 2003, the flow averaged 22.1 MGD and the influent temperature ranged from 15-24°C. The diurnal dry weather flow has a typical pattern for a service area of this size with peaking factors for minimum flow of 0.42 and maximum flow of 1.36 during dry weather conditions (Figure B-4). Neither primary effluent ammonia nor phosphorus loads have changed noticeably during the last three years (Figure B-6).

Table B-5 summarizes the average aeration basin influent composition and COD fractions, as well as values for the two intensive testing phases. The VFA to P ratio was measured at 8.1 mg VFA /mg P during intensive sampling. A ratio of 6–8 mg VFA /mg P would be considered favorable for EBPR. The average VFA/P ratio for the 2000-2003 data set of 4.3 includes periods before and after conversion of the primary sludge thickener to a thickener/fermenter; aeration basin influent VFA concentrations have been higher since this modification.

**Table B-5. Aeration Basin Influent Wastewater Characterization.**

| ABI Wastewater Parameter | Unit | 2000 - 2003 | Phase 1A | Phase 2B |
|---|---|---|---|---|
| $BOD_5$ | mg/L | 135[1] | 111 | 180 |
| $SBOD_5$ | mg/L | - | - | 100 |
| COD | mg/L | - | - | 270 |
| FFCOD | mg/L | - | - | 130 |
| VFA | mg acetic acid/L | 26 | - | 54 |
| TP | mg P/L | 6.1 | 6.5 | 6.6 |
| $PO_4$-P | mg P/L | 3.4 | 3.6 | 3.8 |
| Total Alkalinity (as $CaCO_3$) | mg/L | 240 | 250 | 240 |
| PH | | 8.0 | 7.9 | 7.9 |
| TKN | mg N/L | 33[1] | - | 34 |
| $NH_4$-N | mg N/L | 25 | 24 | 20 |
| TSS | mg/L | 130 | 130 | 120 |
| Flow | MGD | 22 | 19 | 19 |
| Temperature | °C | 19 | 21 | 22 |
| VFA : P Ratio | - | 4.3 | - | 8.1 |
| FFCOD : P Ratio | - | - | - | 20 |
| COD : P Ratio | - | - | - | 40 |
| $BOD_5$ : P | - | 20 | 17 | 28 |
| FFCOD : COD Ratio | - | - | - | 0.48 |
| VFA : FFCOD Ratio | - | - | 10/1/02 | 0.41 |

1. Estimated Value

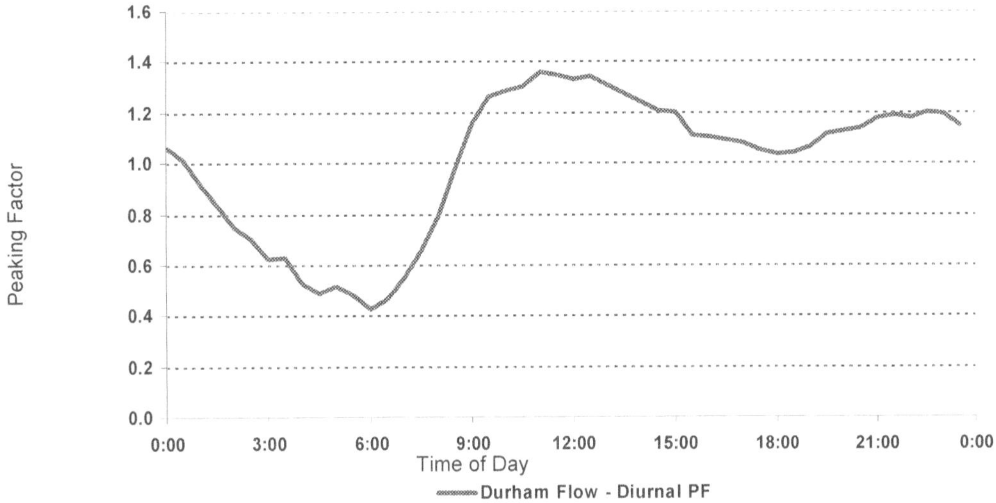


**Figure B-4. Durham Raw influent Diurnal Peaking Factor.**

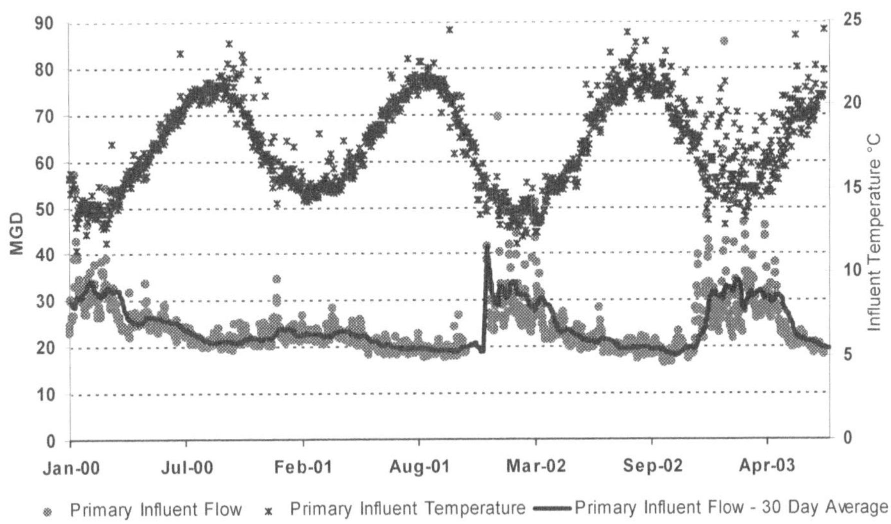


**Figure B-5. Influent Flow, Influent Temperature, and 30-Day average Influent Flow between 2000 – 2003.**

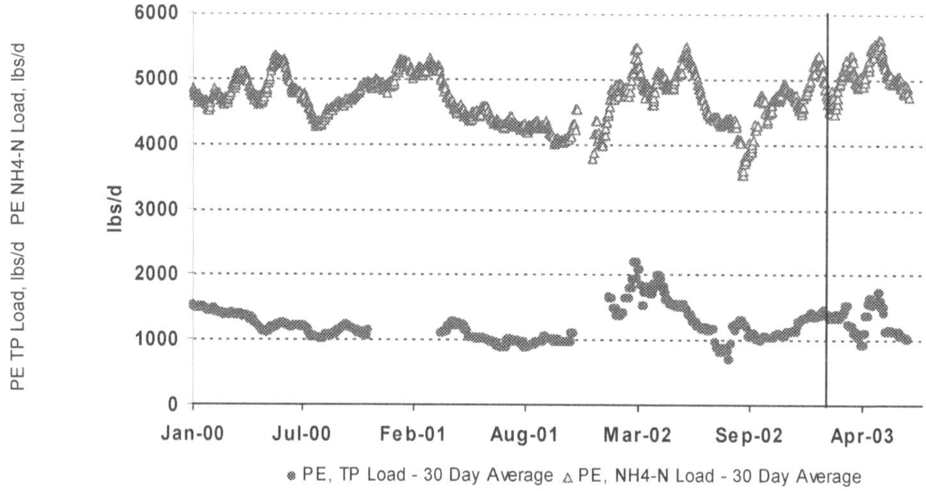


**Figure B-6. Primary Effluent (PE) Phosphorus and Ammonia Load.**

During 2001 and 2002, various shutdowns of primary clarifiers associated with construction of a plant expansion resulted in periodic aeration basin loading peaks. The primary effluent TSS load shown in Figure B-7 reflects this.

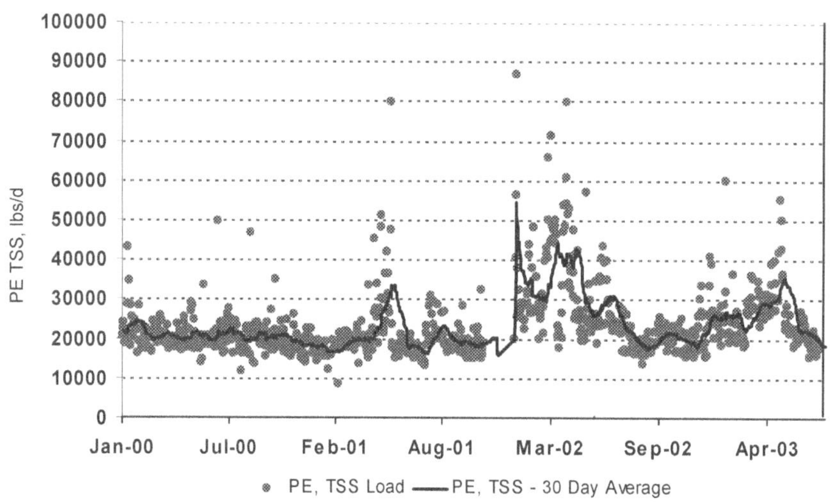


**Figure B-7. Primary Effluent TSS Load.**

The influent loads show some fluctuations during the dry weather permit season. Influent flows and temperatures can vary by 20% lower than to 80% higher than average values, and aeration basin influent ammonia values vary by approximately +/- 40% (Table B-6). Larger swings are seen in aeration basin influent (ABI) total phosphorus and phosphate, where maximum values are up to 150% higher than average values.

**Table B-6. Selected Water Quality Measurement and Process Parameters from 2000 – 2003 (between April 16 and Nov 1)**

| Parameter | Unit | Min | Ave | Max |
|---|---|---|---|---|
| Influent Flow | MGD | 18 | 22 | 38 |
| ABI Temperature | °C | 15 | 19 | 24 |
| ABI $NH_4$-N | mg/L | 15 | 26 | 36 |
| ABI TP | mg/L | 0.7 | 6.1 | 12 |
| ABI $PO_4$-P | mg/L | 0.1 | 3.4 | 8.6 |
| ABI TSS | mg/L | 68 | 130 | 300 |

### B3.2.2 EBPR Performance

Table B-7 summarizes minimum, average, and maximum effluent water quality measurements and process parameters based on the 2000-2003 data set. Water quality measurements are based on final effluent (FE) values measured downstream of tertiary treatment. These values therefore reflect the routine alum addition upstream of the tertiary clarifiers, and periodic alum addition upstream of the secondary clarifiers.

As shown in Table B-7, the plant achieves very low effluent TSS, TP, and ammonia concentrations. Averages values of 0.15 mg/L TP, 1.0 mg/L TSS, and 0.65 mg/L ammonia leave little room for treatment improvement. The SRT is average for plants with similar treatment processes and performance objectives, while hydraulic retention times are slightly lower.

As described earlier, Durham operates four trains with separated sludge systems. Performance does vary somewhat between the trains. Table B-8 summarizes the phosphorus removal performance of all four trains. The table shows that all four trains meet the target secondary effluent phosphorus concentration of 0.5 mg/L over half the time. $80^{th}$ percentile concentrations, however range from 0.28 mg/L to 1.19 mg/L.

Figure B-8 through Figure B-10 illustrate plant performance based on data from Train 2. Figure B-8 shows the Train 2 secondary effluent phosphate and suspended solids. Each permit season shows elevated secondary effluent phosphate concentration during the seasonal startup period and occasionally during the season. The overall probability distribution of Train 2 secondary effluent phosphate is shown in Figure B-9. Train 2 secondary effluent phosphate concentrations of less then 0.5 are accomplished 79% of the time and concentration of less then 0.05 mg/L 50% of the time. Secondary effluent ammonia is shown in Figure B-10.

It should be noted that the plant has been under construction between 2000 and 2003. Throughout the three-year construction period, various process components have been taken in and out of service, adding additional pressure to the system. The facility expansion included the addition of Train 4 in 2000, a centrate storage tank in 2001, a new primary sludge fermenter in 2002, and a new digester complex and primary effluent pump station in 2002.

**Table B-7. Selected Water Quality Measurement and Process Parameters from 2000 through 2003 (between April 16 and Nov 1).**

| Parameter | Unit | Min | Ave | Max |
|---|---|---|---|---|
| FE $NH_4$-N | mg N/L | 0.01 | 1.4 | 21 |
| FE $NO_3$-N | mg N/L | 1.3 | 7.8 | 17 |
| FE TN | mg N/L | | | |
| FE TP | mg P/L | 0.12 | 0.51 | 3.95 |
| FE $PO_4$-P | mg P/L | 0.01 | 0.26 | 4.0 |
| FE $BOD_5$ | mg/L | | | |
| FE TSS | mg/L | 1.0 | 6.5 | 58 |
| FE Total Alk (as $CaCO_3$) | mg/L | 55 | 140 | 360 |
| AB TP Load[1] | Lbs/ 1000 lbs VSS | 0.4 | 3.9 | 7.4 |
| SRT[2] | Days | 5.9 | 11 | 26 |
| MLSS | mg/L | 1880 | 3440 | 5840 |
| MLVSS | mg/L | 1720 | 2540 | 4800 |
| MLVSS/MLTSS | | 64 | 75 | 80 |
| ANR HRT | Hrs | 0.4 | 0.7 | 0.9 |
| ANX HRT | Hrs | 0.7 | 1.2 | 1.5 |
| AER HRT | Hrs | 3.0 | 5.3 | 6.9 |
| TP – Removal (Secondary) | % | 39 | 85 | 100 |
| TN Removal | % | 53 | 70 | 83 |

**Table B-8. EBPR Reliability comparison of Durham AWWTP Train 1 – 4 Between 2000 and 2003 (between April 16 and Nov 1).**

| ABI Wastewater Parameter | Unit | Train 1 | Train 2 | Train 3 | Train 4 |
|---|---|---|---|---|---|
| Sample Population | N/A | 648 | 755 | 643 | 518 |
| Average | mg/L | 0.24 | 0.37 | 0.63 | 0.53 |
| 50th Percentile Secondary Effluent soluble P | mg/L | 0.06 | 0.05 | 0.21 | 0.31 |
| 80th Percentile Secondary Effluent Soluble P | mg/L | 0.28 | 0.54 | 1.2 | 0.73 |
| 98th Percentile Secondary Effluent Soluble P | mg/L | 1.9 | 2.7 | 3.3 | 2.6 |
| Maximum Secondary Effluent Soluble P | mg/L | 4.0 | 4.5 | 6.7 | 4.3 |

[1] 30 Day average
[2] 7-Day running average of recorded daily SRT values

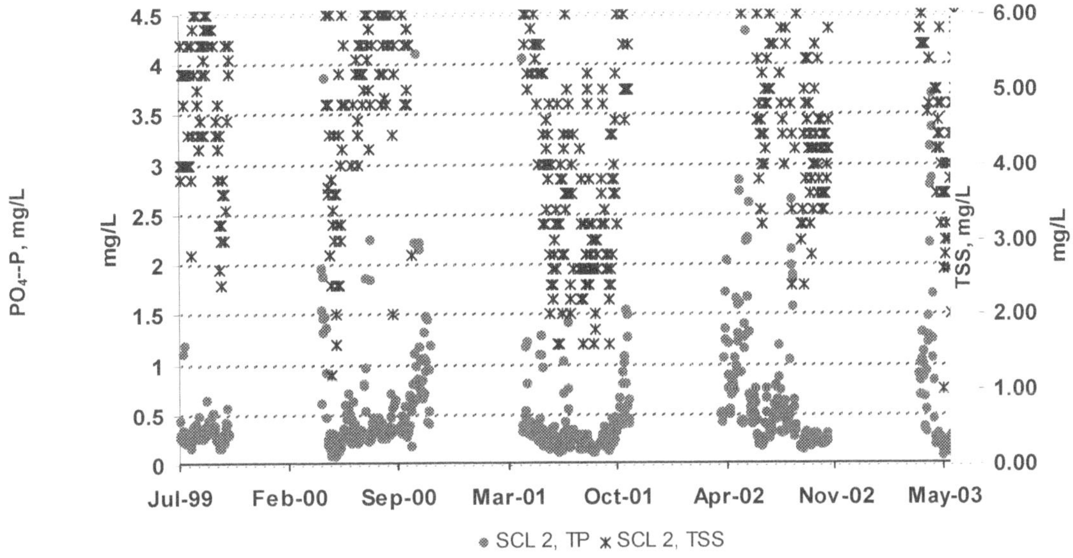


Figure B-8. Secondary Clarifier 2 Effluent Phosphate and Total Suspended Solids (Dry weather permit season data only) Between April 16 and Nov 1.

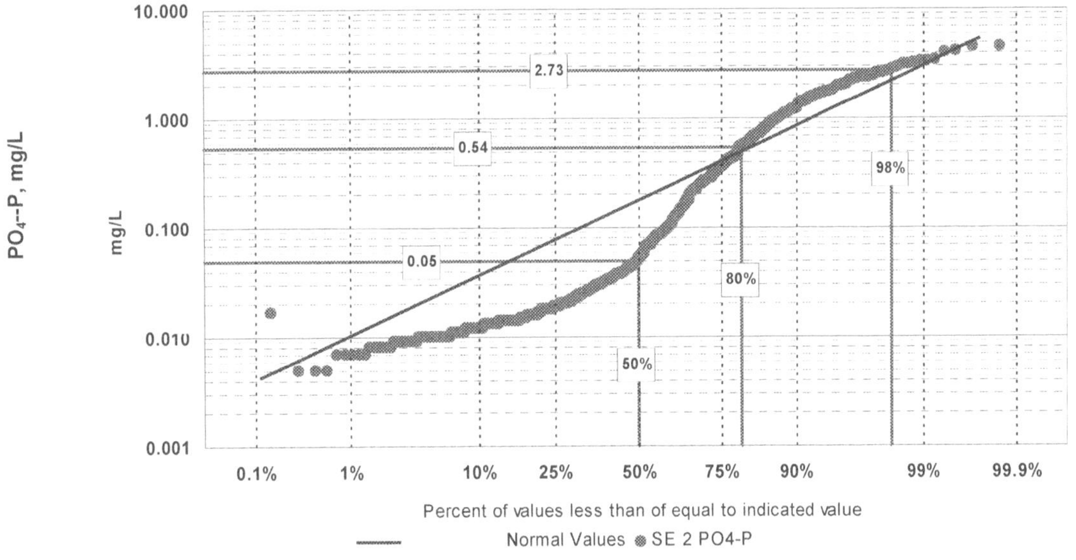


Figure B-9. Secondary Clarifier 2 Effluent Phosphate Probability Distribution between April 16 and Nov 1.

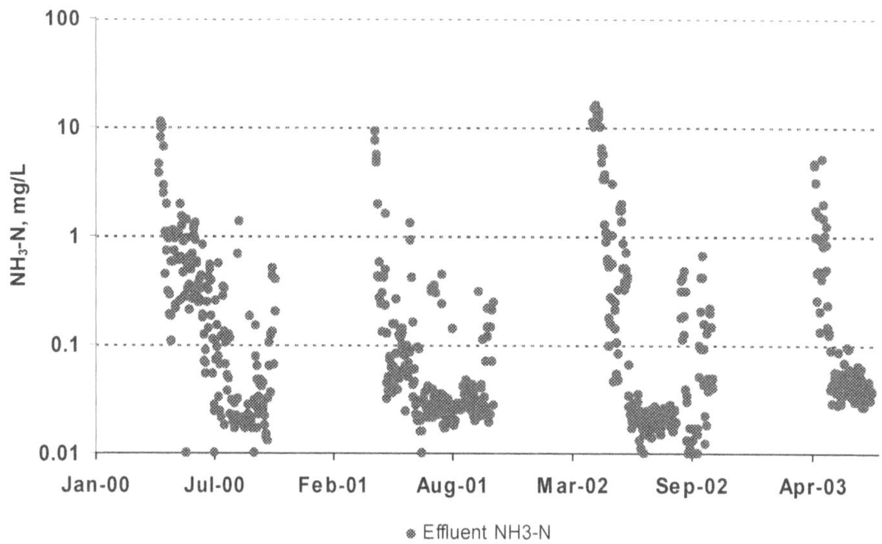


**Figure B-10. Logarithmic Plot of Effluent Ammonia Concentrations Between April 16 and Nov 1.**

### 3.2.2.1 Phase 1 and 2 Testing

During the Phase 1 testing plant performance was comparable with previous years. Figure B-11 shows the plant performance during the summer of 2002, including the Phase 1 intensive sampling period in September 2002. The overall effluent quality remained very good, however, Train 2 experienced elevated effluent phosphate concentrations for short periods of time. Because the facility's phosphorus permit is based on a monthly median concentration, it is common for maintenance and repair activities to be scheduled during the second half of the month. These activities can impact EBPR performance. The District also conducted full-scale stress testing of Train 4 in the fall of 2002; during September 2002, the loading to Train 2 was reduced by approximately 20%.

Data for the complete phosphorus removal season of 2003 was not available at the time of report preparation, however results shown in Figure B-12 show excursions in EBPR performance in May and July of 2003.

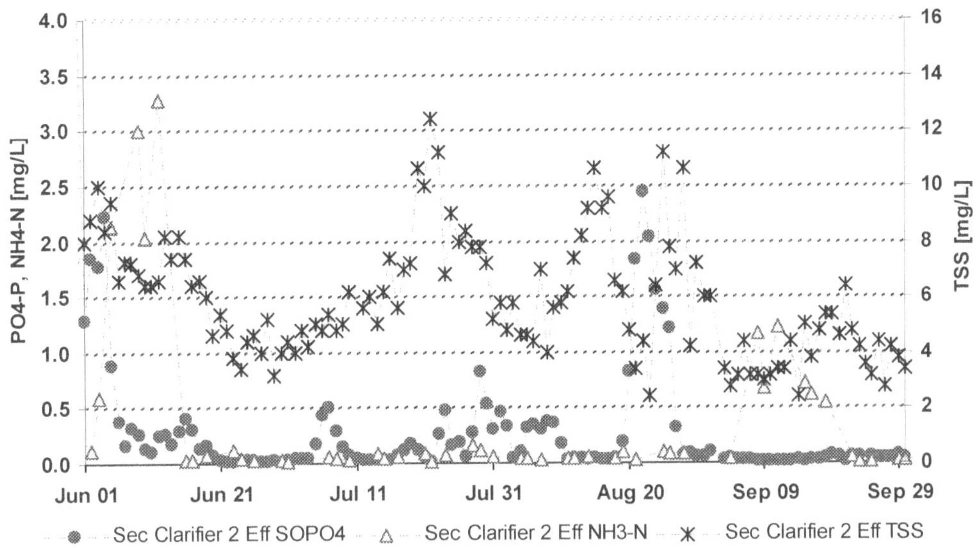


**Figure B-11. Train 2 Secondary Effluent Phosphate, Ammonia and TSS Prior to and During the Phase 1 Testing.**

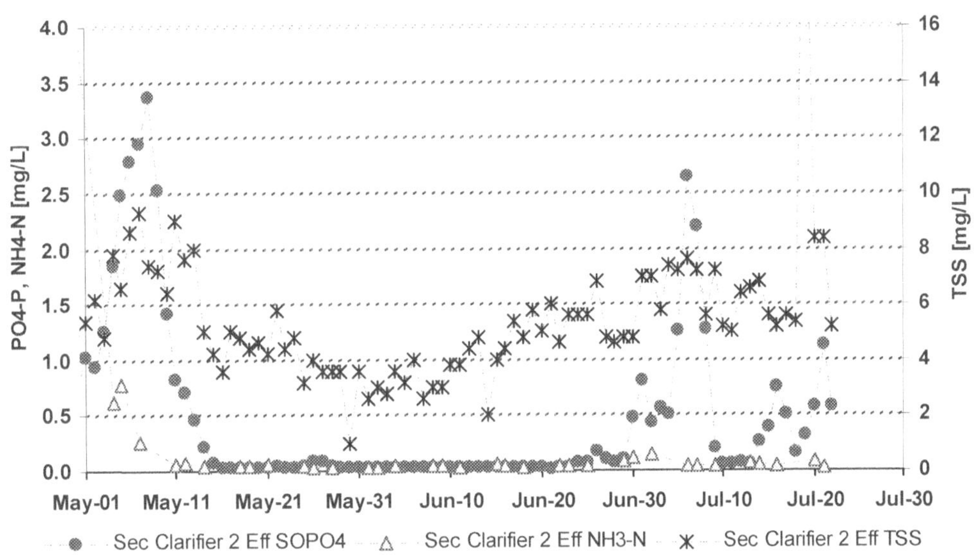


**Figure B-12. Train 2 Secondary Effluent Phosphate, Ammonia and TSS Prior to and During the Phase 2 Testing.**

### B3.2.3 Relationship Between Performance and Process Loading

During the summer permit season, the plant receives a relatively consistent influent load . On average the aeration basin phosphorus load amounts to approximately 4 lb P/d per 1000 lb aeration basin MLVSS, which is lower than other EBPR facilities studied.[3]

As is expected, EBPR performance is strongly impacted by influent VFA concentrations. Approximately 70% of the total aeration basin influent load is comprised of VFAs generated in the collection system (measured in the primary effluent); the remaining 30% of the load is contributed by the fermenter. Figure B-13 shows both primary effluent and fermenter overflow VFA during 2003.

At Durham the process is changed from $BOD_5$ removal to BNR in early April. At that time the primary effluent VFA concentration ranges between 10 mg/L and 20 mg/L in early April when BNR operation starts, and continuously increases through the nutrient removal season.

While the fermenter only contributes to 30% of the total aeration basin influent VFA load, the EBPR process has proven to be very sensitive to fermenter VFA output. Fermenter upsets or shutdowns for other reason are typically accompanied with reduction or loss of EBPR performance. Figure B-13 also indicates that running higher fermenter SRTs in at the beginning of the season could increase VFA production. This could be especially effective because the hydraulic retention time in the anaerobic zone is very short early in the permit season (Figure B-14) allowing only minimal VFA generation in the anaerobic zone.

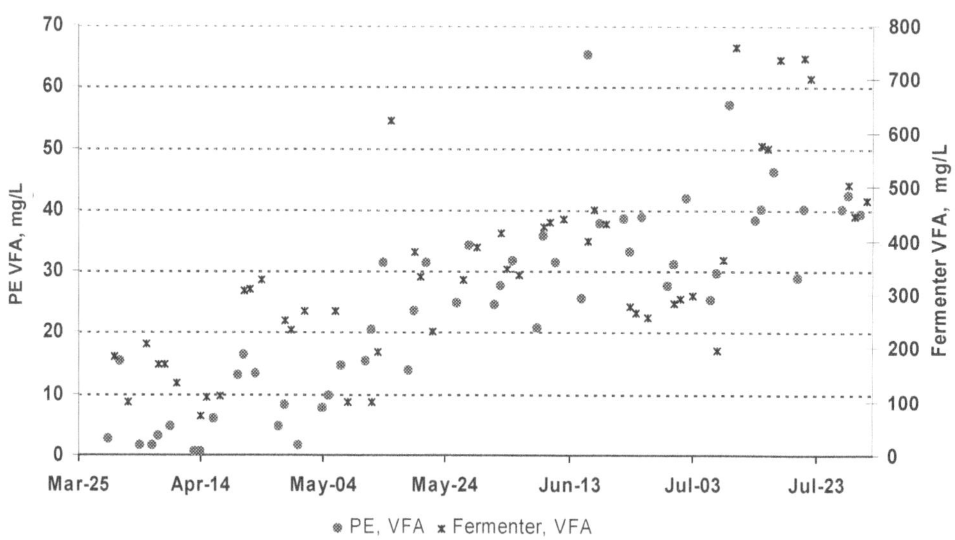


**Figure B-13. Durham AWWTP Primary Effluent VFA and Fermenter Overflow VFA.**

---

[3] However, the MLVSS measurements may be artificially increased due to lime addition. Some of the lime added to the primary effluent results in the formation of carbonates which are partially measured as the solids are exposed to 500°C heat (X-CO3 + Heat → X-O + $CO_2$). This artificially increases the VSS content and lowers the measured load.

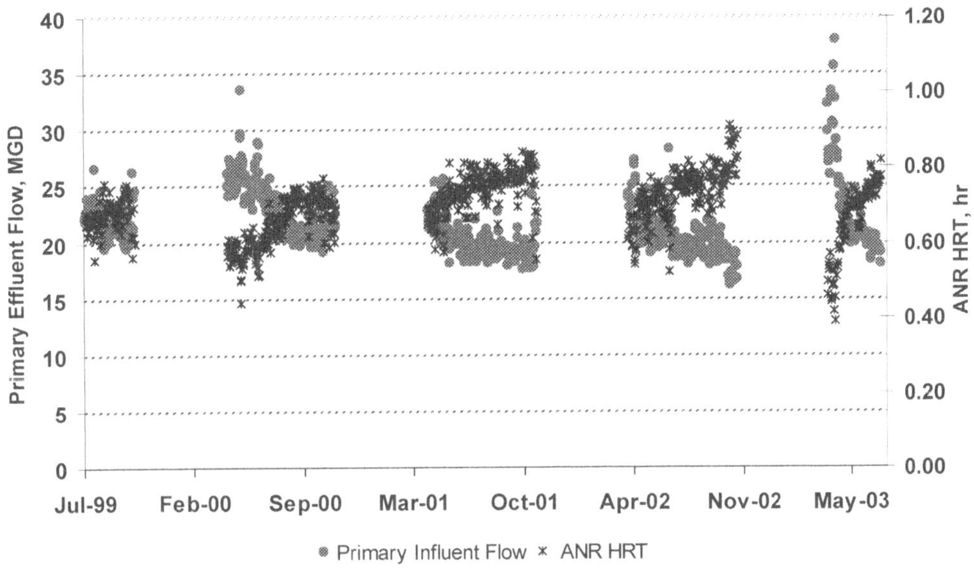


**Figure B-14. Primary Effluent Flow vs. Anaerobic Hydraulic Retention Time.**

It was found that operational stability decreases as $BOD_5$/TP ratios fall below 25 lb $BOD_5$/lb TP. This is similar to observations at the other facilities. Figure B-15 clearly shows that in 2002, when the $BOD_5$/TP ratio averaged less then 20 lb $BOD_5$/lb TP, EBPR was very unstable and produced higher effluent phosphate concentrations. During 2000, in comparison, when the $BOD_5$/TP ratio averaged close to 25 lb $BOD_5$/ lb TP, the process operation was very stable and produced low effluent phosphate concentrations.

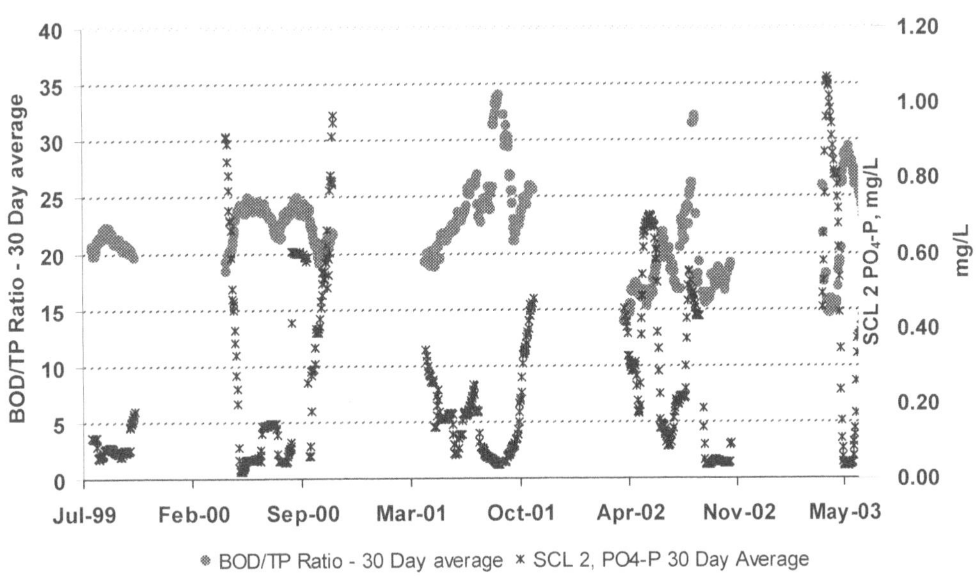


**Figure B-15. Relationship of $BOD_5$/TP Ratio and Secondary Effluent Phosphate for Train 2 Between April 16 and November 1 (PE BOD Estimated from Influent BOD).**

Figure B-16 shows the dewatering centrate phosphate, ammonia and soluble magnesium concentrations. The phosphate concentration ranges from approximately 30 mg/L when EBPR is practiced to 300 mg/L without EBPR. The lowest phosphorus concentrations are measured in the summer and the highest during the winter when EBPR in not in operation. There are two reasons for this seasonal change in dewatering recycle phosphates: the seasonal operation of EBPR and the seasonal addition of metals. EBPR not only increases the phosphorus sent to the digester with the WAS, but also cations, specifically magnesium. These metals are released in the digester (as shown by the magnesium profile in Figure B-16) where they react with and precipitate phosphorus. During the summer permit season, large amounts of alum, calcium and ferric are also added to the process. These metals also end up in the digester and can precipitate phosphorus released in the digester.

The recycle ammonia load is relatively low compared to other facilities—about 13% of the total aeration basin ammonia load. The primary influent total phosphorus included all recycle phosphorus from solids processing and filter backwash until fall of 2001. After 2001, the newly constructed centrate storage tank was added to provide load equalization, and the dewatering centrate recycle point was redirected to the wetwell of the primary effluent pump station. This operational change becomes more apparent in Figure B-17, which shows the primary influent and primary effluent ammonia. Prior to October 2001, dewatering recycle ammonia was sent to plant drain and returned to the influent sewer upstream of the influent sampling point.

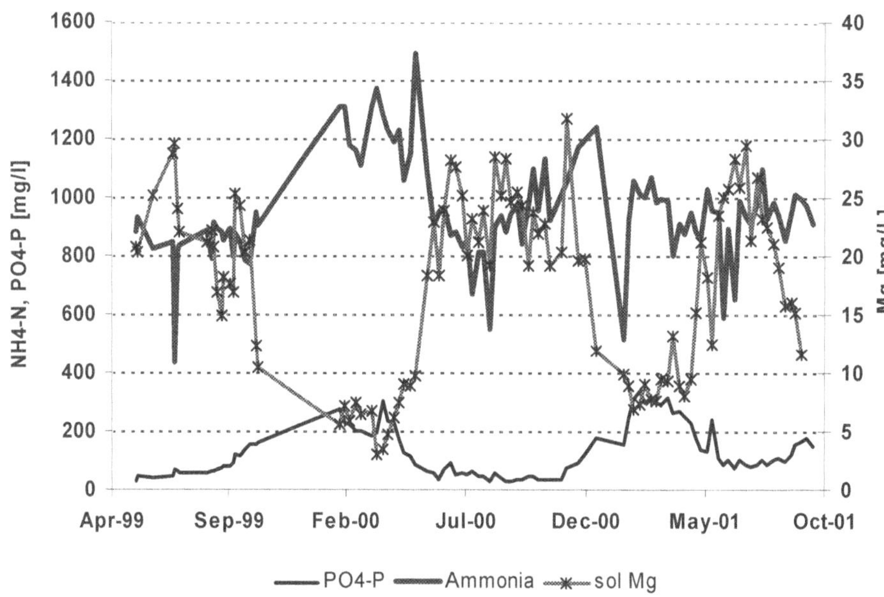


**Figure B-16. Durham AWWTP Dewatering Centrate Phosphate, Ammonia and Soluble Magnesium.**

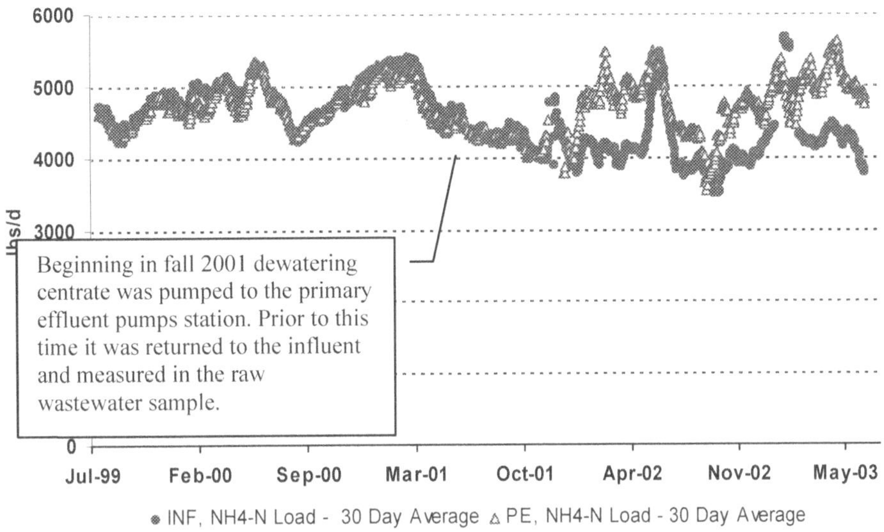


**Figure B-17. 30-day Averages of Raw Influent and Primary Effluent Ammonia from between 2000 and 2003.**

## B.3.3 Additional EBPR Process Evaluations

### B.3.3.1 Aeration Basin Phosphate Profiles

The aeration basin phosphate profile was measured during each sampling phase of the project. Charts of the profiles (Figure B-19, Figure B-20, and Figure B-21) show four lines:

- The individual daily profiles (thin shaded gray lines in background with diamond points).
- Average phosphate profiles calculated from the daily profiles (thick line with circular points).
- Flow corrected average phosphate profile (thick line with triangular points). The flow corrected concentration is calculated as the theoretical phosphorus concentration based on the influent flow. This allows for a direct comparison of the concentration values without the influences of dilution from recycle flows.
- Phosphorus release rates (columns). The release rates (lb/d P / 1000 lb VSS) are calculated as daily phosphorus release compared to the aeration basin volatile solids content.

In order to understand the phosphate profile, the flow scheme must first be investigated. Flows are relatively complicated during $A^2O$ operation since two recycle streams must be accounted for (RAS and ANX recycle). Figure B-18 illustrates the internal recycles used during the profile testing.

#### B.3.3.1.1 Phase 1 Profiles

Aeration basin profiles were measured during September and October 2002 on two different days. On each sampling day, profiles were measured every two hours between 6 am and 6 pm to investigate the diurnal changes in phosphorus uptake and release. Diurnal profiles were taken from Train 2 and Train 4 in mid-September 2002, and again from Train 4 at the end of October 2002. The three aeration basin phosphate profile charts (Figure B-19 - Figure B-21) show:

- Four individual daily profiles (thin lines with diamond points)
- Average phosphate profiles calculated from the daily profiles (thick line with circular points)

- Flow corrected average phosphate profile (thick line with triangular points)
- Phosphorus release rates (columns)

The internal recycle rates at Durham remained fairly constant and were generally consistent with Figure B-19. Compared to other facilities, the measured phosphorus release rates (Table B-9) are relatively small ranging from 7 – 16 lb P released per 1000 lb VSS. For comparison, at the Nansemond facility in Norfolk, VA, the measured release rates ranged from 20 - 70 lb P released per 1000 lb VSS. Furthermore, the amount of phosphorus released compared to the total aeration basin phosphorus inventory (Release of P Content)was significantly less than at other facilities despite the fact the aeration basin influent VFA were highest at Durham.

**Table B-9. Process Control and Water Quality Measurement Results During Intensive Testing (two-week average).**

| Raw Wastewater Parameter | Unit | Train 2 September 2002 | Train 4 September 2002 | Train 4 October 2002 |
|---|---|---|---|---|
| Total Phosphorus Release | mg/L | 11 | 10 | 15 |
| Anoxic Phosphorus Uptake | mg/L | - | - | - |
| Phosphorus Release Rate | Lbs P per 1000 lbs VSS per day | 12 | 7.1 | 16 |
| Phosphorus Uptake Rate | Lbs P per 1000 lbs VSS per day | 15 | 7.0 | 20 |
| Total Aeration Basin Phosphorus Inventory | Lbs P | 2560 | 2390 | 3070 |
| Release of P-Content | Lbs P/d | 520 | 270 | 790 |
| Total Phosphorus Release Rate | Lbs P per 1000 lbs VSS per hr | 27 | 16 | 40 |

Figure B-19 through Figure B-21 show the aeration basin phosphorus profiles. No release was measured in the second anaerobic cell. No samples were taken from the anoxic cells.

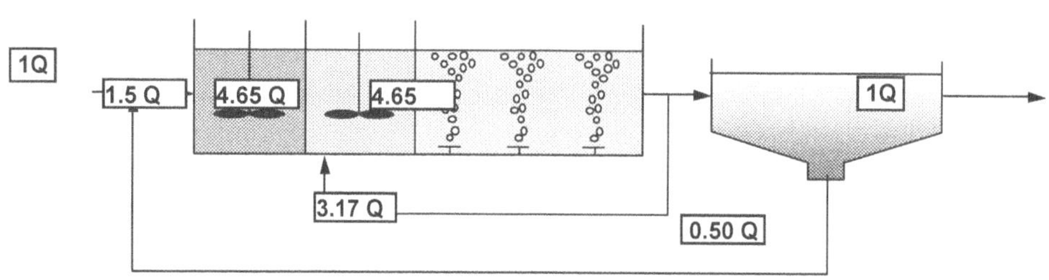


**Figure B-18. Example of RAS and Internal Recycle Flows at Durham during Aeration Basin Profile Testing.**

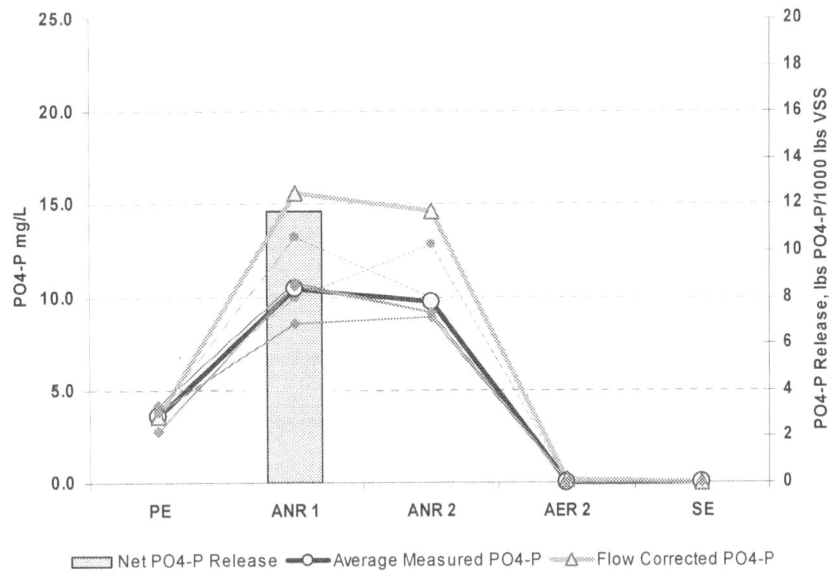


**Figure B-19. Train 2 Phosphate Profiles First Diurnal Sampling September 2002.**

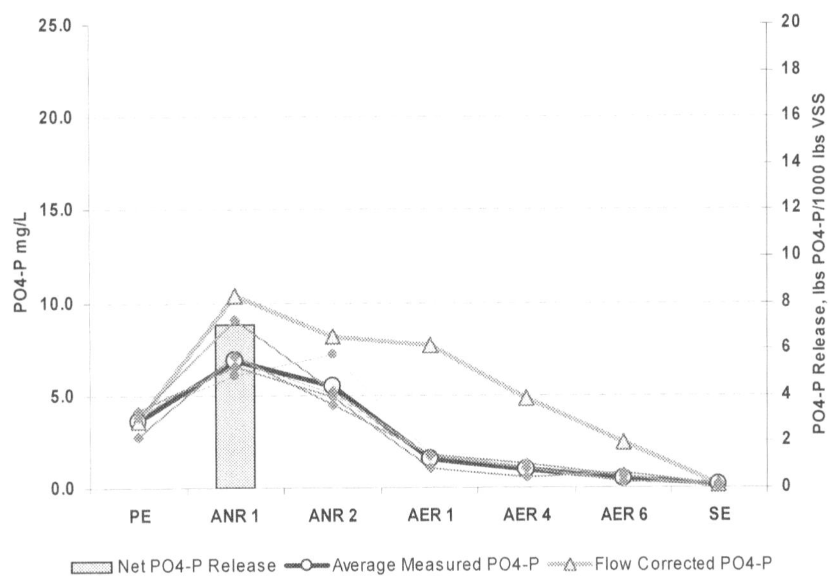


**Figure B-20. Train 4 Phosphate Profiles During First Diurnal Sampling September 2002.**

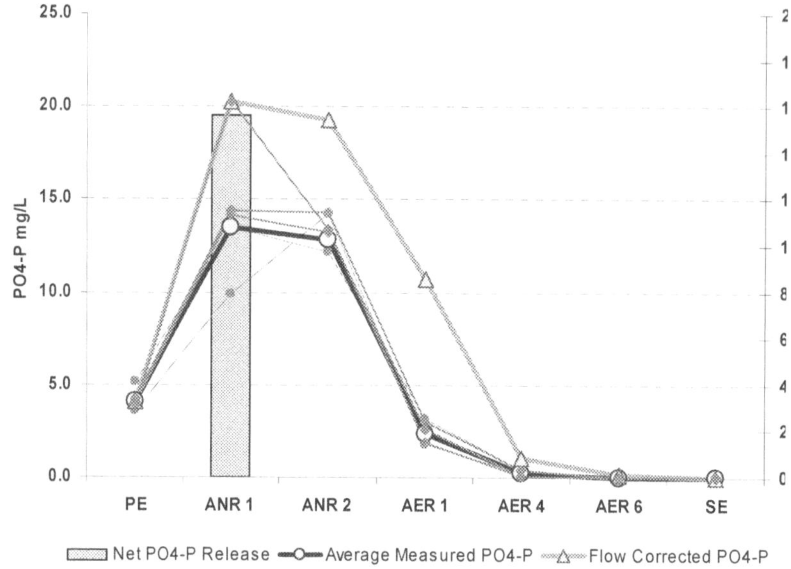


**Figure B-21. Train 4 Phosphate Profiles During Second Diurnal Sampling October 2002.**

### B.3.3.1.2 Phase 2 Profiles

Figure B-22 shows the phosphate profile measured during the last two weeks of September 2003. The results are comparable with those from 2002.

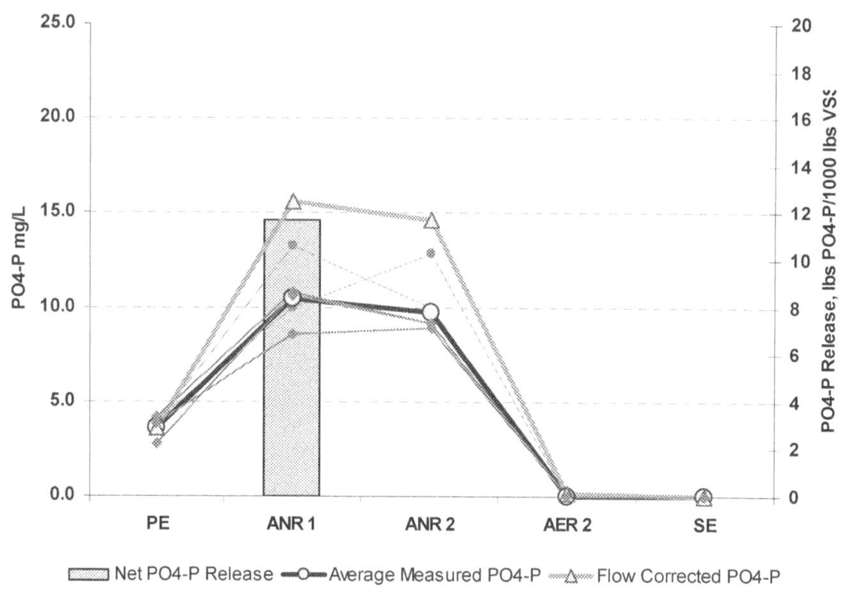


**Figure B-22. Train 2 Phosphate Profiles from September 2003 (Phase 2).**

All of the phosphate profiles show a lack of phosphorus release in the second anaerobic cell. Even if all primary effluent VFA have been consumed prior to the second anaerobic cell, one would still expect continued release driven by VFAs generated in the anaerobic zone. It was suspected that the high pH of (7.9 – 8.2) in the anaerobic zone might inhibit the fermentation reaction. Bench tests conducted in October 2003 were unable to prove whether fermentation was inhibited by lime addition. It is also possible that phosphorus limiting conditions (related to the low phosphorus inventory and high chemical feed) precluded further release in the second anaerobic cell.

**B3.3.2 Uptake and Release Tests (URT)**

URTs provide a quantification of the PAO population in the basin. By adding VFA to mixed liquor, the measured phosphorus release is limited by the PAO population in the basin and by the amount of stored phosphorus. URTs can be used for:

- Determining the amount of PAO activity from phosphorus release rate. The release rate (mg P/g VSS/h) is indicative of PAO activity.
- Determining the potential competition for VFA in the anaerobic zone. The phosphorus release ratio (mg P released/mg VFA removed) indicates PAO activity. A phosphorus release ratio of 0.5 mg P released/mg VFA removed indicates PAO activity. Lower ratios are indicative of competition from other VFA consuming organisms such as GAOs.
- Determining key kinetic parameters for process simulation model calibration, such as $\Delta$P/Acetic acid and magnesium and potassium uptake and release rates.
- Provides information regarding the phosphorus uptake kinetics, which represents the time required in the aerobic zone to remove the phosphorus.

Two URT tests were conducted at the Durham plant according to the modified testing plant in May and August of 2003. The testing procedure is outlined in Appendix J.

Table B-10 summarizes the results from the URT at Durham during May 2003 (Figure B-23). The test was conducted during a period where EBPR was not operating well in effort to learn more about the reasons for the process failure. Mixed liquor from Train 3 and Train 4 were both tested and Train 3 showed slightly better uptake and release results. The cause of the performance problems was not identified, and the process recovered within a matter of weeks.

In both trains, the phosphorus release ratio was approximately 0.3 mg P per mg COD. In a system with a strong PAO population, the ratio would be expected to be approximately 0.5 mg P/mg COD. Lower phosphorus release ratios typically point to the presence of either nitrates or GAOs, which consume VFAs without a corresponding release in phosphorus.

The release rate measured in the uptake and release tests was 8.0 mg P/g VSS/hr for Train 3 and 7.4 mg P/g VSS/hr (Figure B-23). The uptake Mg:P ratio was smaller than the release indicating that during the uptake phase less phosphorus was accumulated per unit of magnesium uptake.

**Table B-10. Uptake and Release Tests Results from May 2003.**

| Test Parameter | Unit | Train 3 | Train 4 |
|---|---|---|---|
| Phosphorus Release Rate | mg P /g VSS /hr | 8.0 | 7.4 |
| Phosphorus Uptake Rate | mg P /g VSS/ hr | 2.3 | 2.3 |
| Release Rate/ Uptake Rate | - | 3.5 | 3.3 |
| ΔP/Acetic Acid | mg $PO_4$-P/ mg COD | 0.29 | 0.32 |
| ΔP/ΔMg - Release | mg $PO_4$-P / mg Mg | 5.5 | 5.8 |
| ΔP/ΔMg - Uptake | mg $PO_4$-P / mg Mg | 3.9 | 4.7 |
| MLVSS | mg/L | 2170 | 2592 |

A second set of uptake and release tests was conducted in August of 2003. Table B-11 summarizes the result from those tests. Mixed liquor from Trains 1 and 2 were tested; Train 2 was not performing well at the time, as evidenced by the low phosphorus release ratio. Train 1 delivered better results than measured during the April tests, with a phosphorus release ratio of 0.36 mg P/mg COD.

The release rate was 10.1 mg P/g VSS/hr for Train 1 and 4.2 mg P/g VSS/hr for Train 2 (Figure B-24 and Figure B-25). The uptake Mg:P ratio was almost equal to the release Mg:P ratio for both Train 1 and Train 2. Because sodium acetate was used for VFA addition, the pH did not increase much during the anaerobic retention time, preventing calcium phosphate precipitating. There was also much less calcium in the sample. Calcium concentrations dropped from 33 mg/L to 30 mg/L during 45 minutes of anaerobic reaction time compared to the change from 50 mg/L to 30 mg/L during the earlier tests.

**Table B-11. Uptake and Release Tests Results from August 26th 2003.**

| Test Parameter | Unit | Train 1 | Train 2 |
|---|---|---|---|
| Phosphorus Release Rate | mg P /g VSS /hr | 10 | 4.2 |
| Phosphorus Uptake Rate | mg P /g VSS/ hr | 3.6 | 1.5 |
| Release Rate/ Uptake Rate | - | 2.8 | 2.9 |
| ΔP/Acetic Acid | mg $PO_4$-P/ mg COD | 0.36 | 0.16 |
| ΔP/ΔMg - Release | mg $PO_4$-P / mg Mg | 3.3 | 2.8 |
| ΔP/ΔMg - Uptake | mg $PO_4$-P / mg Mg | 3.1 | 2.1 |
| MLVSS | mg/L | 3030 | 2600 |

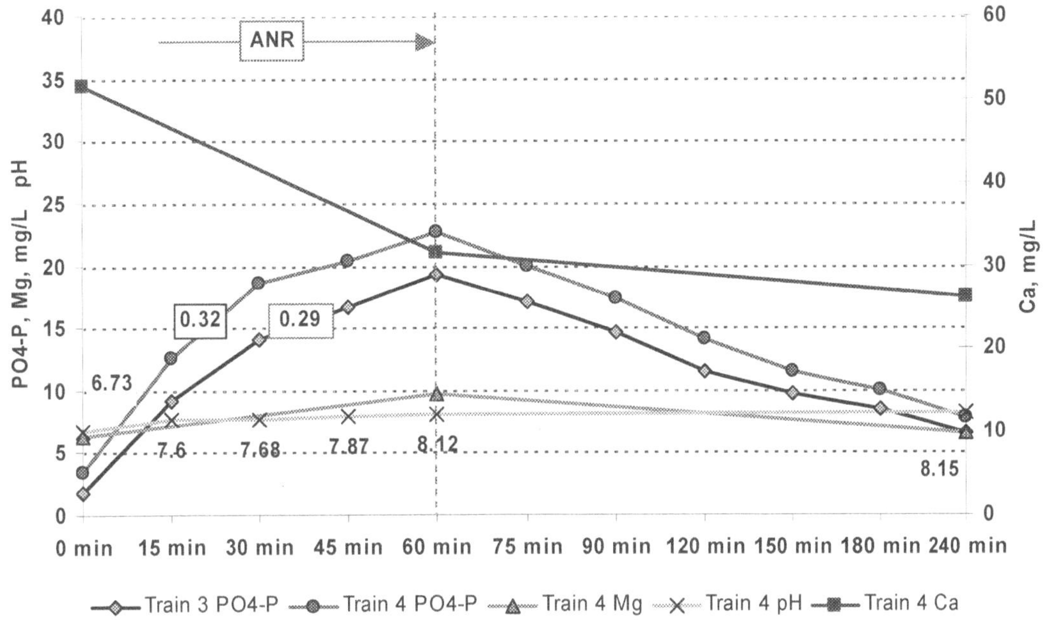


**Figure B-23. Phosphorus Uptake and Release Test Results from May 2003.**

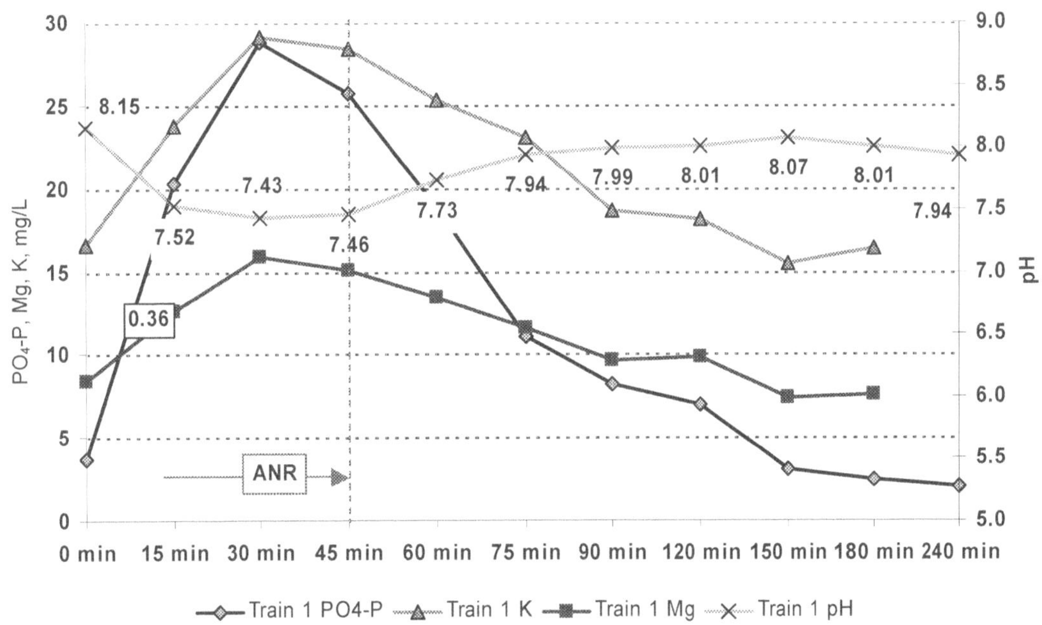


**Figure B-24. Phosphorus Uptake and Release Test Results from August 26th 2003—Using Mixed Liquor from Train 1.**

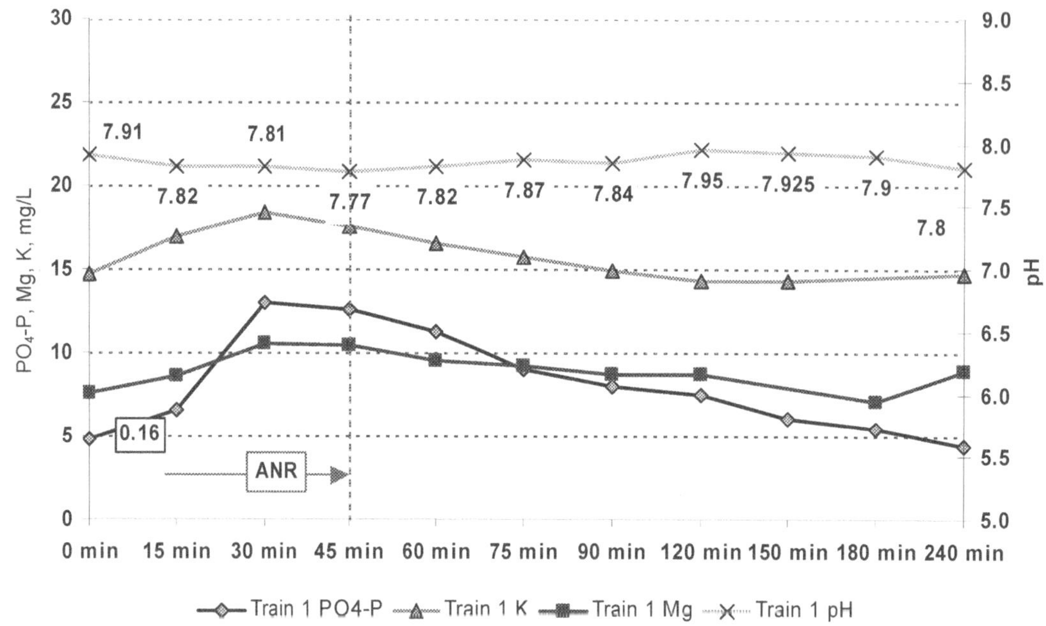


**Figure B-25. Phosphorus Uptake and Release Test Results from August 26th 2003—Using Mixed Liquor from Train 2.**

Comparison of the URT does indicate process inhibition during the May testing. As described earlier, this typically indicates the presence of GAOs or nitrates. However, the earlier uptake and release test was conducted using mechanical mixing rather than nitrogen gas mixing. It is possible that, under these conditions, oxygen trapped in the headspace of the reactor was introduced by the mechanical mixing.

Figure B-26 shows the results of additional uptake and release tests conducted at Durham as part of a separate BNR stress testing project. After VFA consumption, a measurable phosphorus uptake was observed in the test that was mechanically mixed, suggesting that the small amount of air trapped between the water surface and the bottle cap was sufficient to result in phosphorus uptake. Figure B-26 shows that for the lower VFA spike dosages (20 mg/L and 40 mg/L) a reduction of phosphate was measured during the anaerobic period of the URT test.

Based on these results, it is not possible to determine the cause of the lower phosphorus release rates observed during the URT at Durham.

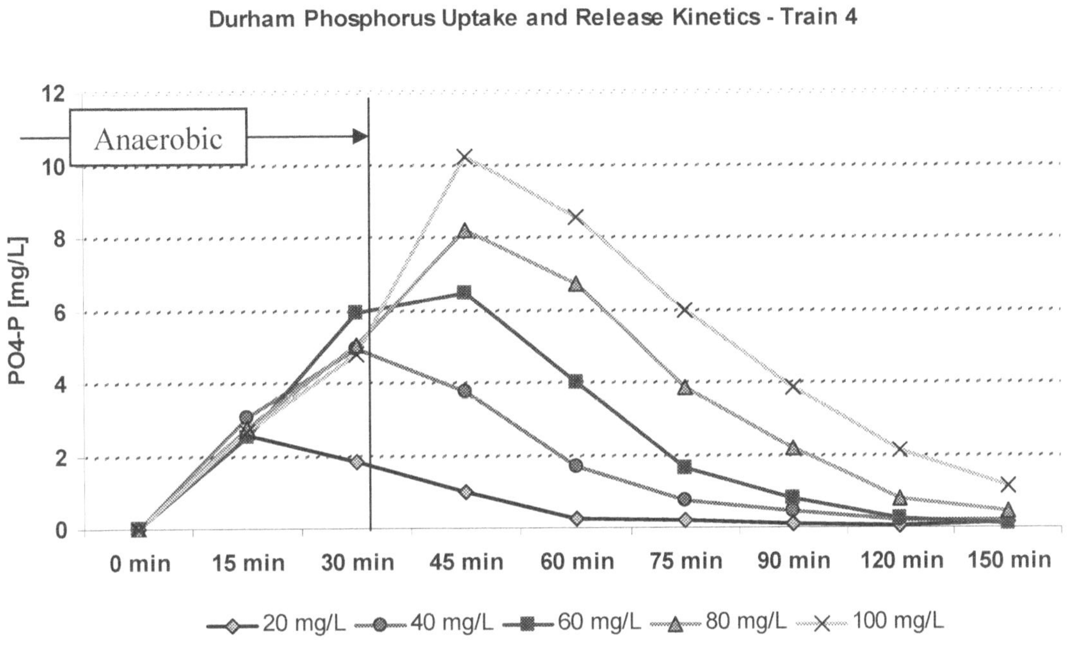


**Figure B-26. URT Curves from Tests Conducted in September of 2002 as Part of the Ongoing BNR Stress Testing.**

### B.3.3.3 Phosphorus and Metal Analysis

Lime addition plays a significant role in the phosphorus and metal mass balance at Durham. During the 2002 BNR season, three tons per day of lime were added to the primary effluent, increasing the primary effluent alkalinity 250 – 300 mg/L, the pH to 8 and soluble calcium to 30 – 40 mg/L. The impact of metals addition was evaluated through both metals profile testing and a mass balance analysis.

#### B.3.3.3.1 Metals Profiles

Metals concentrations were analyzed along with phosphorus and other constituents during the diurnal profile sampling. Figure B-27, Figure B-28, and Figure B-29 show calcium, magnesium, and potassium profiles, respectively. The high lime addition is believed to contribute to an increase in the measured calcium concentration between aeration basin influent and aeration basin effluent (Figure B-27). The calcium increase between the primary effluent and the first anaerobic zone could only be explained by dissolving lime. Because not all lime dissolves between the primary clarifier and the aeration basin, the measured dissolved calcium in the anaerobic zone only represents a fraction of the total calcium added to the process. Based on findings from other research one would expect similar uptake and release P:Mg and P:K ratios (Table B-12). At Durham, the measured uptake ratios were higher than the release ratios, suggesting that more phosphorus was accumulated per unit of magnesium uptake than the release. This discrepancy could be due to the magnesium content in lime.

Table B-12. Metal and Phosphorus Uptake and Release Rates at Durham AWWTP.

| Metal Phosphate Ratio | Release by Weight | Uptake by Weight | Release - Molar | Uptake - Molar |
|---|---|---|---|---|
| P/Mg | 1.7 | 1.9 | 1.4 | 1.5 |
| P/K | 0.78 | 0.93 | 0.98 | 1.2 |

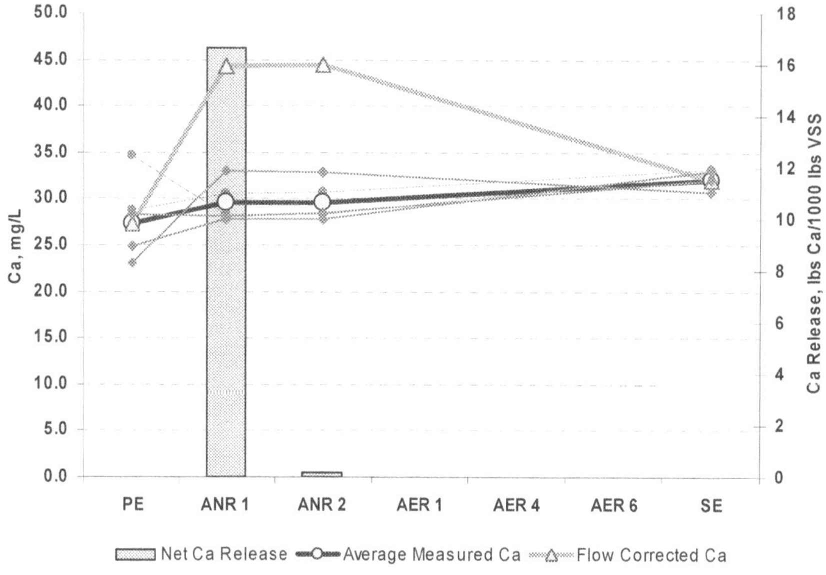


**Figure B-27. Measured Soluble Calcium Profiles During Diurnal Sampling October 2002.**

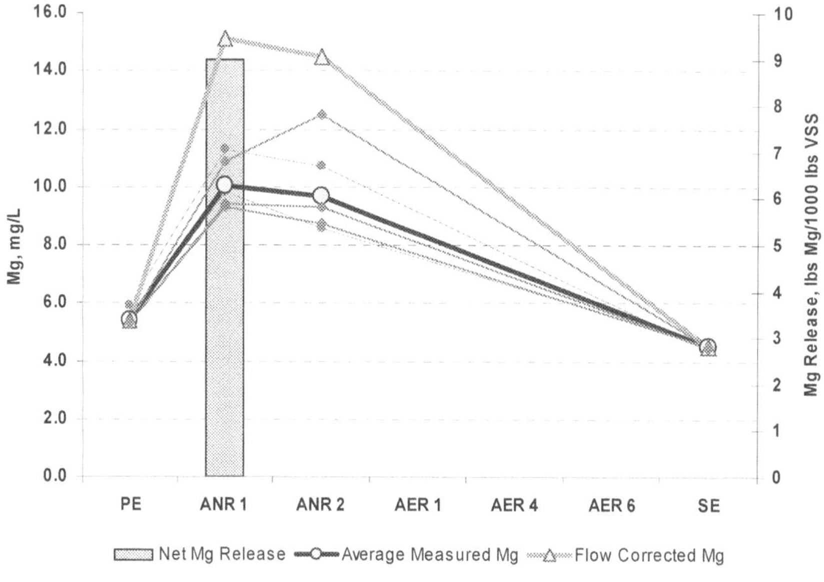


**Figure B-28. Measured Soluble Magnesium Profiles During Diurnal Sampling October 2002.**

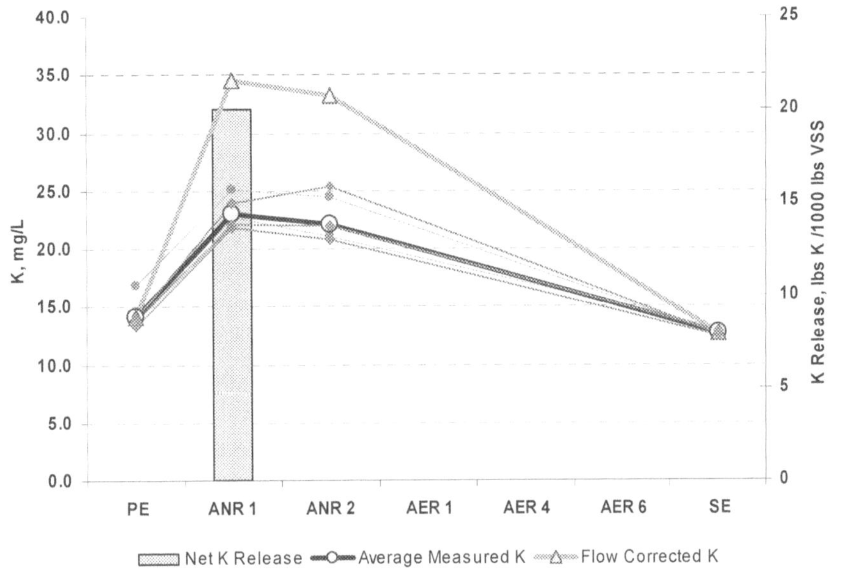


**Figure B-29. Measured Soluble Potassium Profiles During Diurnal Sampling October 2002.**

### B.3.3.3.2 Mass Balance

Following the phosphorus mass flow through the treatment process provides useful EBPR performance information, including a determination of the recycle phosphorus load. Based on a mass balance at Durham (Figure B-30), the combined recycle load equals 24% of the influent phosphorus load, the majority is phosphate recycle with the dewatering centrate and fermenter overflow (14%). The remaining 10% are recycled with the filter backwash. However, this part of the recycle load does not impact EBPR since the phosphorus is chemically bound (aluminum phosphates). Figure B-30 illustrates the phosphorus mass flow at Durham.

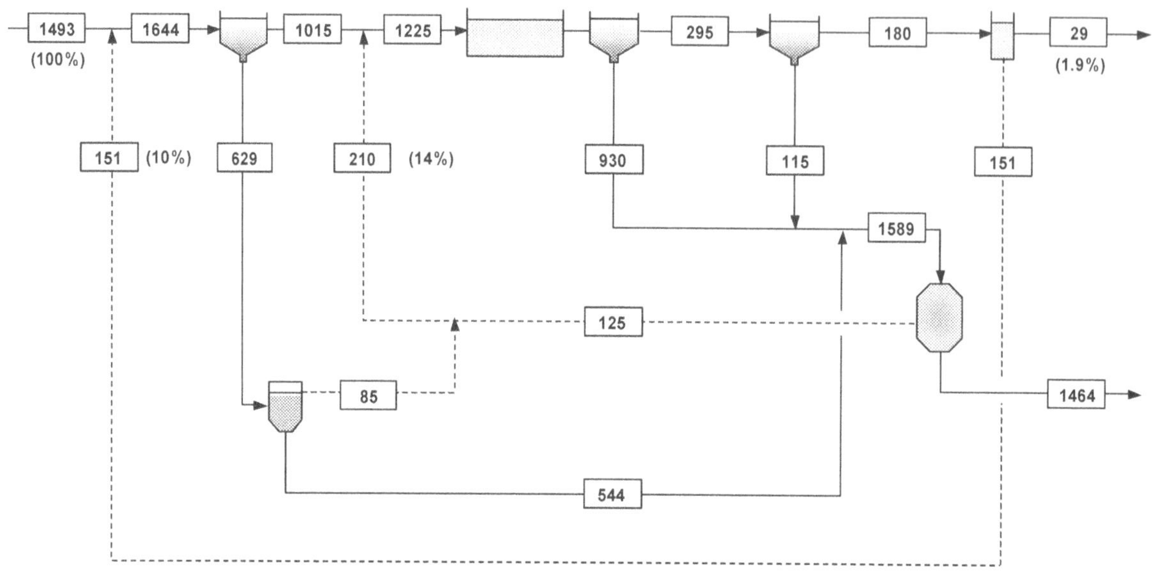


Figure B-30. Durham BNR Season Phosphorus Mass Balance (all numbers in lb P/d).

Mass balances for the metals could not be completed for Durham due to insufficient data. However Table B-3 in section B-B.2.3 shows the chemical use. Together with the low recycle phosphorus and metal content in digested sludge (Table B-13) it is safe to assume that the metal addition (especially alum) plays a major role in the overall plant performance.

Table B-13. Digested Sludge Metals.

| ABI Wastewater Parameter | Unit | 2000 - 2003 |
|---|---|---|
| Aluminum | mg/kg | 39,000 |
| Calcium | mg/kg | 22,500 |
| Potassium | mg/kg | 10,800 |
| Iron | mg/kg | 16,700 |
| Magnesium | mg/kg | 6,410 |
| Phosphorus | mg/kg | 1,530 |

## B.3.4 Microscopic Examination of Digested Sludge

Digested sludge samples were examined at 200x magnification for the presence of struvite crystals. The presence and size of such crystals can serve as an indicator of EBPR activity because increased EBPR results in increased magnesium transfer to the digester, promoting struvite crystallization.

**The first set of images was taken at the beginning of the 2003 EBPR season ( FigureB-31) and the second towards the end of the season ( Figure B-32). In the first sample, the crystals were very small and only a few could be found. The images in Figure B-33 show struvite crystals that washed out of digested sludge that began plugging recirculation pumps. This occurred after the mechanical digester mixing was turned off to minimize or control digester foaming. Without mixing,**

**previously suspended struvite crystals settled to the bottom of the digester making it difficult for the recirculation pumps to move the sludge.**

Figure B-34 shows the same crystals magnified at 50x.

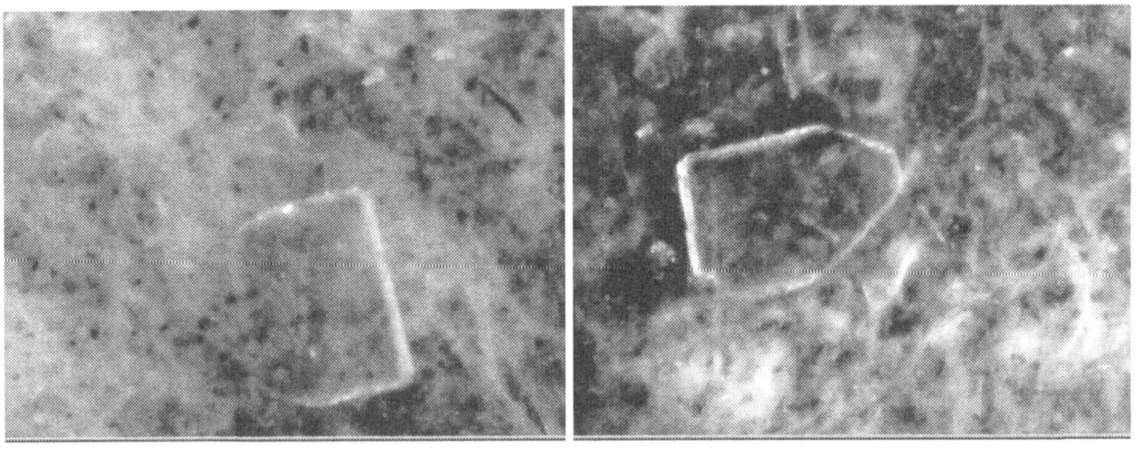

**FigureB-31. Struvite Crystals Found in Digested Sludge Sample at the Beginning of the 2003 EBPR Season.**

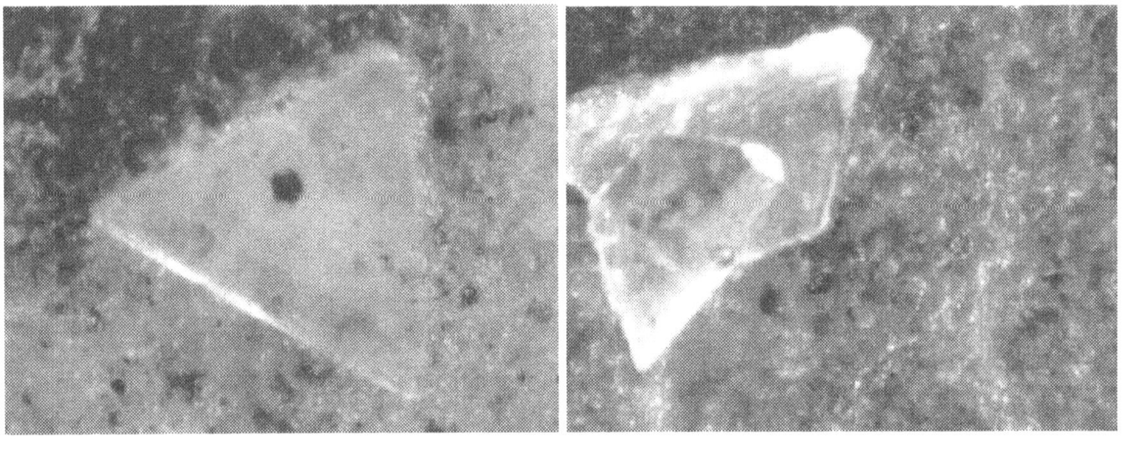

**Figure B-32. Struvite Crystals Found in Digested Sludge Sample at the End of the 2003 EBPR Season .**

Figure B-33. Struvite Crystals Washed Out of Digested Sludge After Mixing was Tuned Off for Several Hours in August of 2003 (200x magnified).

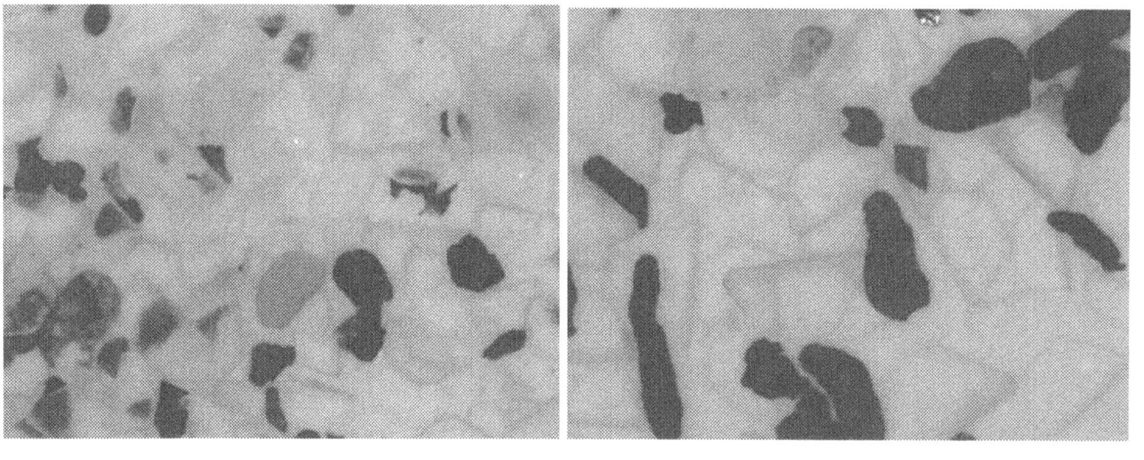

Figure B-34. Struvite Crystals Washed Out of Digested Sludge After Mixing was Tuned of for Several Hours in August of 2003 (50x magnified).

### B.3.5 PAO and GAO Investigations

Mixed liquor samples from the Durham plant were examined to observe the presence of polyphosphate granules (indicative of EBPR) and to look for the presence of GAOs. GAOs typically consist of large cocci (2-3 µm) arranged in tetrad clusters within the floc. Samples were collected in August 2003.

Both microscopic study and molecular techniques were used to examine the presence of GAOs in the mixed liquor from the plant. Light microscopy observation was used to observe the tetrad-forming GAOs, and molecular probing using specific oligonucleotide probes targeting GAOs was completed to identify both tetrad-forming and non-tetrad-forming GAOs. In addition Neisser and DAPI staining were applied to observe the presence and relative abundance of PAOs.

Figure B-35 and Figure B-36 show the observation of PAOs and GAOs using a Neisser stain. Figure B-37 and Figure B-38 shows the results of DAPI stain. The GAOs have Neisser positive stained cell walls, but are not positive for cell inclusions as are PAOs. The Neisser positive stained PAOs are dark blue to purple in color, and usually group in clusters within the flocs. The DAPI-stained PAOs are yellow in color, which is characteristic of polyphosphate granules observed under the fluorescence microscope. Other non-polyphosphate-accumulating bacteria including GAOs show up in blue with the DAPI stain.

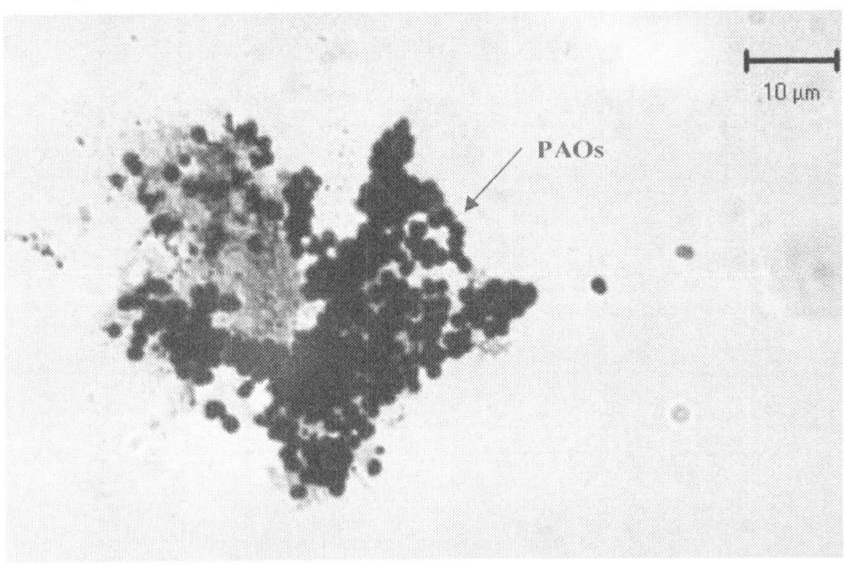


**Figure B-35. Neisser Stain Observation of PAOs of Mixed Liquor Sample from Durham WWTP, Train 1. August 26, 2003.**

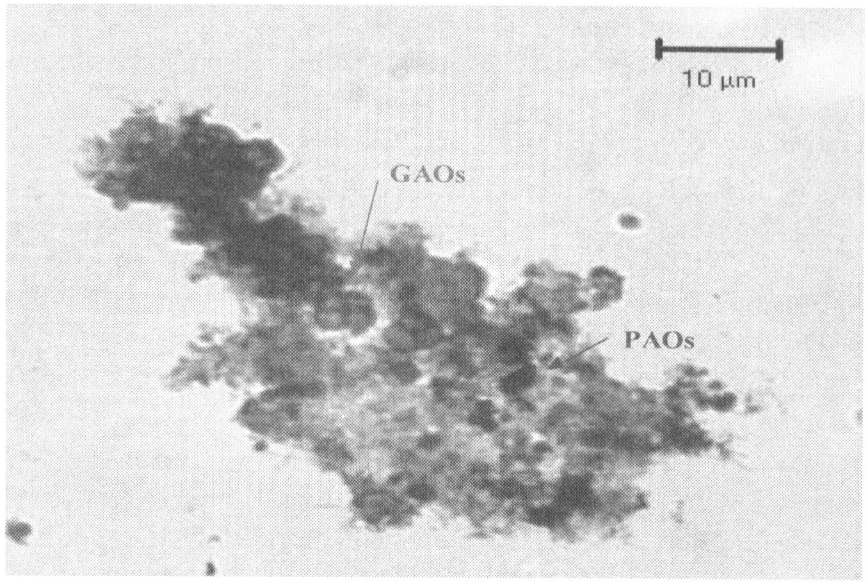


Figure B-36. Neisser Stain Observation of PAOs and GAOs in the Mixed Liquor Sample from the Durham WWTP, Train 2, August 26, 2003.

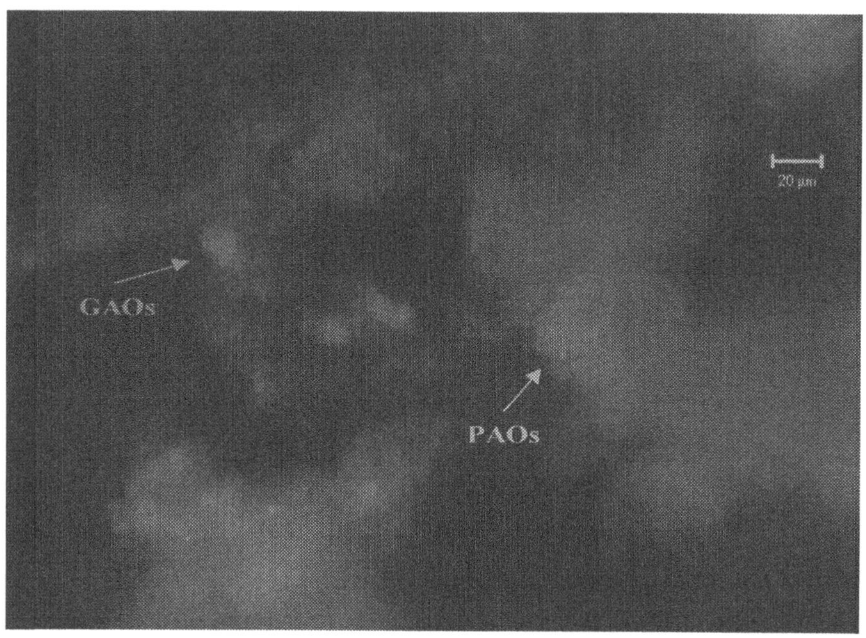


Figure B-37. DAPI Stain Observation of PAOs and Tetrad-Forming GAOs in the Mixed Liquor Sample from Durham WWTP, Train 1, August 26, 2003.

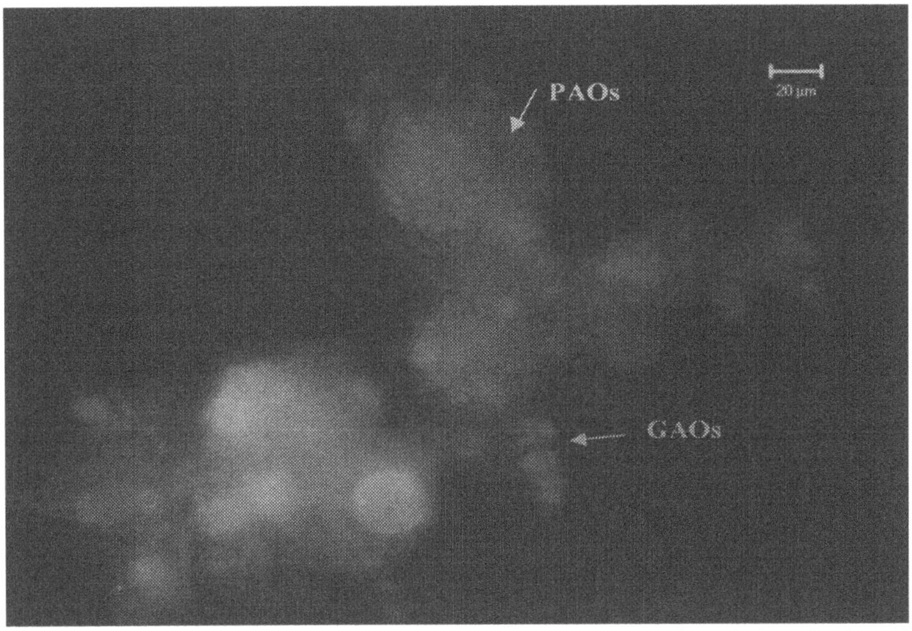


**Figure B-38. DAPI Stain Observation of PAOs and GAOs in the Mixed Liquor Sample from Durham WWTP Plant, Train 2, August 26, 2003.**

Table B-14 compares the observation of PAOs and GAOs for mixed liquor samples taken from Train 1 and Train 2. Tetrad-forming GAOs were easily found in both sludge samples, although they were slightly more abundant in the Train 2 sample than in the Train 1 sample. Both Neisser stain and DAPI stain confirmed the abundant presence of PAOs in both mixed liquor samples. The dominance of PAOs in the samples indicated that the growth of poly-P accumulating populations was favored and they were responsible for the phosphorus removal performance. There were some tetrad-forming GAOs that were easily found in the samples. Their presence could reduce the phosphorus removal rate and phosphorus removal efficiency.

Examination of the mixed liquor samples via molecular methods (fluorescence in situ hybridization (FISH)) using 16S rDNA-based probes targeting *Accumulibacter* (represent PAOs) (Crocetti et al. 00) and some non tetrad-forming GAOs- *Competibacter* (GAOQ989 andGB-G2, Crocetti *et al*. 00 and Kong et al. 00) showed that 10-20% of the Eubacteria populations was *Accumulibacter*-PAOs. No *Competibacter*-GAOs was observed. It should be noted that FISH methods only targeted certain PAO and GAO species for which the probes are available. DAPI and Neisser stain targeted all cells that are capable of accumulating poly-P, therefore should reveal more PAO populations than FISH method.

Table B-14. Summary of PAOs and Tetrad-Forming GAOs Observation for Durham WWTP.

| Date of Sample | GAO observation (Tetrad-forming) | GAO observation (Non-tetrad-forming) | PAO observation | Comments |
|---|---|---|---|---|
| Durham 8-26-03, Side 1 | some, 3 out of 10 fields observed GAOs, some grouped together | Not observed. | abundant, big flocs of PAOs | PAOs dominant , some GAOs were easily observed |
| Durham 8-26-03, Side 2 | easily found , some,5 out of 10 fields observed GAOs, some grouped together | Not observed. | abundant, big flocs of PAOs | PAOs dominant , some GAOs were easily observed |

The phosphorus removal performance at the plant was evaluated for the time period when the samples were taken for GAOs evaluation. The results of this evaluation are shown in Table B-15. The relatively higher abundance of GAOs observed in Train 2 may only partially contribute to the decreased phosphorus removal efficiency as indicated by the lower value of phosphorus uptake rate and phosphorus release ratio in side 2. Since the difference in tetrad-forming GAO abundance in the mixed liquor from Train 1 and Train 2 was not significant, there are other factors that led to the observed difference in the P removal performance.

Table B-15. Comparison of GAOs Abundance and P removal Performance.

| Date of Sample | GAOs abundance | P Release Rate gP/gVSS.hr | P Uptake Rate gP/gVSS.hr | P Release Ratio gΔ P/gΔAcetic Acid | Release Rate/ Uptake Rate |
|---|---|---|---|---|---|
| DP 8-26-03, side 1 | Some Easily found | 10 | 3.6 | 0.36 | 2.8 |
| DP 8-26-03, side 2 | Some Easily found | 4.2 | 1.5 | 0.16 | 2.9 |

The results of GAO tests indicated that the conditions in the basins allowed the growth of GAOs during the period of study. The presence of GAOs, however, did not necessarily hamper the phosphorus removal to a great extent provided there was a sufficient population of PAOs to sustain adequate phosphorus removal. Many factors were shown to have an effect on the competition between the PAOs and GAOs including feed substrate, COD/P ratio, pH, SRT, temperature and DO. More complex substrate favors the growth of GAOs. So any change of influent wastewater characteristics may have an influence on the stimulated growth of GAOs. P-limiting condition (COD/P >50) favors the GAOs. Lower pH (<7.3) could favor the growth of GAOs. Higher temperature (>20°C) and longer SRT (>7-10 days) were also known to favor the growth of GAOs.

Other factor such as the availability of other electron acceptors in the anaerobic zone may also have effect on P removal process. The amount of nitrate introduced to the anaerobic zone could affect the intracellular poly-p storage form and the stability of the PAO population and performance.

## B.4 Conclusions

The study showed that phosphorus removal varies considerably among the four treatment trains. While 50$^{th}$ percentile secondary effluent concentrations are low in all trains (ranging from 0.06 mg/L to 0.31 mg/), 80$^{th}$ percentile concentrations range from 0.28 mg/L to 1.19 mg/L. Aeration basin influent VFAs are higher than in most plants, due in part to the seasonal operation of a primary sludge fermenter. The average aeration basin influent BOD:P and VFA:P ratios are 20:1 and 4.3:1, respectively. EBPR was observed to be more unstable when BOD:P ratios dropped below 20:1, and the process was very stable at BOD:P ratios over 25:1.

Despite the high VFA concentrations, phosphorus release rates at Durham are low. Influent phosphorus concentrations are relatively low, and the total aeration basin inventory is also lower than at other facilities. Phosphorus release rates measured during the uptake and release tests were also lower than at other facilities. Some GAOs were observed in microscopic testing, but no GAOs were observed using molecular methods. The cause of the low phosphorus release at Durham was not determined.

APPENDIX C

# LOWER REEDY ENHANCED BIOLOGICAL PHOSPHORUS REMOVAL (EBPR) EVALUATION REPORT

## C.1 Summary of Findings

The Lower Reedy Wastewater Treatment Plant (WWTP) is located in Greenville, SC, and is operated by the Western Carolina Regional Sewer Authority (WCRSA). The plant has a maximum month design capacity of 8.6 million gallons per day (mgd) (18.75 mgd peak day) and treats a combination of domestic and industrial wastewater. The layout of the basins allows the facility to be operated in a variety of EBPR modes ($A^2O$, VIP, UCT, or modified UNC Charlotte). During the time of this study, the facility was operated in the VIP mode using only two of the three available secondary treatment trains. The facility adds lime for alkalinity control and alum for phosphorus polishing, and plans to add supplemental volatile fatty acid (VFA) feed in the near future.

Overall the aeration basin influent characteristics would be considered favorable for EBPR. Water temperatures range from 15-30°C, and the average BOD to P ratio is approximately 24 mg BOD/mg P. The aeration basin influent BOD averaged 198 mg/L vs. 8.5 mg/L of phosphorus. Ratios considered favorable for EBPR are 20 mg BOD/mg P or higher. 50% of the aeration basin influent phosphorus is being returned from solids processing, and the dewatering recycle contributes an average of 8% to the ammonia load. Including the dewatering recycle the aeration basin influent ammonia concentration average 25 mg/L and TKN 34 mg/L.

The operational strategy at Lower Reedy is to achieve effluent ammonia of less than 1.0 mg/L and effluent phosphorus of less than 0.5 mg/L. From 2000 through 2003, the average effluent total phosphorus concentration was 0.67 mg/L and the plant produced an effluent phosphorus concentration of equal to or less then 2 mg/L 91% of the time. At the same time, the plant produced average effluent ammonia of 0.62 mg/L and average total nitrogen of 4.7 mg/L. The effluent ammonia was very consistent until the summer of 2003, during which a steady increase in effluent ammonia was observed. The system achieved overall phosphorus and nitrogen removal of 88.2% and 81%, respectively.

| | |
|---|---|
| Reliability | Average effluent phosphate of 0.75 mg/L exceeds operational target of 0.5 mg/L, but complies with discharge permit. Overall process reliability has improved since alum addition added in 2002. |
| Operational Criteria | Average aerobic solids retention time (SRT) is 5 days. Anaerobic hydraulic retention time (HRT) is 1 hour, anoxic HRT typically 1.5 hours, and aerobic HRT 5.2 hours. |
| Influent Organic Characterization | Aeration basin influent BOD:P ratio is approximately 24:1 and VFA:P ratio is 7.2:1, with approximately half of the phosphorus load contributed by the solids recycle stream. Process appears somewhat less stable at BOD:P ratios below 25:1. |
| P Removal Capacity | P release is high, suggesting potential for high EBPR. However, effluent concentrations observed in profiles and in uptake and release tests varied and were not directly correlated with phosphorus release. P removal may be limited by aerobic detention time. |
| Measures Showing PAO Growth | Strong phosphorus release observed in some phosphorus profiles, and low GAO population observed in microscopic testing. |
| P and Metals Mass Balance | Mass balance suggests high recycle P, contributing 54% of the aeration basin influent phosphorus loads. |
| Secondary Release | Some phosphorus profiles showed release in the anoxic zones, possibly suggesting secondary release. These observations were generally associated with high initial phosphorus release, suggesting early consumption of available VFAs. |

## C.2 Introduction

The analysis of this facility was completed as part of a project sponsored by the Water Environment Research Foundation (WERF). The project objective is to develop information for evaluating the ability of municipal wastewater treatment plants to efficiently and cost-effectively remove phosphorus to low levels through processes that involve EBPR. This was completed in large part through detailed studies at several treatment facilities that employ EBPR. The goals of the study were to:

- Determine process reliability.
- Determine the biological phosphorus removal efficiency.
- Determine the biological phosphorus removal capacity.
- Determine how chemical addition affects Bio-P removal.

### C.2.1 Facility Introduction

Figure C-1. Lower Reedy WWTP, Western Carolina Regional Sewer Authority, Greenville, SC.

The Lower Reedy Wastewater Treatment Plant (WWTP) (Figure C-1 and Figure C-2) is located in Greenville, SC, and is operated by the Western Carolina Regional Sewer Authority (WCRSA). The plant has a maximum month design capacity of 8.6 mgd (18.75 mgd peak day) and treats a combination of domestic and industrial wastewater. The average plant flow from 2000 through 2003 was 4.16 mgd. The Lower Reedy plant was upgraded to biological nutrient removal including EBPR in May 2000. The plant was configured such that different biological nutrient removal (BNR) process designs (A2O, VIP, UCT, or Modified UNC) could be operated providing a great level of flexibility to plant operators. When required, lime can be added to augment influent alkalinity and alum can be added to the secondary clarifiers for chemical phosphorus polishing. The effluent permit requirements are summarized in Table C-1. Two of WCRSA's facilities - Lower Reedy and Mauldin Road - have a combined permit to discharge treated effluent into Lake Greenwood via Reedy Creek. This permit structure provides WCRSA some flexibility regarding the distribution of the permitted effluent nutrient load between the two plants. The operational strategy at Lower Reedy is to achieve effluent ammonia of less than 1.0 mg/L and effluent phosphorus of less than 0.5 mg/L.

Table C-1. Lower Reedy Permit Requirements.

| Parameter | Units | Monthly | Weekly |
|---|---|---|---|
| BOD | mg/L | 30 | 45 |
| TSS | mg/L | 30 | 45 |
| $NH_4$-N – May-Oct | mg N/L | 5.4 | 8.1 |
| $NH_4$-N – Nov-Apr | mg N/L | 8.4 | 12.6 |
| Total Phosphorus | mg/L | 2.0 | - |

Figure C-2. Flow Equalization (top left), Aeration Basin 2 (top right), Chlorine Contact Tank (bottom left), Anaerobic Digester (bottom right).

## C.2.2 Facility Process Design

### C.2.2.1 Liquid Stream

The Lower Reedy facility has three identical activated sludge trains, each consisting of five equally sized unaerated (anaerobic (ANR) and anoxic (ANX)) basins and one aerated aerobic (AER) basin. Only two of the three trains have been in operation at any time during the last three years. The system is currently operating in VIP mode, as shown in Figure C-3.

Denitrified recycle from the unaerated basins is combined with the primary effluent upstream of the activated sludge basins and split to the first anaerobic zone in each train. One variable speed denitrified recycle pump (also called anoxic recycle pump) is used per train. Each pump is rated at 1,750 gpm (2.5 mgd). At the time of testing the anoxic pumps were operating at fixed rates of 1,680 and 1,770 gpm for the two trains in operation.

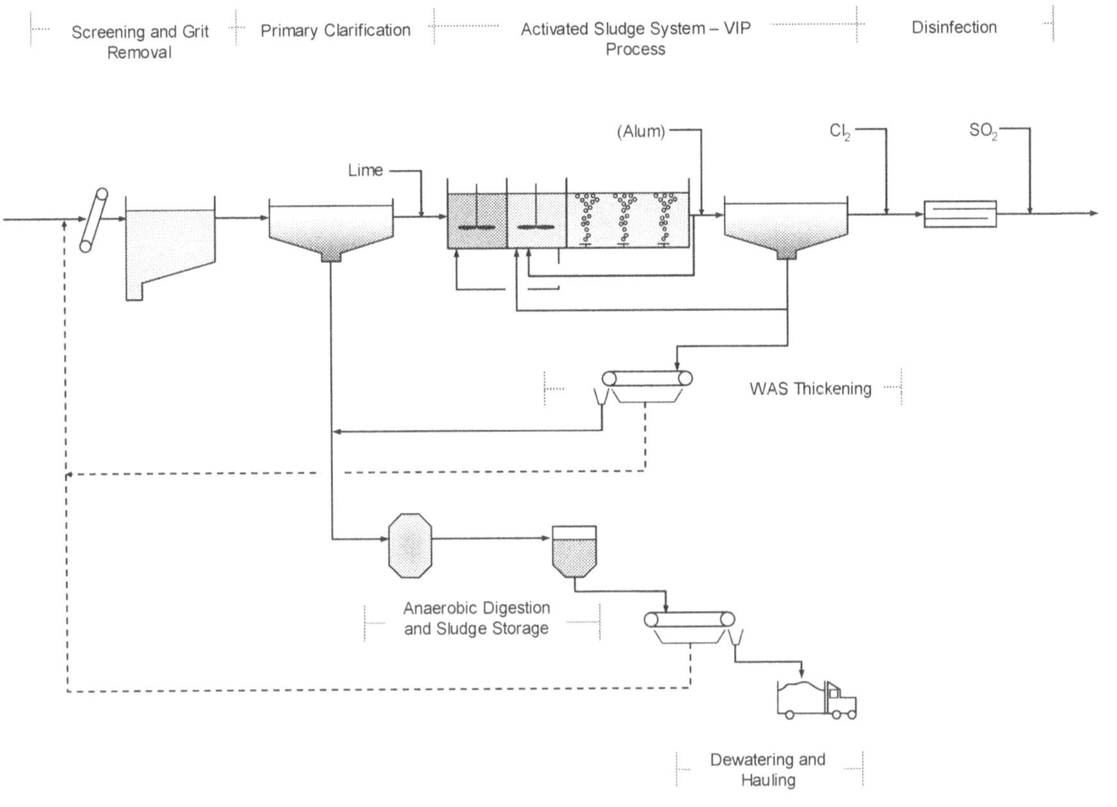


**Figure C-3. Process Schematic for Lower Reedy WWTP Liquid and Solids Treatment.**

Aerobic recycle from each of the aeration basins is returned to the unaerated basins to create anoxic zones. Five anoxic recycle variable speed pumps (rated at 7,000 gpm each) are available to return mixed liquor from the end of the aerobic to the head of the anoxic zone. One pump is used with each train.

Return activated sludge (RAS) is pumped separately into the first anoxic basin of each activated sludge train, and is flow placed at 65% of the effluent flow.

One unique aspect of the facility is the long hydraulic retention time. At the design flow and with all three trains in operation, the anaerobic zone provides 2.6 hours of hydraulic retention time, the anoxic zone 4 hours and the aerobic zone 9.6 hours. Operators can use an upstream equalization basin to shave peak flows and help maintain relatively constant flow to the activated sludge basins.

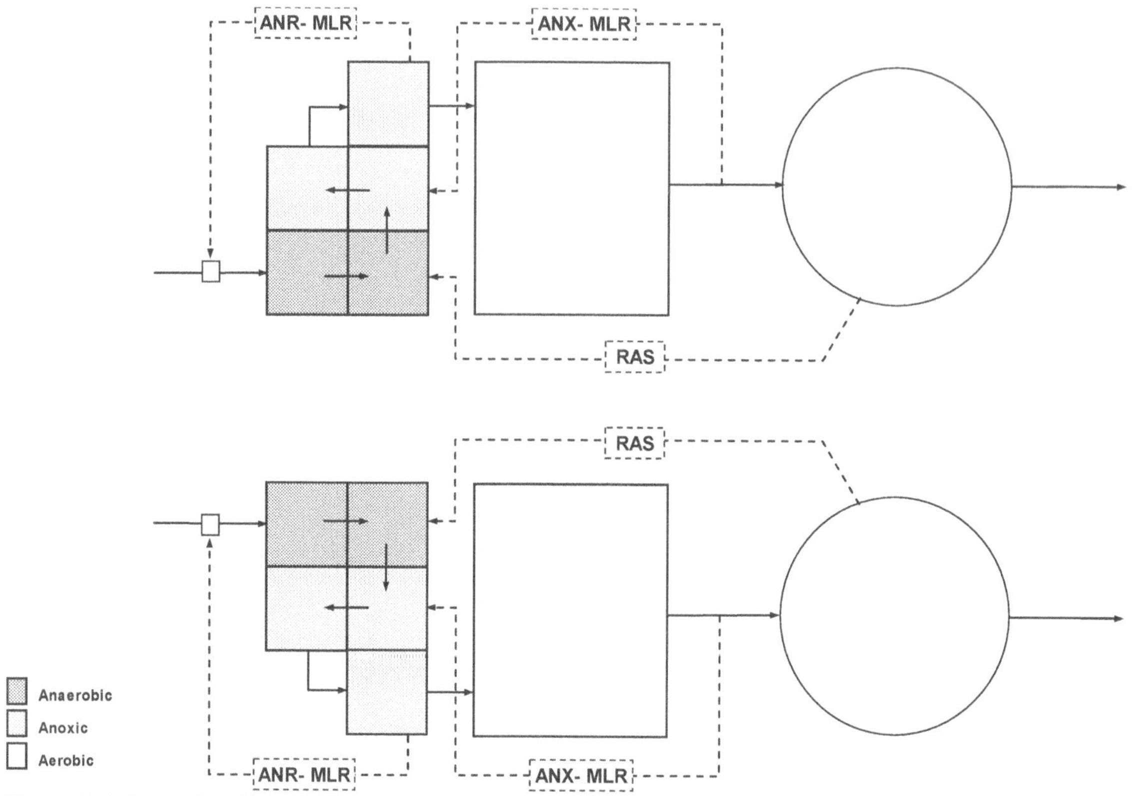


**Figure C-4. Secondary Treatment Layout at Lower Reedy WWTP (shows the two operating activated sludge trains).**

**Table C-2. Key Process Parameters.**

| Component | Unit | Design | Average |
|---|---|---|---|
| Anaerobic HRT | hrs | 1.8 – 3 | 3.1 |
| Anoxic HRT | hrs | 2.7 – 4.5 | 4.5 |
| Aerobic HRT | hrs | 7 – 12 | 11.3 |
| Total HRT | hrs | 11.5 – 19.5 | 19.4 |
| RAS Rate | % | 50 % | 52% |
| Anaerobic Mixed Liquor Recycle Rate | % | 50 % | 55% |
| Anoxic Mixed Liquor Recycle Rate | % | 150 % | 116% |
| SRT | days | 7-12 | 9.4 |
| MLSS | mg/L | 2000 – 3500 | 3220 |
| Solids Recycle Return | | Upstream of Primary Clarifier | |
| Flow | MGD | 8.6 | 4.8 |

### C2.2.2 Solids Stream

Primary sludge is thickened in the primary clarifier and waste activated sludge (WAS) is thickened using a gravity belt thickener. WAS thickeners are typically run 12 hours/day, starting at approximately 3 am. Thickened sludge is sent to a digester feed tank for flow equalization and mixing, and then fed continuously to the anaerobic digester. The dewatering filtrate is returned to the primary influent. Digested sludge is dewatered during weekdays between 7 a.m. and 3 p.m. The sludge management program is scheduled to change to land application of thickened digested sludge in the near future.

**Table C-3. Solids Processing Summary.**

| Component | Technology | Comments |
|---|---|---|
| Primary Sludge Thickening | Gravity | Thickened in Primary Clarifier |
| WAS Thickening | Gravity Belt Thickener | Operated 8/7 |
| Primary Sludge Stabilization<br>WAS Stabilization | Anaerobic Digestion | TSS = 2% |
| Dewatering | Belt Filter Press | Liquid land application planned for new future |
| Centrate Storage | None | |

### C2.2.3 Chemical Addition

The Lower Reedy plant adds lime for alkalinity supplement. Since August 2002, alum can be added if needed. Table C-4 summarized the chemical use from 2001 through 2003, which shows high lime use despite an influent alkalinity of 122 mg/L. Aeration basin effluent alkalinity ranges from 100 mg/L to 300 mg/L. On demand acetic acid feed to the anaerobic zone is planned for the near future to supplement influent volatile fatty acids (VFAs) when necessary.

**Table C-4. Chemical Feed Summary**

| Chemical | Unit | 2001 | 2002 | 2003 |
|---|---|---|---|---|
| Lime | Tons/yr | 544 | 288 | 288 |
| Alum | Tons/yr | - | - | 57.2 |
| | | | | |
| Lime | lb/MGD | 69 | 344 | 548 |
| Alum | lb/MGD | - | - | 112 |

### C2.2.4 Process Control

The operational strategy is to maintain the sludge inventory at a constant value. Currently, the sludge inventory is maintained to produce a solids retention time (SRT) of about 7-12 days depending on water temperature. Wasting is controlled on a daily basis by measuring the RAS (WAS) concentration while wasting and stopping the wastage once the target mass is wasted to maintain the target SRT.

RAS is controlled at a constant flow and adjusted to account for wasting (i.e. the underflow from the clarifiers is kept constant as the WAS flow varies during the day).

Dissolved oxygen (DO) is controlled to maintain a target concentration of 2.5 mg/L in the aeration basin. Actual values are typically between 3 and 4 mg/L. The aeration basin is equipped with coarse bubble aeration.

## C.3 Process Performance Evaluation

### C.3.1 Approach

The process evaluation entailed two data collection activities: a review of historical data and the conduct of special sampling to supplement routinely recorded water quality and process measurements. The following data were collected (see Table C-5):

**Routine Data Review**

The reviewed data represent all available water quality and process monitoring information collected from January 2000 through August 2003. Data were analyzed to determine general process performance and reliability of meeting effluent quality goals, and to evaluate the effect of process operating parameters on performance. These routine data are referred to as the **2000-2003 data set**.

**Special Data Collection**

Project team members conducted field tests and interviewed plant staff during each phase of the project (Table C-5). During Phase 1 the focus of site visits was on supplementing detailed survey information and data. Aeration basin phosphate and ammonia profiles were measured and preliminary phosphorus uptake and release tests were conducted. Results of the Phase 1 uptake and release tests are not presented in this report because test methodology was modified for Phase 2 to permit better interpretation of the results. Wastewater characteristic samples were collected for one day during Phase 1. During Phase 2, data collection for each intensive sampling period extended over two weeks and generally included collection of six samples for each water quality parameter.

During the intensive sampling periods, the plant staff was asked to take additional samples that were either not taken or only taken infrequently as part of the routine sampling plan. The sampling plan included influent VFA, BOD, COD, readily biodegradable COD as measured by the flocculated filtered COD test (FFCOD), and soluble BOD (sBOD) as well as phosphorus and nitrogen species. Influent, effluent, and digested sludge phosphorus and metals samples were included in the sampling plan to allow performance of P and metals mass balances.

The results of the special data collection efforts are included in the following data sets:

- **Phase 1 data set** This data set was collected in Year 1 of the project during September 2002.
- **Phase 2A data set** This data set was collected in Year 2 of the project during the first intensive sampling period in June 2003.
- **Phase 2B data set** This data set was collected in Year 2 of the project during the second intensive sampling period in October 2003.
- **Phase 2F data set** This data set was collected in Year 2 of the project during the field visit in June 2003.

**Table C-5. Summary of Field Tests and Sampling.**

| Field Testing Segment | Month | Data Set ID | Purpose Conducted |
|---|---|---|---|
| Historical Data Review | January 2000 – August 2003 | 2000 - 2003 | Influent, Effluent Averages<br>Process Parameter Averages |
| Phase 1 – Field Tests | September 2002 (One day) | Phase 1 | Aeration Basin Profiles<br>Uptake and Release Test<br>Operator Interviews |
| Phase 2 – Field Tests | June 2003 | Phase 2F | Uptake and Release Test<br>Operator Interviews<br>Data Collection |
| Phase 2 – Intensive Sampling Period 1 | June 2003 | Phase 2A | Influent Characterization<br>Aeration Basin Profiles<br>Phosphorus Mass Balance<br>Metal Mass Balance |
| Phase 2 – Intensive Sampling Period 2 | October 2003 | Phase 2B | Influent Characterization<br>Aeration Basin Profiles<br>Phosphorus Mass Balance<br>Metal Mass Balance |

## C.3.2 Overall Process Characterization and Performance

This section of the report summarizes the aeration basin influent characteristics and secondary treatment process performance based on routine data gathering and additional data collection described in Section C.3.1. It also discusses relationships between primary effluent loading and process performance.

### C3.2.1 Aeration Basin Influent Characterization

The raw wastewater composition at Lower Reedy shows very typical municipal composition. However, the service area does include several industrial contributors. The ratio of municipal to industrial wastewater is approximately 80:20. Since the facility startup, the flow has increased by 24% from 4.6 to 5.7 mgd, which equals and average annual increase of 6.4%. Daily flows range from 0.8 mgd to 14.8 mgd.

Overall the aeration basin influent characteristics would be considered favorable for EBPR. Water temperatures range from 15 to 30°C (Figure C-5), and the average BOD to P ratio is approximately 24 mg BOD/mg P. The aeration basin influent BOD averaged 198 mg/L vs. 8.5 mg/L of total phosphorus. Ratios considered favorable for EBPR are 20 mg BOD/mg P or higher. 50% of the aeration basin influent phosphorus is being returned from solids processing, and the dewatering recycle contributes an average of 8% to the ammonia load. Including the dewatering recycle, the aeration basin influent ammonia concentration average 25 mg/L and TKN 34 mg/L. Table C-7 summarizes the aeration basin influent compositions. Figure C-6 and Figure C-7 show the key primary effluent loads.

Table C-6. Aeration Basin Influent Wastewater Characterization.

| ABI Wastewater Parameter | Unit | 2000 - 2003 | Phase 1[1] | Phase 2A | Phase 2B |
|---|---|---|---|---|---|
| BOD | mg/L | 199 | 190 | 160 | 180 |
| SBOD | mg/L | - | - | - | 97 |
| COD | mg/L | - | 430 | - | 420 |
| ffCOD | mg/L | - | 140 | - | 200 |
| VFA | mg/L | 63 | - | 90 | 48 |
| TP | mg P/L | 8.8 | 8.8 | 6.6 | 12 |
| $PO_4$-P | mg P/L | 5.6 | 6.6 | 4.7 | 7.7 |
| Total Alkalinity (as $CaCO_3$) | mg/L | 150 | 170 | 130 | 170 |
| pH | | 6.9 | 7.0 | 6.1 | 6.3 |
| TKN | mg N/L | 34 | 34 | 25 | 37 |
| $NH_4$-N | mg N/L | 25 | 26 | 21 | 25 |
| TSS | mg/L | 94 | 78 | 44 | 170 |
| Flow | MGD | 5.4 | 4.6 | 5.7 | 5.0 |
| Temperature | °C | 21 | 22 | 20 | 20 |
| VFA : P Ratio | - | 7.2 | - | 14 | 3.9 |
| ffCOD : P Ratio | - | - | 16 | - | 15 |
| COD : P Ratio | - | - | 49 | - | 34 |
| BOD:P Ratio | - | 25 | 21 | 25 | 15 |
| ffCOD : COD Ratio | - | - | 0.33 | - | 0.46 |
| VFA : ffCOD Ratio | - | - | - | - | 0.29 |

[1] Single composite sample

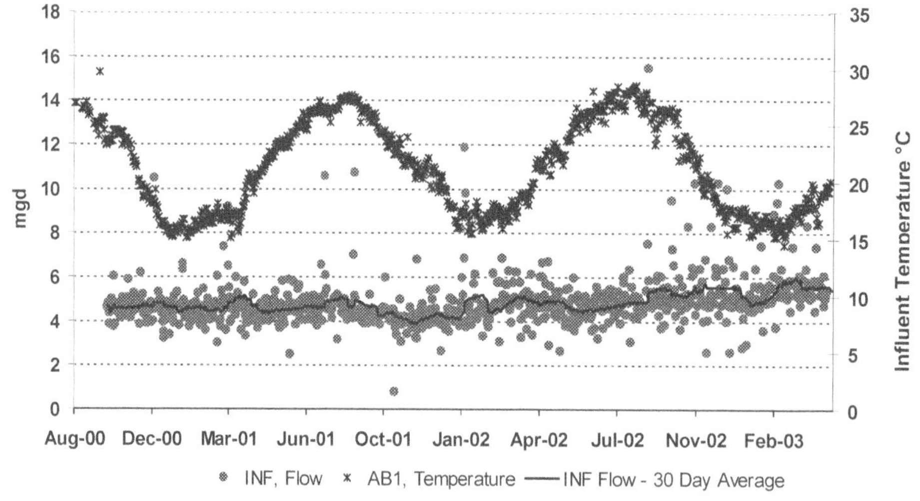


**Figure C-5. Influent Flow, Influent Temperature, and 30-day Average Influent Flow Between 2000 and 2003.**

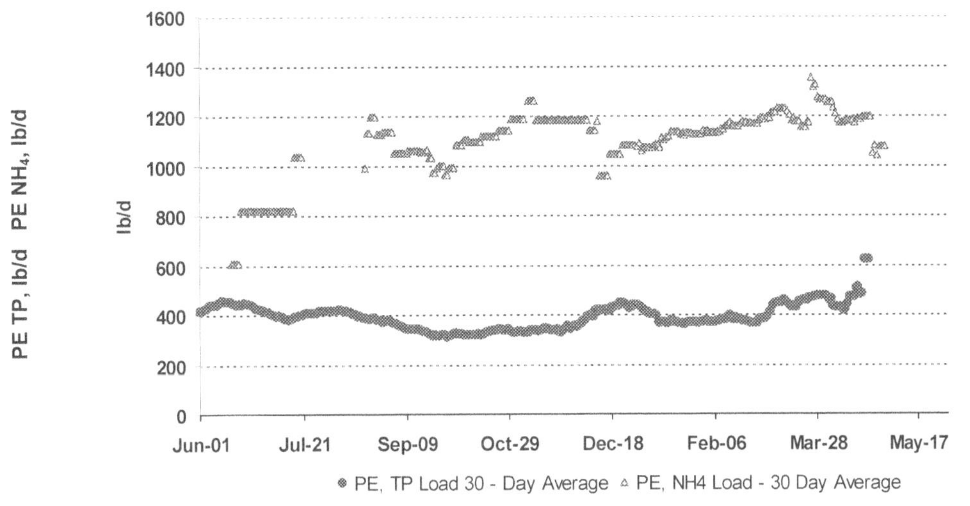


**Figure C-6. Primary Effluent Phosphorus and Ammonia Load.**

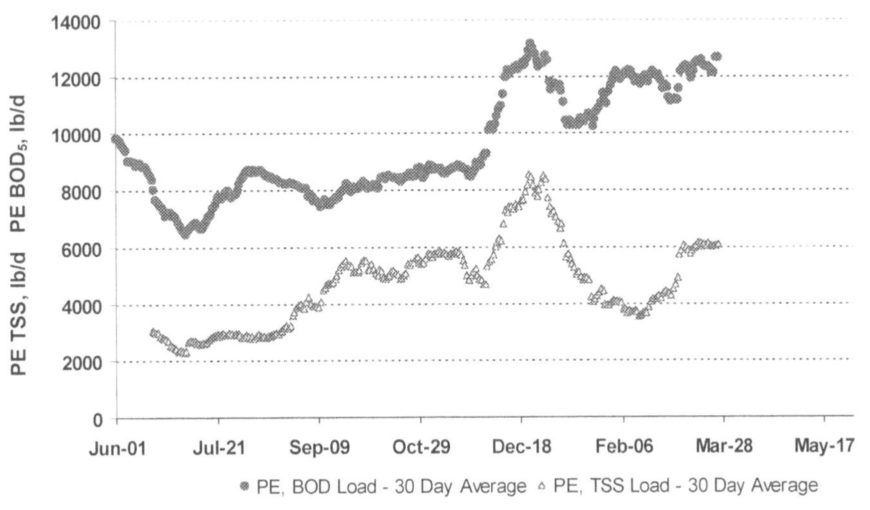


**Figure C-7. Primary Effluent BOD and Suspended Solids Load.**

The aeration basin influent (ABI) shows large fluctuations in flow and concentration. Concentrations of aeration basin influent phosphate concentrations vary the most widely (Table C-7).

Table C-7. Selected Water Quality Measurement and Process Parameters from 2000 – 2003.

| Parameter | Unit | Min | Ave | Max |
|---|---|---|---|---|
| Influent Flow | MGD | 0.77 | 4.8 | 14.5 |
| ABI Temperature | °C | 15 | 21 | 30 |
| ABI BOD | mg/L | 86 | 200 | 500 |
| TKN | mg N/L | 7.2 | 34 | 56 |
| ABI $NH_4$-N | mg N/L | 10 | 25 | 49 |
| ABI TP | mg P/L | 1.5 | 8.9 | 35 |
| ABI $PO_4$-P | mg P/L | 1.4 | 5.9 | 18 |
| ABI TSS | mg/L | 23 | 95 | 370 |

### C3.2.2 EBPR Performance

Table C-8 summarizes the average effluent quality for the various sampling periods. Overall the plant performs very well. The three-year average effluent total phosphorus concentration was 0.67 mg/L, and the plant produced an effluent phosphorus concentration of equal to or less then 2 mg/L 91 percent of the time (Figure C-8 and Figure C-11). At the same time, the plant produced average effluent ammonia of 0.62 mg/L and average total nitrogen of 4.7 mg/L. The effluent ammonia was very consistent until the summer of 2003, at which time a steady increase in effluent ammonia was observed (Figure C-9). The system achieved overall phosphorus and nitrogen removal of 88.2% and 81%, respectively.

Since 2003, the plant has had the ability to add alum to the secondary clarifier for additional phosphorus removal. Through July 2003 alum use averaged 113 pounds per million gallons treated (Table C-4), or approximately 14 mg/L.

The primary clarifiers remove an average of 34% of the influent BOD and 38% of influent TSS. The poor TSS removal is attributed to the fact that primary clarifiers are used to thickened primary sludge. The solids removal performance correlates well with the blanket depth. Figure C-11 shows the 30-day averages for primary clarifier blanket depth and primary effluent BOD load. According to record data the blanket depth ranges between 6 and 10 ft. The data also show that the higher the sludge blanket the less efficient the primary solids retention.

The mixed liquor sludge volume index averages 134 mL/g indicating well settling sludge. While there are some spikes, the average effluent TSS of 4.7 mg/L also indicates good performance of the secondary clarifiers.

**Table C-8. Process Performance Measurement Results During Intensive Testing (two-week average).**

| Process Performance Parameter | Unit | 2000 – 2003 | Phase 1 | Phase 2A | Phase 2B |
|---|---|---|---|---|---|
| ABE $PO_4$-P | mg P/L | 0.7 | 0.08 | - | 1.7 |
| SE $PO_4$-P | mg P/L | 0.75 | 0.11 | 0.77 | 1.38 |
| SE TP | mg P/L | 0.90 | 0.22 | 0.83 | 1.59 |
| SE TN | mg N/L | 6.12 | 5.2 | 9.45 | 8.23 |
| Secondary P-Removal | % | 88.2 | 96.5 | 85.9 | 75.2 |
| Secondary N-Removal | % | 81.0 | 84.5 | 68.5 | 74.7 |
| RAS $PO_4$-P | mg P/L | - | - | 2.1 | 0.9 |
| RAS Nitrate | mg N/L | 1.6 | 2.0 | 1.15 | 1.0 |
| RAS DO | mg/L | - | - | 0.05 | 0.1 |
| SRT | days | 9.4 | 9.8 | 9.3 | 12.6 |
| RAS Rate | % Q | 51.2 | 52 | 50.1 | 48.7 |
| ML Recycle to Anaerobic Zone | % Q | 67 | 60 | 43.3 | 50.2 |
| ML Recycle to Anoxic Zone | % Q | 119 | 124 | 96.1 | 115 |
| Plant Flow | MGD | 4.62 | 4.5 | 5.81 | 4.77 |
| Anaerobic HRT | hours | 3.2 | 2.9 | 2.66 | 3.2 |
| Anoxic HRT | hours | 4.8 | 4.1 | 4.0 | 4.9 |
| Aerobic HRT | hours | 11.6 | 9.9 | 9.64 | 11.8 |
| MLSS | mg/L | 3220 | | | |
| MLVSS | mg/L | 2200 | 1890 | 2400 | 1980 |
| MLSS/MLVSS | | 77.5 | | | |
| SVI | ML/g | 134 | | | |
| TN - % Removal | % | 81.0 | | | |

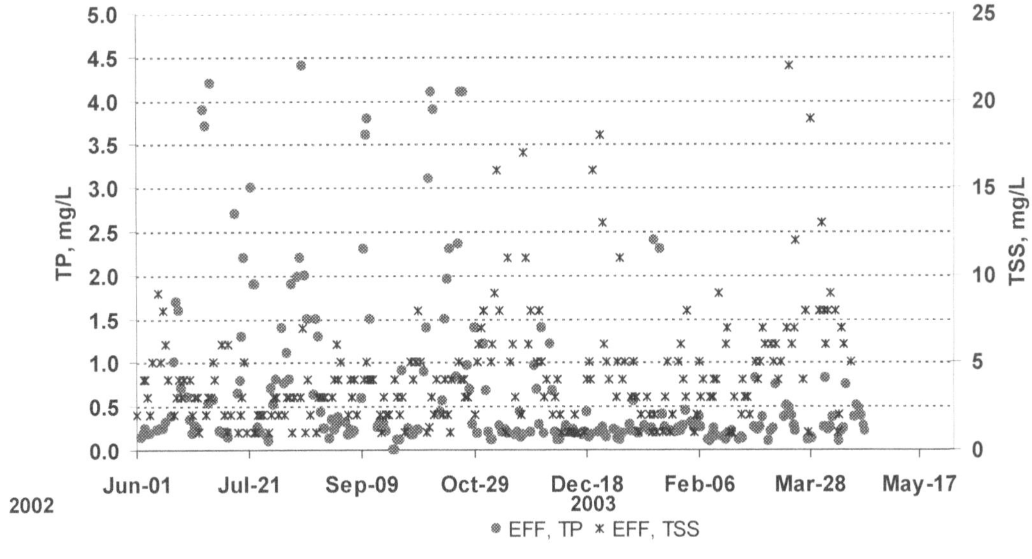


**Figure C-8. Effluent Total Phosphorus and Total Suspended Solids Between Summer of 2002 and Spring of**

2003.

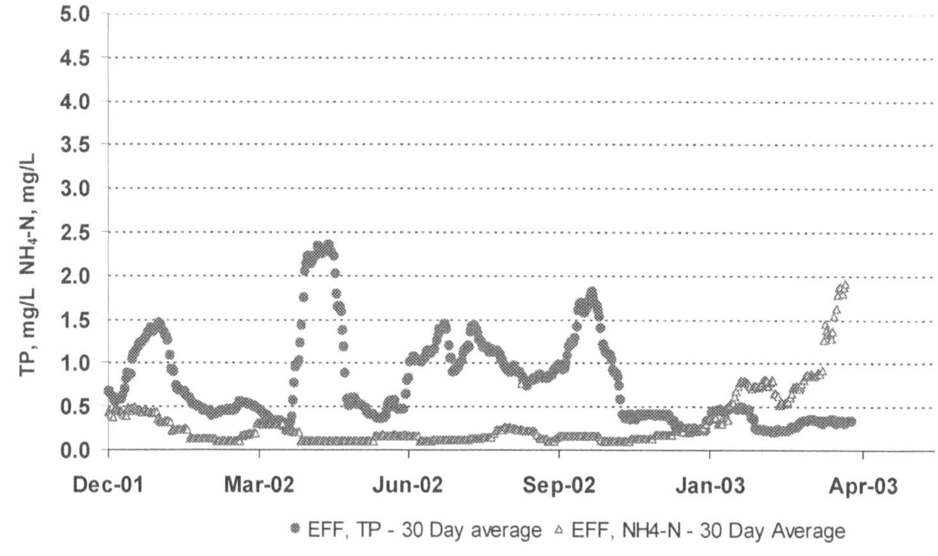


**Figure C-9. 30-day Average for Effluent Total Phosphorus and Ammonia.**

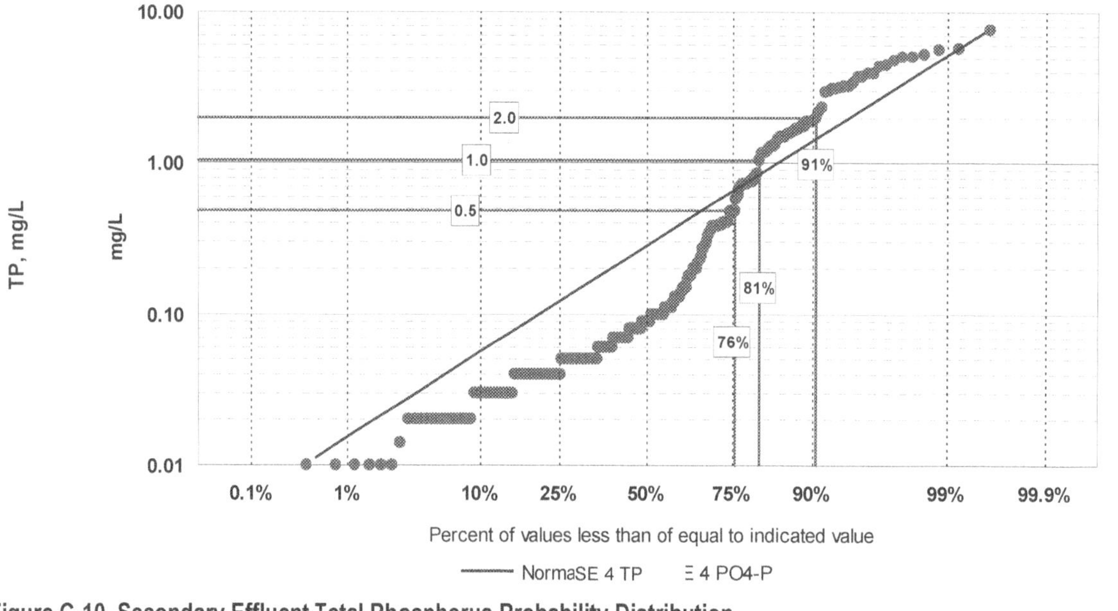


**Figure C-10. Secondary Effluent Total Phosphorus Probability Distribution.**

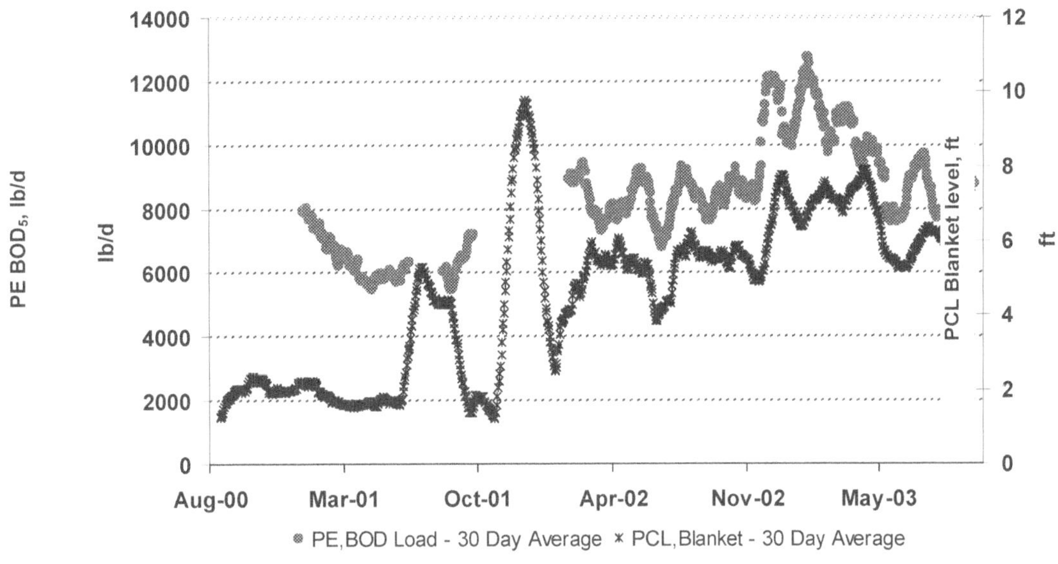


**Figure C-11. Primary Clarifier Blanket Level vs. Primary Effluent BOD.**

### *C.3.2.2.1 Phase 1 Testing*

During Phase 1, Lower Reedy delivered good and consistent nitrification results but phosphorus removal was plagued by frequent EBPR performance problems. While the plant maintained permit compliance without chemical addition, effluent total phosphorus would at times exceed 4 mg/L (Figure C-12). On the day of the site visit the plant was producing low P and low N effluent. Secondary effluent phosphate was 0.11 mg/L and total nitrogen was measured at 5.2 mg/L.

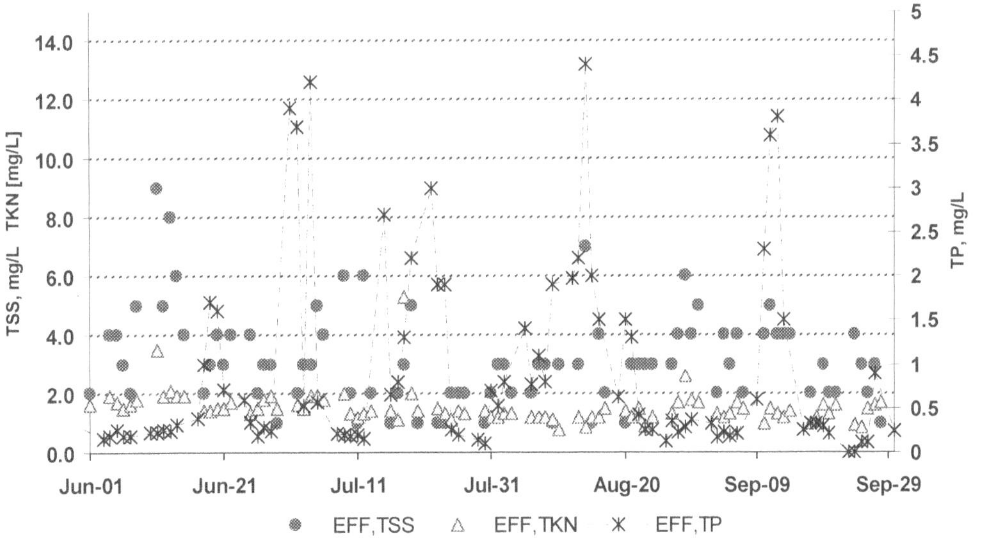


**Figure C-12. Phase 1 Secondary Effluent Phosphorus, TKN, and Suspended Solids.**

### C3.2.3 Phase 2 Testing

Performance during Phase 2 was opposite that of Phase 1, with high effluent ammonia and TKN concentrations but low phosphorus concentrations. What is unusual about the high effluent TKN is the major contribution from soluble organic nitrogen. The performance shown in Figure C-13 and Figure C-14 suggest that the increased organic nitrogen is due to an internal source, most likely solids processing.

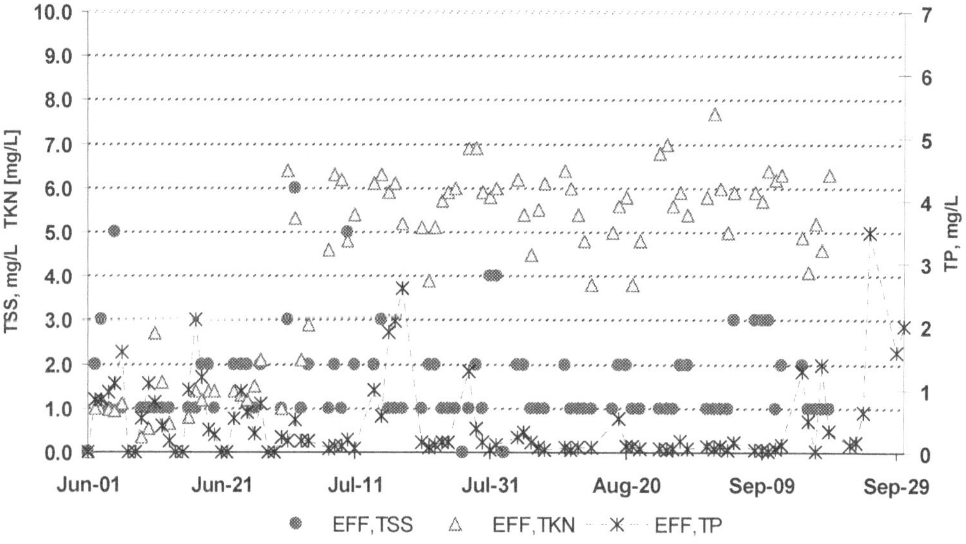


**Figure C-13. Phase 2 Secondary Effluent Phosphorus, TKN, and Suspended Solids.**

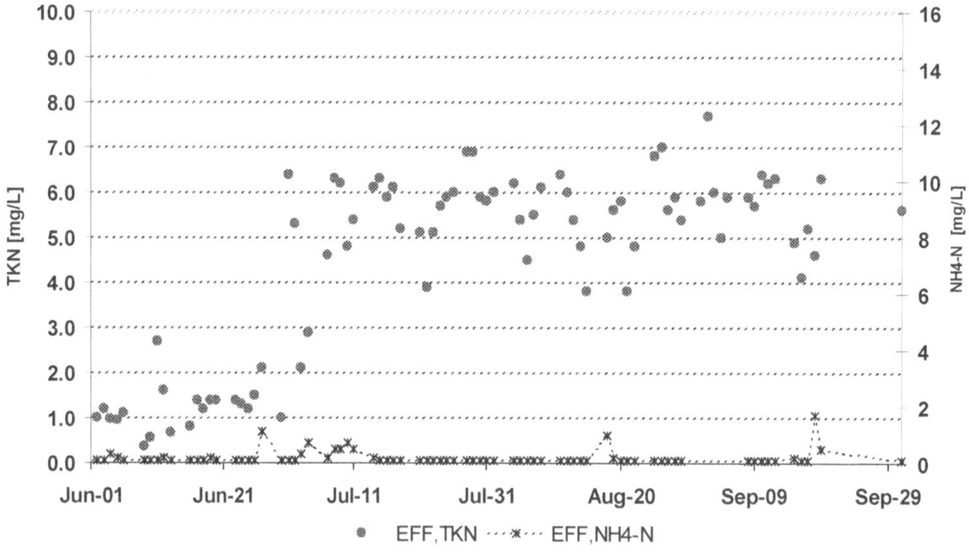


**Figure C-14. Effluent Ammonia and TKN During Phase 2.**

### C3.2.4 Relationship Between Performance and Process Loading

Since the BNR startup, the facility has only operated two of the three trains at a time. Based on the hydraulic retention time, the aerated time is about 30% - 50% longer than other similar facilities.

The aeration basin BOD and TSS load has increased disproportional to the influent flow. This appears to be due to the decrease in primary clarifier solids retention mentioned in the previous section (Figure C-11).

The aeration basin phosphorus load averages 5.0 lb P/lb VSS/d. Compared to other investigated facilities, this is considered to be an average load. No correlation was found between aeration basin phosphorus load and effluent phosphorus concentration.

The BOD/P ratio does show some cause and effect relationship with effluent phosphorus (Figure C-15). When the BOD/P ratio drops below 25 lb BOD/lb P, effluent phosphorus performance is less stable.

Since the spring of 2003, influent and primary effluent data indicate an increased solids processing recycle load (Figure C-18) that appears to not adversely affect nitrogen or phosphorus removal.

It is likely that primary clarifier operation impacts EBPR at Lower Reedy. At blanket depths of 5-10' one could expect significant VFA concentrations in the thickened primary sludge and pH of less then 6.0. The trapped VFAs offer a potential to optimize EBPR operation, however the high VFA concentrations can also cause operational problems. In its current operation operators have no control over how much of the primary sludge blanket VFA are elutriated and enter the aeration basin. Significant upswings in VFA concentration can cause an increase in effluent phosphorus. When significantly more phosphorus is released such that the uptake capacity of the aerobic zone is exceeded, the effluent phosphorus concentration increases.

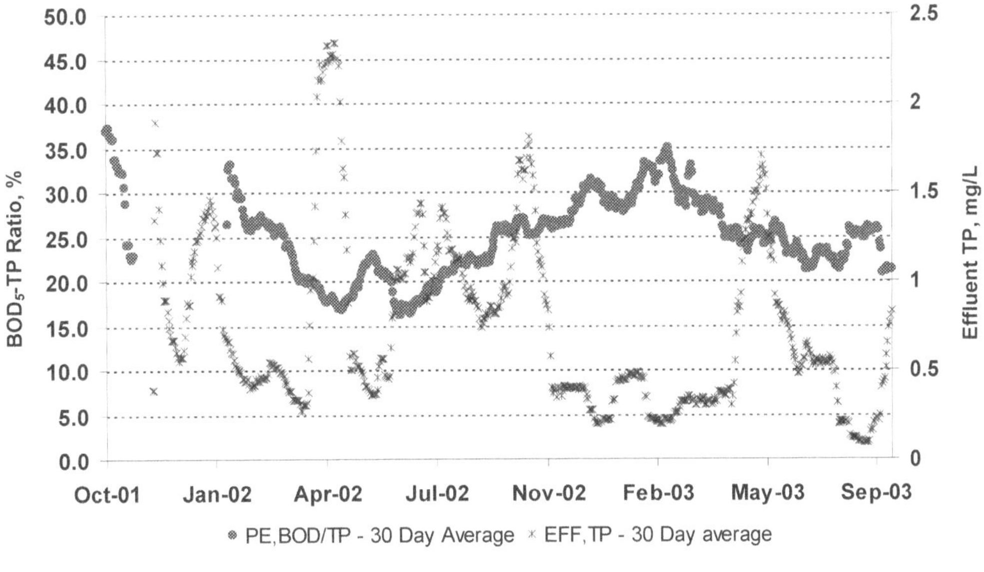


**Figure C-15. 30-day Averages of Primary Effluent BOD/TP Ratio and Effluent Total Phosphorus.**

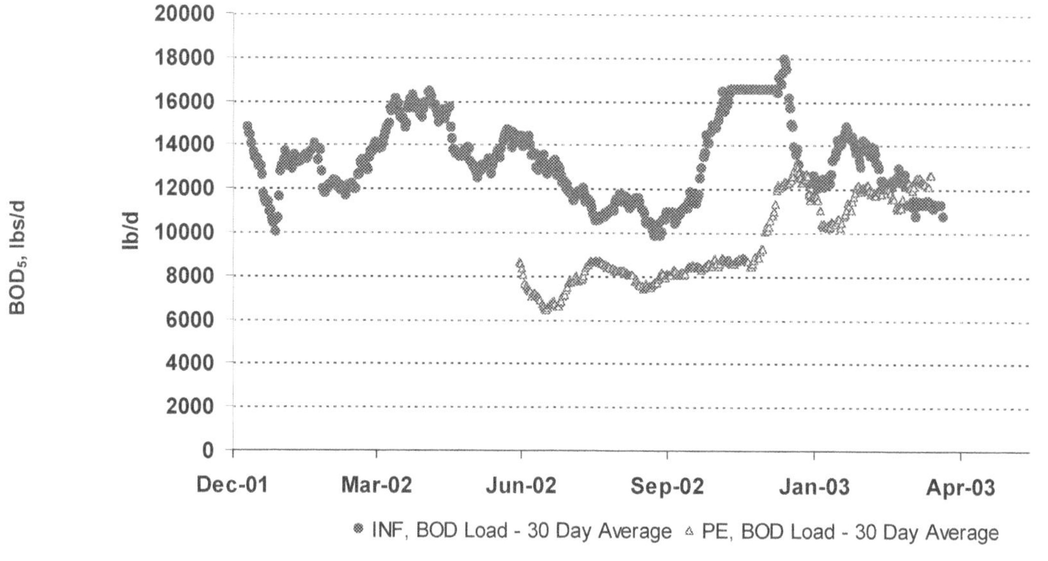


Figure C-16. 30-day Averages for Raw Influent and Primary Effluent BOD Between 2001 and 2003.

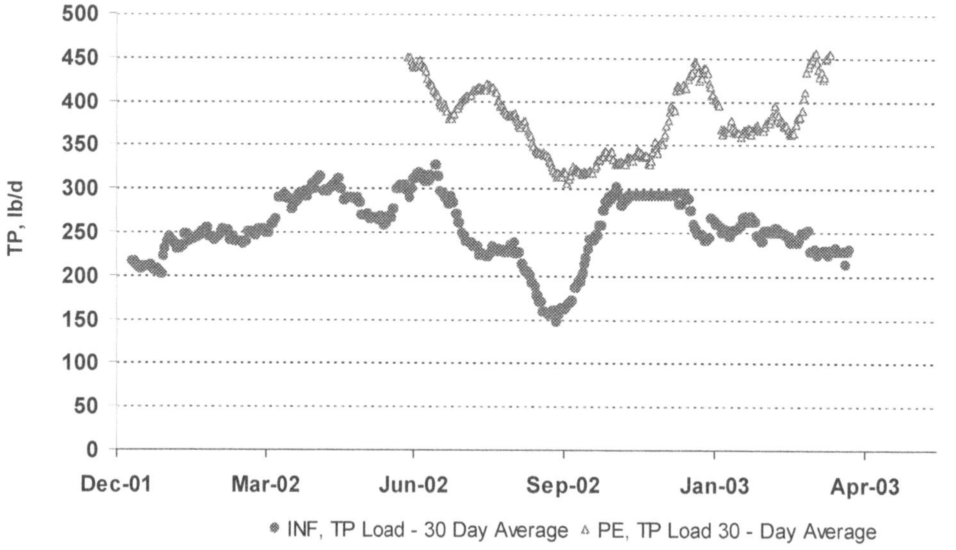


Figure C-17. Influent and Primary Effluent 30 Day Average Total Phosphorus Load Between 2001 and 2003.

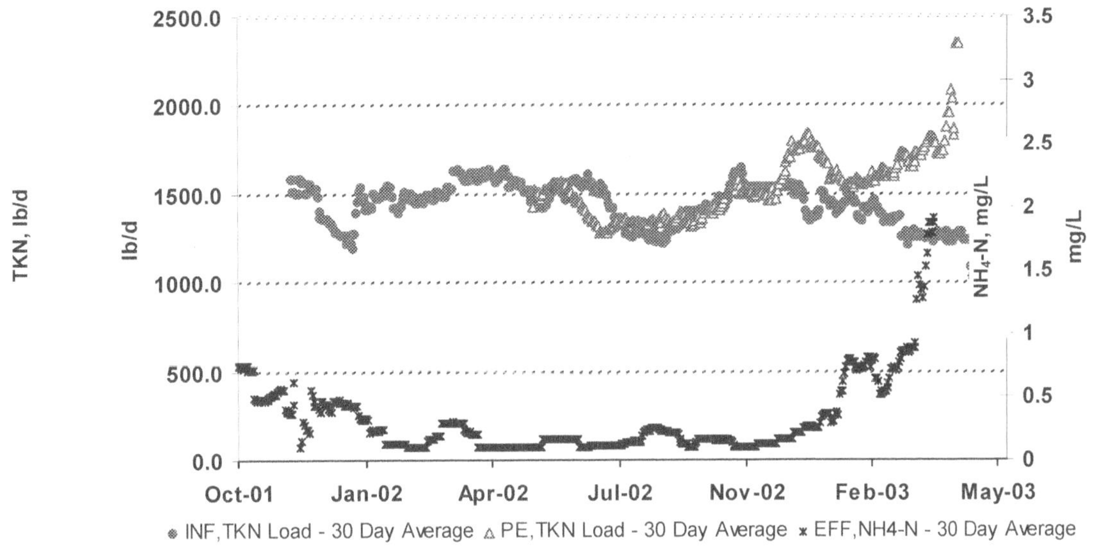


Figure C-18. 30-day Averages of Raw Influent and Primary Effluent TKN load and Effluent Ammonia Concentration Between 2001 and 2003.

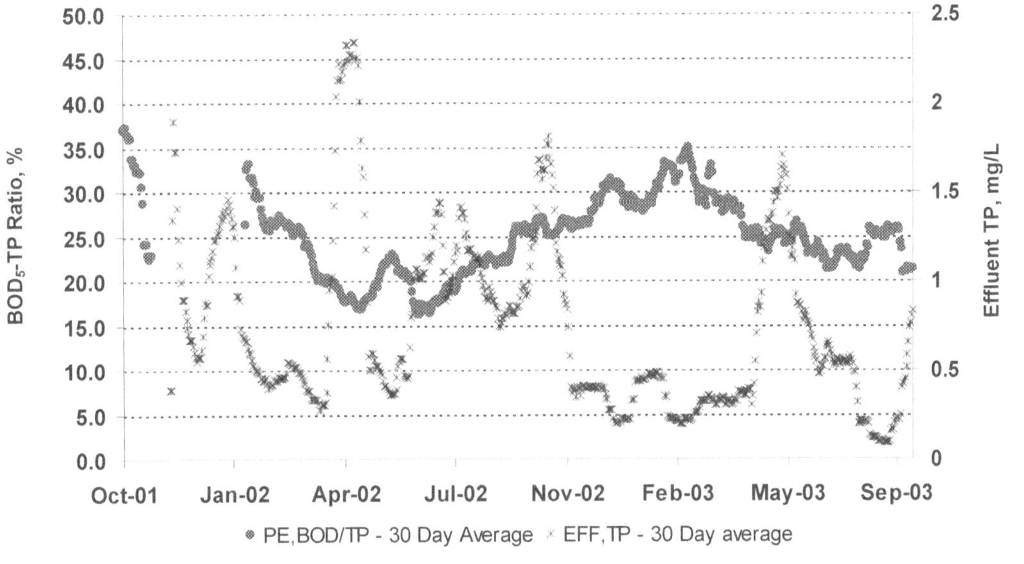


Figure C-19. 30-day Averages of Primary Effluent BOD/TP Ratio and Effluent Total Phosphorus.

## C.3.3 Additional EBPR Process Evaluation

### C3.3.1 Aeration Basin Phosphate Profiles

The aeration basin phosphate profile was measured during each sampling phase. Charts of the profiles (Figure C-21 through Figure C-26) show four lines:

- The individual daily profiles (thin shaded gray lines in background with diamond points).
- Average phosphate profiles calculated from the daily profiles (thick line with circular points).

- Flow corrected average phosphate profile (thick line with triangular points). The flow corrected concentration is calculated as the theoretical phosphorus concentration based on the influent flow. This allows for a direct comparison of the concentration values without the influence of dilution from recycle flows.
- Phosphorus release rates (columns). The release rates (lb P / 1000 lb MLVSS / d) are calculated as daily phosphorus release compared to the aeration basin volatile solids content.

In order to understand the phosphate profile, the flow scheme must first be investigated. The flows are relatively complicated for the Lower Reedy WWTP mode of operation since three recycle streams must be accounted for: RAS, anaerobic recycle (ANR), and anoxic recycle (ANX). Figure C-20 illustrates the internal recycle flows for the three periods when the aeration basin profiles where measured.

#### *C.3.3.1.1 Phase 1 Profiles*

In addition to the phosphate profiles measured as part of this study, Lower Reedy routinely measured the phosphorus profiles in 2002 (Phase 1). This allowed the comparison of the profiles measured during the field tests and intensive sampling weeks with those from the routine data. Figure C-20 shows a representative example of the internal recycle rates typical for the VIP operation at Lower Reedy.

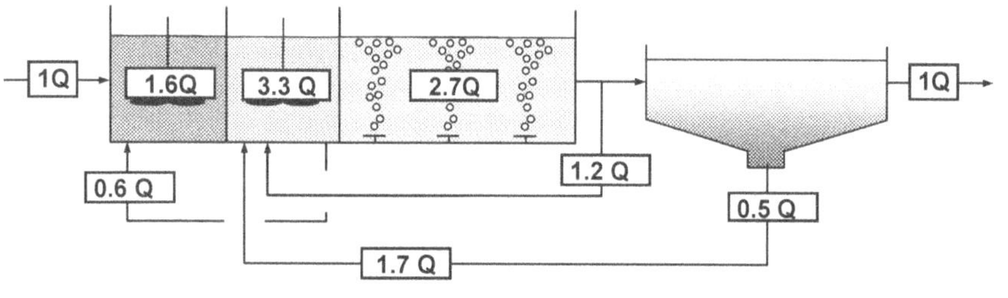


**Figure C-20. Example of RAS and Internal Recycle Flows at Lower Reedy WWTP.**

During the Phase 1 field test, the profiles revealed only a marginal phosphorus release of 10.7 mg/L (flow normalized).The measured phosphorus release was 8.3 lb P / lb MLVSS / day and the phosphorus uptake 16.3 lb P / lb MLVSS / day. The initial phosphorus release rate, which is a more suitable indicator for the PAO activity, was determined as 12.8 lb P/ lb MLVSS / hr (307 lb P/ lb MLVSS / d). The highest initial phosphorus release rate measured at Lower Reedy was 52.7 lb P/ lb MLVSS / hr (1,265 lb P / lb MLVSS / d).

Unlike what was observed at other VIP and UCT plants, the profile measured during the Phase 1 field test did not show phosphorus release in the anoxic zone.

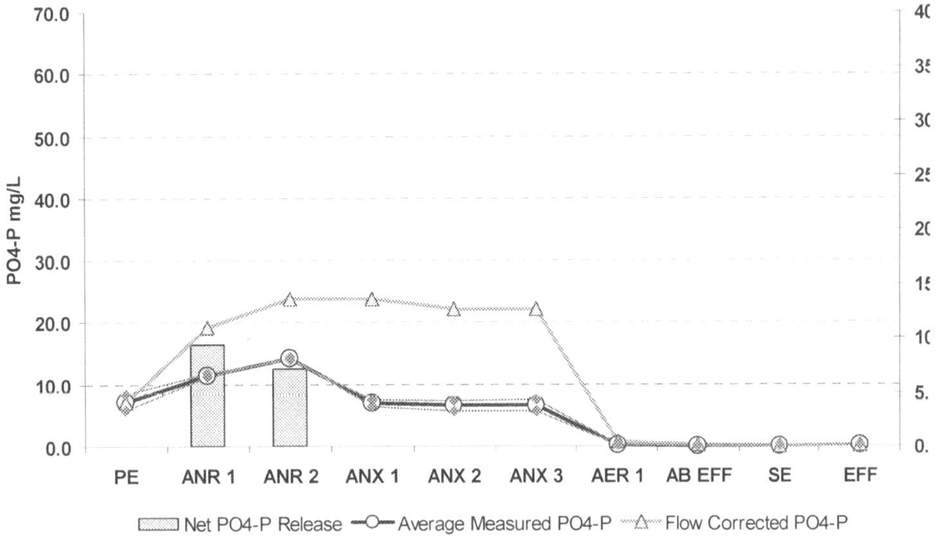


**Figure C-21. Phosphate Profiles Measured During Phase 1 Field Testing.**

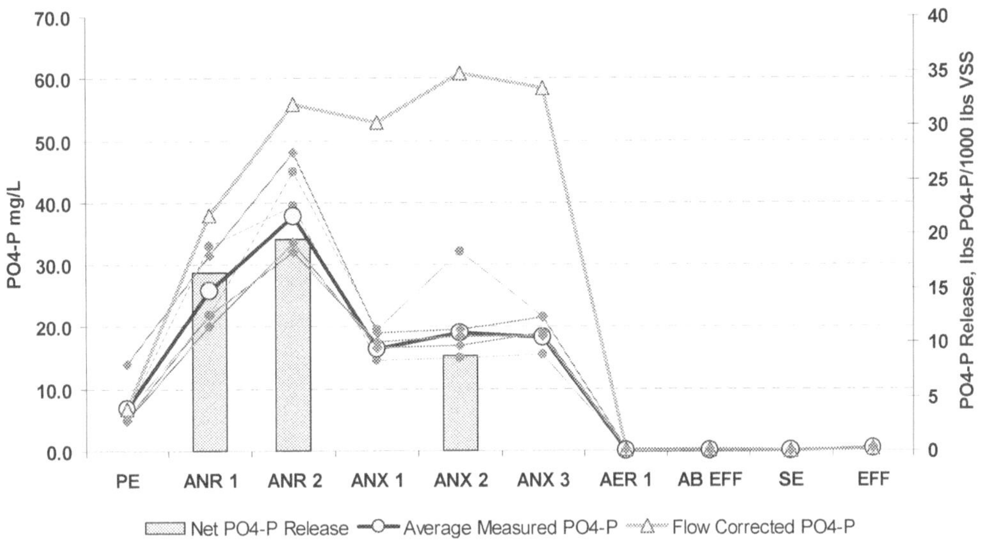


**Figure C-22. Phosphate Profiles from February - March 2002.**

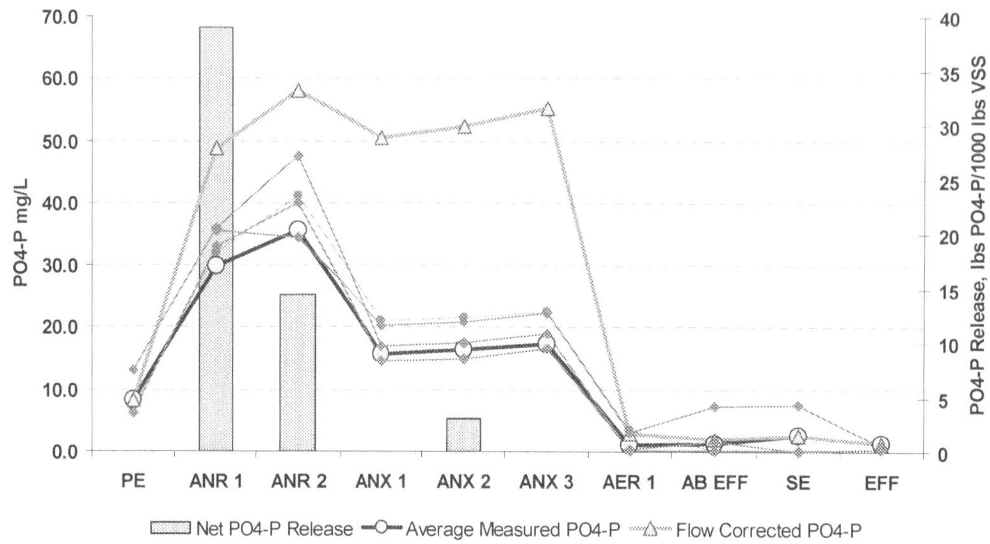


**Figure C-23. Phosphate Profiles from April - May 2002.**

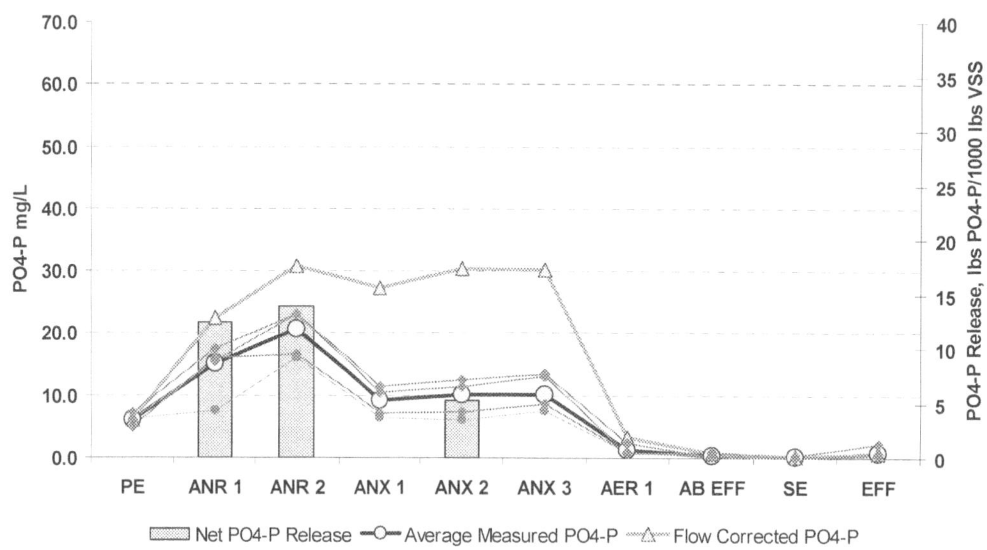


**Figure C-24. Phosphate Profiles from August - September 2002.**

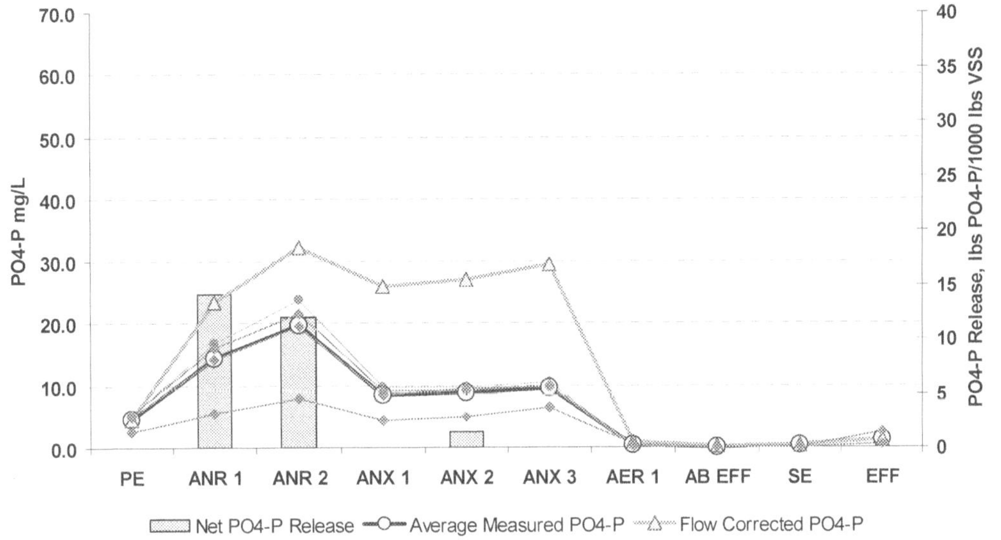


**Figure C-25. Phosphate Profiles from November 2002.**

### *C.3.3.1.2 Phase 2 Profiles*

During the Phase 2 intensive sampling the plant performed very well. The profiles shown in Figure C-26 show a relatively high release and very low effluent phosphate. The total phosphorus release rate was 33.7 lb P per 1000 lb MLVSS / d with an initial phosphorus release rate of 22.4 lb P / 1000 lb MLVSS / hr (538 lb P / 1000 lb MLVSS / d). The measured release of 33 mg/L implies (based on a theoretical ratio of 0.5 mg $PO_4$-P released per mg acetate consumed) that a total of 66 mg/L of VFAs were consumed. The VFA concentration measured in the primary effluent was 48 mg/L, indicating that on average 17 mg/L of VFA were generated in the anaerobic zone.

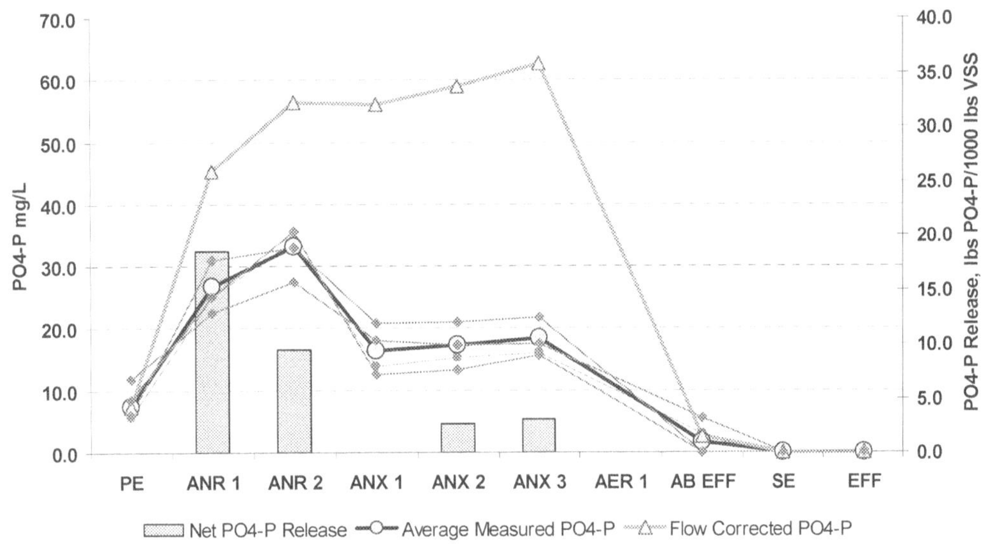


**Figure C-26. Phosphate Profiles During Second Intensive Sampling Period.**

#### *C.3.3.1.3 Profile Summary*

Six phosphorus profiles were collected at Lower Reedy during the course of the project. Table C-9 summarizes the results from all profiles shown in the previous sections. The average flow normalized phosphate profiles for all presented periods are shown in Figure C-27.

The data did not demonstrate any apparent relationships between phosphorus removal and the measured parameters. The initial release rate, which is an indicator for VFA supply and PAO activity, did not correlate with either the effluent phosphorus concentration or net phosphorus removal. Figure C-28 shows that for the periods during which phosphorus profiles were collected, no relationship exists between the phosphorus release and net phosphorus removal.

While no relationship was found between phosphorus removal and the parameters shown in Table C-9, the results suggest that the EBPR process is well buffered. Whether the changes in release are related to the VFA supply at the time or to the PAO population strength was not conclusively determined based on profile testing. The uptake and release tests presented in the next section could indicate that the differences in release are linked to the PAO population.

**Table C-9. Process Performance Measurements Results During Intensive Testing (two-week average).**

| Process Performance Parameter | Unit | Phase 1 | March 02 | May 02 | Aug 02 | Nov 02 | Phase 2B |
|---|---|---|---|---|---|---|---|
| Net Phosphorus Removal | Lb/d | 169 | 164 | 200 | 166 | 139 | 143 |
| Total Phosphorus Release | mg/L | 11 | 41 | 36 | 19 | 20 | 39 |
| Anoxic Phosphorus Uptake | mg/L | 3.2 | 7.1 | 10.0 | 5.2 | 7.5 | 1.8 |
| Phosphorus Release Rate | Lbs P per 1000 lbs VSS per day | 8.3 | 22 | 28 | 32 | 28 | 34 |
| Phosphorus Uptake Rate | Lbs P per 1000 lbs VSS per day | 16 | 32 | 40 | 51 | 41 | 40 |
| Total Aeration Basin Phosphorus Inventory | Lbs P | 1460 | 1823 | 1460 | 730 | 865 | 1160 |
| Release of P-Content | Lbs P/d | 232 | 781 | 777 | 413 | 466 | 750 |
| Initial Phosphorus Release Rate | Lbs P per 1000 lbs VSS per hr | 12.8 | 19.8 | 52.7 | 17.4 | 20.6 | 22.4 |

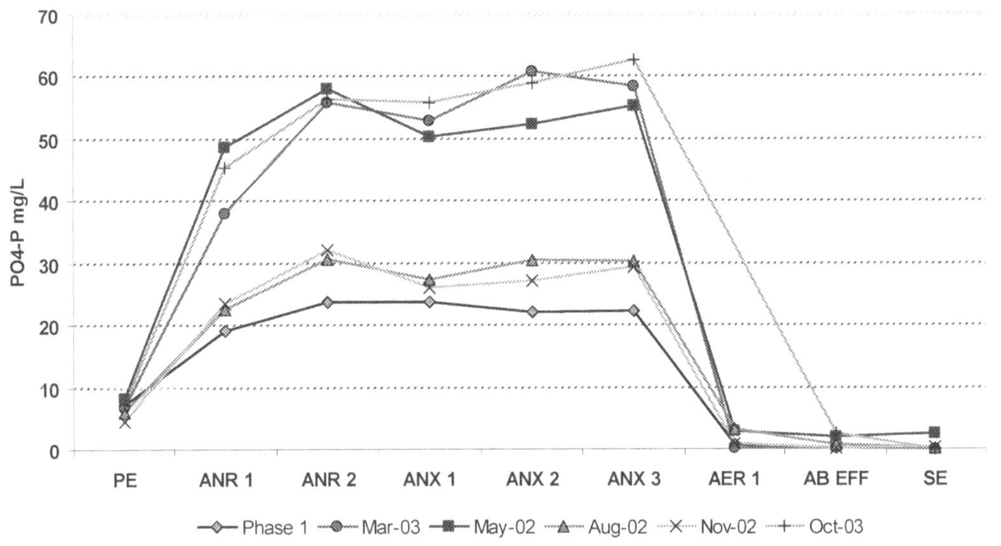


**Figure C-27. Average Flow Normalized Phosphorus Profiles from Phase 1 and 2.**

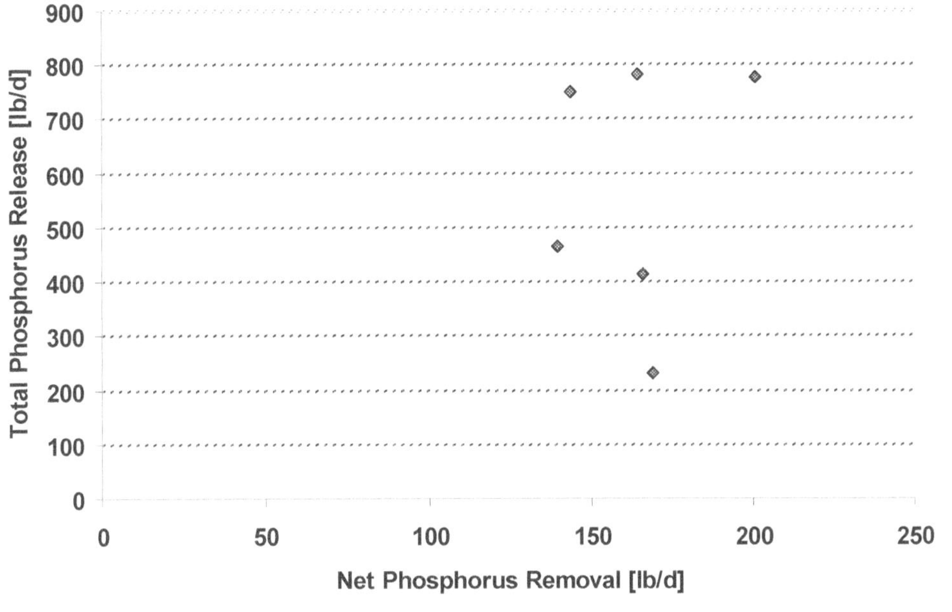


**Figure C-28. Relationship Between Phosphorus Release and Net Phosphorus Removal at Lower Reedy During Phase 1 and 2.**

### C3.3.2 Uptake and Release Tests (URT)

URTs provide a quantification of the PAO population in the basin. By adding VFA to the influent, phosphorus release is limited by the PAO population in the basin and by the amount of stored phosphorus. Theoretically, the phosphorus release under anaerobic conditions is 0.5 mg P/mg VFA. Lower reported values could indicate competition from other bacteria in the system such as GAOs.

URTs can be used for:

- Determining the amount of PAO activity from phosphorus release rate. The release rate is indicative of PAO activity.
- Determining the potential for competition in the anaerobic zone for VFA. A phosphorus release ratio of 0.5 mg P released/mg VFA removed indicates PAO activity. Lower ratios are indicative of competition from other VFA consuming organisms such as GAOs.
- Determining key kinetic parameters for process simulation model calibration, such as the phosphorus release ratio (ΔP/Acetic acid) and magnesium and potassium uptake and release rates.

Table C-10 summarizes the results from the URT at Lower Reedy during Phase 2. Figure C-29 and Figure C-30 show the phosphate and metal release curves.

When comparing the uptake and release test result with the phosphorus release and uptake measured during respective intense sampling periods, one can find that the uptake and release test is consistent with the release determined by the aeration basin profiles.

Comparing both testing phases, the release rates were quite different, 19 mg P /g VSS/hr for Phasc 2A vs. 31 mg P /g VSS/hr for Phase 2B, while the uptake rates were similar (5.7 mg P /g VSS/hr and 6.1 mg P /g VSS/hr, respectively). This shows again that the phosphorus uptake and phosphorus release are not directly related. Provided sufficient aerobic retention time, PAOs show the ability to compensate for differences in phosphorus release.

The fact that the release did not level off during the anaerobic test period indicates that VFAs were not limiting during either test. The Phase 2A URT showed a lower phosphorus release ratio by PAOs possibly indicating the presence of competing organisms such as GAOs.

**Table C-10. Uptake and Release Tests Results.**

| Test Parameter | Unit | Phase 2A | Phase 2B |
|---|---|---|---|
| Phosphorus Release Rate | mg P/g VSS /hr | 18.6 | 30.6 |
| Phosphorus Uptake Rate | mg P/g VSS/ hr | 5.7 | 6.1 |
| Release Rate/ Uptake Rate | - | 3.3 | 4.7 |
| ΔP/Acetic Acid | mg $PO_4$-P/ mg COD | 0.36 | 0.66 |
| ΔP/ΔMg – Release | mg $PO_4$-P/ mg Mg | 3.94 | 4.04 |
| ΔP/ΔMg – Uptake | mg $PO_4$-P/ mg Mg | 3.62 | 3.48 |
| ΔP/ΔK – Release | mg $PO_4$-P/ mg K | - | 3.28 |
| ΔP/ΔK – Uptake | mg $PO_4$-P/ mg K | - | 4.22 |
| MLVSS | mg/L | 1700 | 2000 |

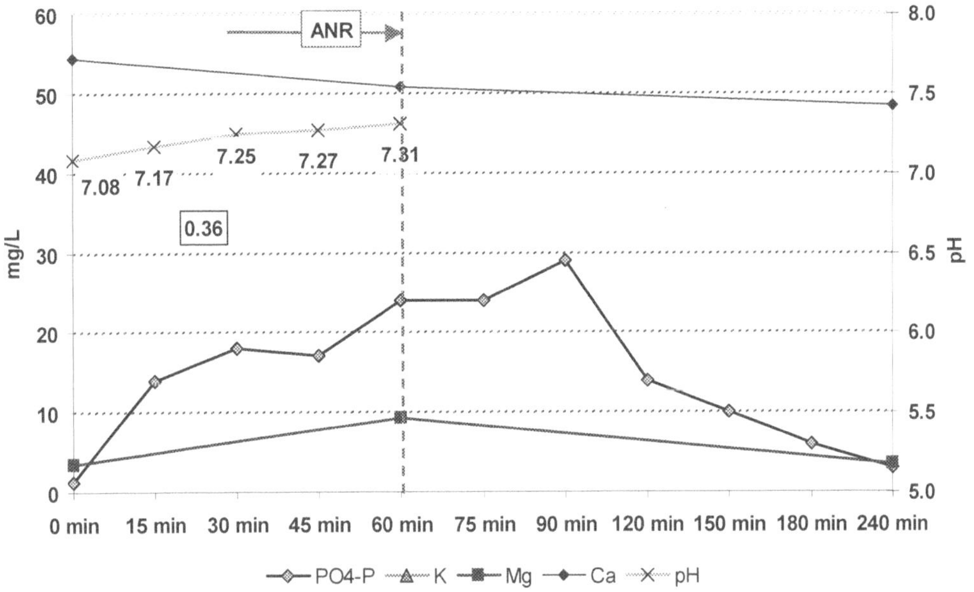


Figure C-29. Phosphorus Uptake and Release Test Results from Phase 2A.

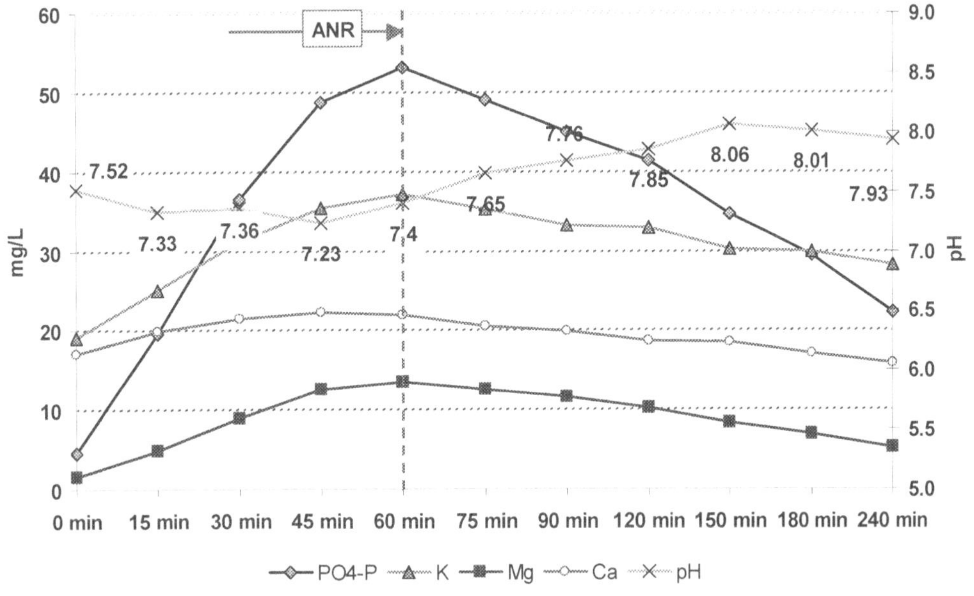


Figure C-30. Phosphorus Uptake and Release Test Results from Phase 2B.

## C3.3.3 Phosphorus and Metal Mass Balance

The purpose of the phosphorus mass balance is to track the average mass flow of phosphorus throughout the system. Figure C-31 shows the phosphorus mass balance for Lower Reedy. On average 89% of the influent phosphorus is removed from the liquid stream. The recycle load is fairly high at 54%.

The profiles from the Phase 2A intensive sampling show magnesium and potassium release and uptake in the range that was observed at other facilities (Figure C-32). The soluble calcium concentrations increase after primary clarification due to the lime addition. Since the calcium uptake and release measured during the URT was not nearly as high as the other metals (see Figure C-30), it appears the calcium profile is associated with lime addition and calcium phosphorus precipitation between the anaerobic and aerobic zone.

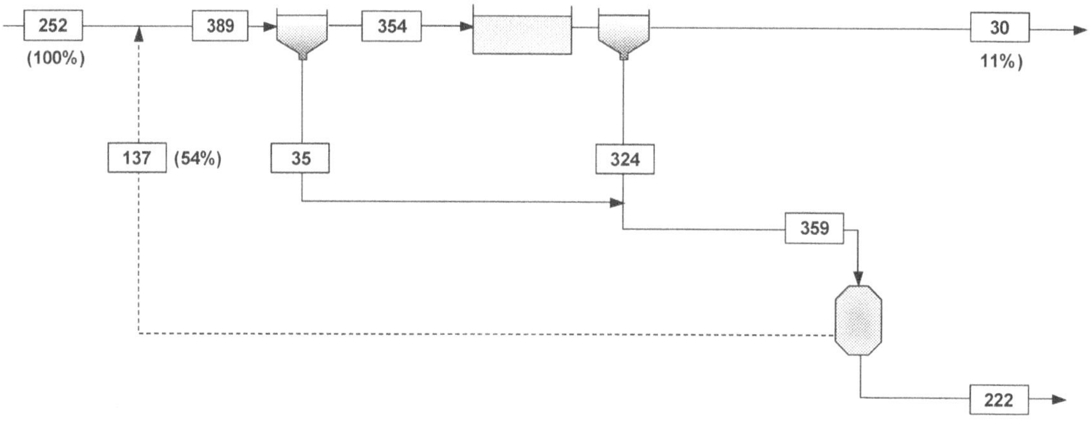


Figure C-31. Phosphorus Mass Balance for Lower Reedy WWTP (2002 annual average).

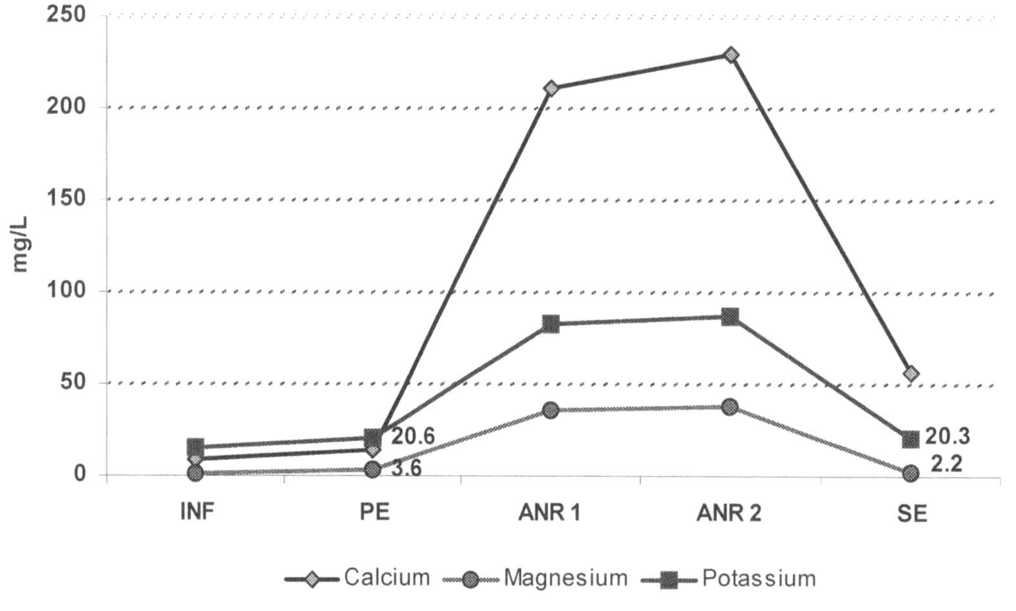


Figure C-32. Lower Reedy Secondary Treatment Metal Profiles.

## C.3.4 Microscopic Examination of Digested Sludge

A digital microscope was used to inspect dewatered solids at 200x magnification. 200x magnification is usually sufficient to detect struvite (magnesium ammonium phosphate) and other crystals that might form during solids stabilization and the dewatering processes. At Lower

Reedy one would expect to see struvite crystals in large numbers because of the high magnesium and phosphorus transfer to the digester.

Pictures in Figure C-33 show crystals, although they were not as easy to find as anticipated. Plant staff reported struvite scaling problems, and observed that scaling problems decrease with increasing lime dosage. The additional metal supply (Ca) with the lime could be responsible for the reduction in struvite scaling by limiting the availability of phosphate through calcium phosphate precipitation.

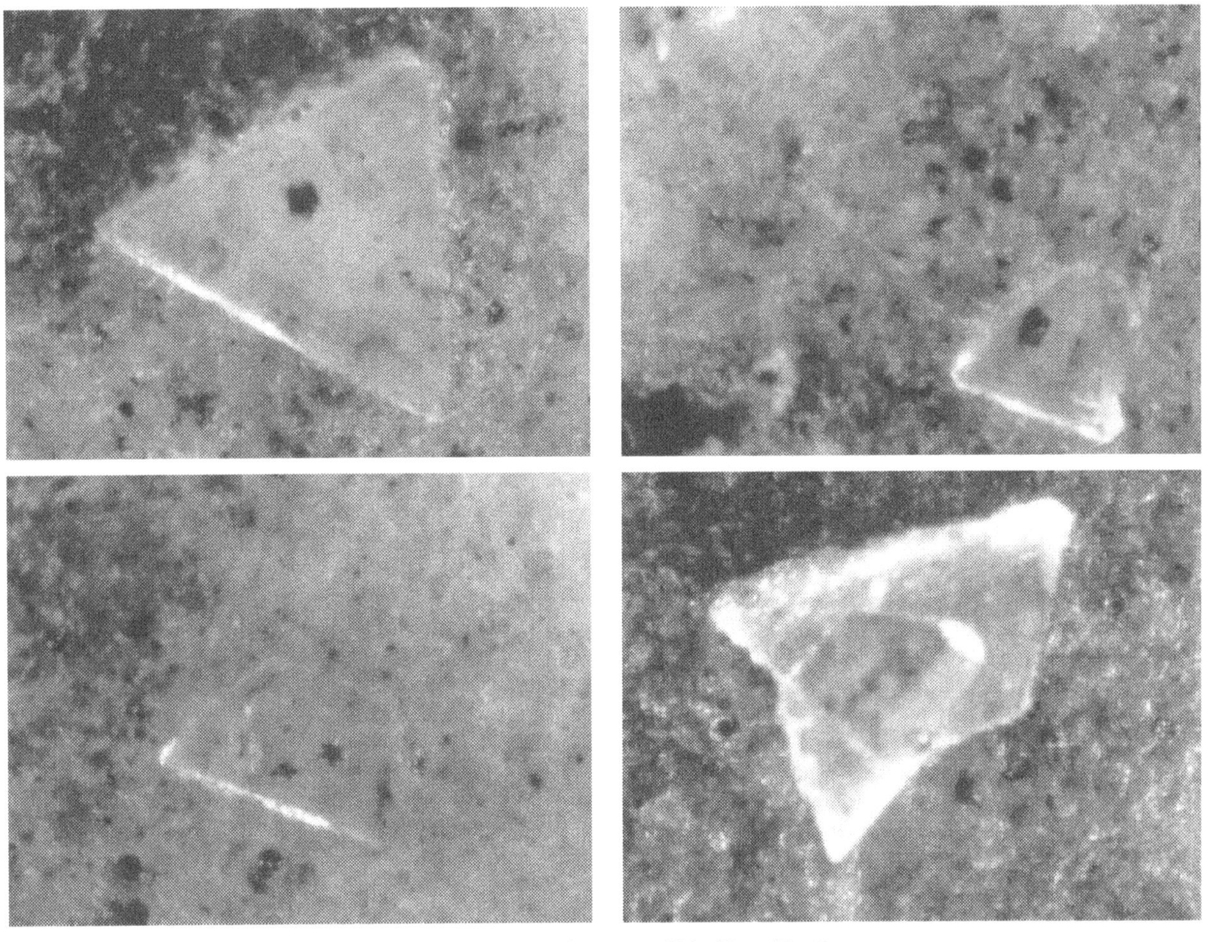

**Figure C-33. Digital Microscope Pictures of Dewatered Sludge at 200x Magnification.**

## C.3.5 PAO and GAO Investigations

A mixed liquor sample from the Lower Reedy plant was examined to observe the presence of polyphosphate granules (indicative of EBPR), and to look for the presence of GAOs, which typically consist of large cocci (2-3 μm) arranged in tetrad clusters within the floc. The sample was collected on September 16.

Both microscopic observations and molecular techniques were used to examine the presence of GAOs in the mixed liquor from the plant. Light microscopy observation was used to observe the tetrad-forming GAOs, and a molecular probing using specific oligonucleotide probes targeting GAOs was completed to identify both tetrad-forming and non-tetrad-forming GAOs. In addition, Neisser and DAPI staining were applied to observe the presence and relative abundance of PAOs.

Figure C-34 shows the observation of PAOs and GAOs using a Neisser stain. Figure C-35 shows the results of DAPI stain. The GAOs have Neisser positive stained cell walls, but are not positive for cell inclusions as are PAOs. The Neisser positive stained PAOs are dark blue to purple in color, and usually group in clusters within the flocs. The DAPI-stained PAOs are yellow in color, which is characteristic of polyphosphate granules observed under the fluorescence microscope. Other non-polyphosphate-accumulating bacteria including GAOs show up in blue with DAPI stain.

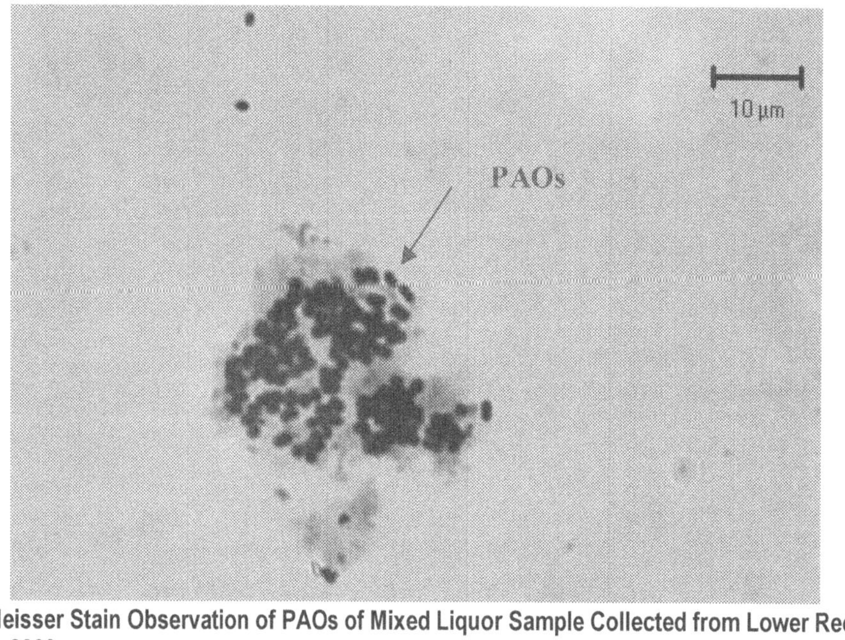


**Figure C-34. Neisser Stain Observation of PAOs of Mixed Liquor Sample Collected from Lower Reedy on September 16, 2003.**

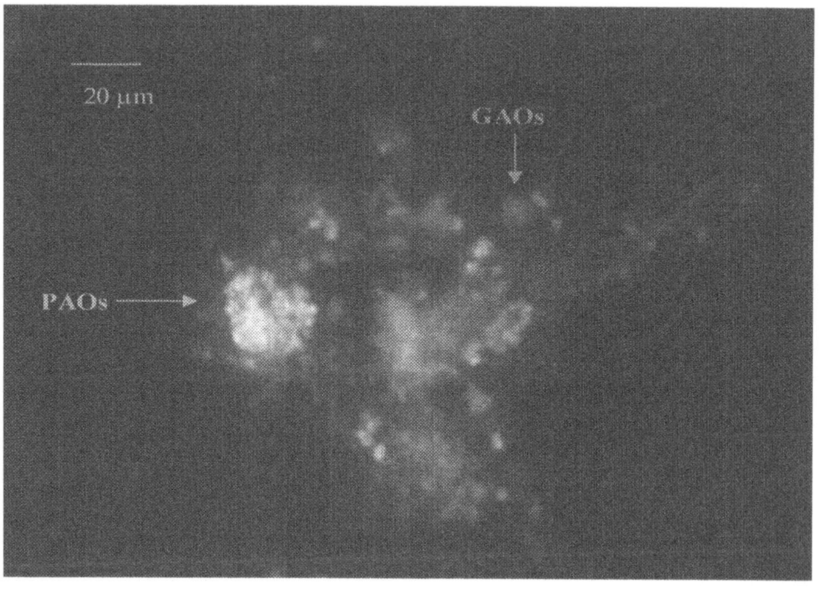


**Figure C-35. DAPI Stain Observation of PAOs and Tetrad-Forming GAOs in a Mixed Liquor Sample Collected from Lower Reedy on September 16. 2003.**

Table C-11 summarizes the observation of PAOs and GAOs for the mixed liquor samples from September 16, 2003. Only a few tetrad-forming GAOs were observed, and PAOs were found to be abundant.

Examination of the mixed liquor samples via molecular methods (fluorescence in situ hybridization (FISH)) using 16S rDNA-based probes targeting *Accumulibacter* (represent PAOs) (Crocetti et al. 2000) and some non tetrad-forming GAOs-*Competibacter* (GAOQ989 andGB-G2, Crocetti et al. 2000 and Kong et al. 2000) showed that only 1-10% of the Eubacteria populations was *Accumulibacter*-GAOs. No *Competibacter*-GAOs was observed. It should be noted that FISH methods only targeted certain PAO and GAO species for which the probes are available. DAPI and Neisser stain were targeting at all cells that are capable of accumulating poly-P, therefore should reveal more PAO populations than FISH method.

**Table C-11. Summary of PAOs and Tetrad-forming GAOs Observation for Lower Reedy WWTP.**

| Date of Sample | GAO observation (tetrad-forming) | GAO observation (*Competibacter*) | PAO observation | Comments |
|---|---|---|---|---|
| **Lower Reedy 9-16-03** | ***A few,*** 1 out of 10 fields observed GAOs, 3-5 tetrapods per field | Not observed. | **Abundant**, cocci shaped cells group together in big flocs | PAOs dominant in big flocs, only a few GAOs were found |

The phosphorus removal performance at the plant was evaluated for the time period when the samples were taken for the GAOs study. The results of phosphorus removal performance evaluation data are shown in Table C-12.

**Table C-12. Summary of P removal Performance at the Lower Reedy WWTP.**

| Date of Sample | GAOs abundance | PAOs abundance | Effluent P mg/L | P Release Rate gP/gVSS.hr | P Uptake Rate gP/gVSS.hr | P Release Ratio gΔ P/gΔ Acetic Acid | Release Rate/ Uptake Rate |
|---|---|---|---|---|---|---|---|
| LR 6-03 | NA | NA | 0.9 | 18.6 | 5.7 | 0.36 | 3.3 |
| *LR 9-16-03 | A few | Abundant | 1.5 | 24.3 | 4.6 | 0.49 | 5.3 |

* The results from laboratory bench tests. Sample was shipped overnight with ice-pack but not in a cooler and was stored in 4°C refrigerator for 6 hours before analysis.
**NA = not available

Evaluation of the phosphorus removal performance showed that the plant has a high ratio of phosphorus release rate to phosphorus uptake rate. Due to the high phosphorus release during the anaerobic phase, the overall phosphorus removal was low (high effluent phosphorus concentration). The GAO study showed that only a few GAOs were observed in the sludge. At the time of sample collection, the effluent phosphate concentration was somewhat variable, with effluent concentrations up to 1.5 mg/L. During this period, the plant also had very high phosphorus release, which could have contributed to higher than average effluent phosphorus concentrations. The presence of a few GAO bacteria did not appear to affect the EBPR performance.

## C.4 Conclusions

The study showed that the plant did not meet its operational target of 0.5 mg/L secondary effluent phosphorus (although Lower Reedy and its companion facility did comply with the combined discharge permit limit). From 2000 through 2003, the average effluent total phosphorus concentration was 0.67 mg/L and the plant produced an effluent phosphorus concentration of equal to or less then 2 mg/L 91 percent of the time.

Process reliability has improved with alum addition capabilities were added in 2002. The aeration basin influent BOD:P ratio averages approximately 24:1. Ratios above 25:1 appeared to produce more stable effluent phosphorus concentrations, indicating that the addition of acetic acid feed could help stabilize process performance. Approximately half of the aeration basin influent phosphorus load is contributed by the solids recycle stream.

The phosphorus release at Lower Reedy is variable, but can be as high as 40 mg/L. This is high compared with other EBPR facilities, suggesting the potential for high EBPR. However, effluent concentrations observed in profiles and in uptake and release tests varied and were not directly correlated with phosphorus release. Phosphorus removal may be limited by the aerobic detention time.

Microscopic observation of mixed liquor collected in September 2003 showed a dominant PAO population, with few GAOs observed. The GAO presence did not appear to affect performance.

APPENDIX D

# McDOWELL CREEK ENHANCED BIOLOGICAL PHOSPHORUS REMOVAL (EBPR) EVALUATION REPORT

## D.1 Summary of Findings

The McDowell Creek Wastewater Treatment Plant is located is located 20 miles North of Charlotte North Caroline and is operated by the Charlotte-Mecklenburg Utility (CMU). McDowell Creek has a 6 MGD design treatment capacity. Average plant flow in 2003 exceeded the design flow, and the plant is scheduled to be expanded to a capacity of 12 MGD in 2007. The plant can be operated in the following process configurations:

- University of Cape Town (UCT)
- Modified Orange Water and Sewer Authority (OWASA)
- Charlotte, North Carolina (CNC)

Even though not required for permit compliance, the operational strategy aims for secondary phosphate concentration of less then 0.2 mg/L. The target effluent total nitrogen concentration is less then 5 mg/L with an ammonia-nitrogen concentration of less then 0.2 mg/L. Other effluent quality targets are average month BOD and TSS of less then 5 mg/L. After an initial startup and optimization period the EBPR process has delivered very consistent low effluent phosphorus concentrations. However, alum addition remains a part of the operation strategy to lower the primary effluent and dewatering recycle phosphate and provide EBPR backup. Since 2000, the McDowell Creek plant has also added an industrial sugar waste to supplement influent VFA and soluble BOD. Dosage rates range from 1000 gpd – 60000 gpd.

The primary effluent BOD/P ratio is approximately 30:1 before addition of the sugar waste, indicating favorable conditions for EBPR. The plant has a year around permit and produced average effluent ammonia concentration of 0.25 mg/L and effluent nitrate of 4.1 mg/L. Because phosphorus removal is not a regulatory requirement, secondary effluent phosphorus concentrations are not routinely measured. Final effluent phosphorus concentrations, however, are under 0.5 mg/L 95% of the time. During the intensive sampling periods, secondary effluent total phosphorus was typically under 0.2 mg/L.

At McDowell Creek the dewatering filtrate is equalized after a portion of the phosphate is precipitated by alum addition. The recycle phosphorus load is approximately 23%, but due to chemical addition in the primary clarifier, the net increase between raw influent and primary clarifier influent soluble phosphorus loading is only 8%.

During the year of 2003 much of North Carolina experienced record rainfalls. The maximum month flow reached 7.3 MGD with a maximum day of 13 MGD. At 7.3 MGD, the plant operated essentially 21% over its rated capacity. During the high flows, effluent total phosphorus generally remained under 0.2 mg/L. This performance is indicative of the stable phosphorus removal observed at the McDowell Creek facility.

The following table.

Table D-1. Summary of Key Findings from this Study.

| | |
|---|---|
| Reliability | Plant achieves effluent Total Phosphorus (TP) of 0.5 mg/L with 95% reliability. Net increase in phosphorus from raw influent to primary influent due to recycle streams is 8%. |
| Operational Criteria | Solids retention time (SRT) averages 7 days. Anaerobic hydraulic retention time (HRT) is maintained at less than 1 hour, anoxic HRT typically 2-3 hours, and aerobic HRT averages 10 hours. |
| Influent Organic Characterization | Aeration basin influent BOD:P ratio averages 30 mg BOD/mg P. Sugar waste is also added to the aeration basins, further increasing BOD:P ratio. |
| Measures Showing PAO Growth | Uptake and release tests conducted in the laboratory showed $\Delta$P/Acetic acid ratios from 0.2 to 0.36. Microscopic examination did show some PAOs, although GAOs were dominant at the time the sample was collected. |
| Factors Hindering PAO Growth | Anaerobic phosphorus release is low compared with other facilities investigated in the study. Release in the first anaerobic zone was measured at approximately 14 lb P/1000 lb VSS/hr during two testing periods. Phosphorus content in the mixed liquor was lower in Phase 2 than Phase 1, with values of 2.3% and 4.3%, respectively. |

## D.2 Introduction

The analysis of this facility was completed as part of a project sponsored by the Water Environment Research Foundation (WERF). The project objective is to develop information for evaluating the ability of municipal wastewater treatment plants to efficiently and cost-effectively remove phosphorus to low levels through processes that involve EBPR. This was completed in large part through detailed studies at several treatment facilities that employ EBPR. The goals of the study were to:

- Determine process reliability.
- Determine the biological phosphorus removal efficiency.
- Determine the biological phosphorus removal capacity.
- Determine how chemical addition affects Bio-P removal.

Definitions of terms used in this analysis are described in Appendix P.

### D.2.1 Facility Introduction

The McDowell Creek Wastewater Treatment Plant is located is located 20 miles North of Charlotte North Caroline and is operated by the Charlotte-Mecklenburg Utility (CMU). McDowell Creek has a 6 MGD design treatment capacity.

Figure D-1. McDowell Creek, Charlotte Mecklenburg Municipal Utility District, Charlotte, NC.

In 2003 the actual plant flow average exceeded 6 MGD to due extreme wet weather flows. The average flow for 2001 and 2002 was 4.7 MGD compared to 6.1 MGD in 2003. The design for a facility expansion is underway and the new secondary treatment trains are scheduled to come online in 2007 and increase the total plant capacity to 12 MGD.

The facility was first placed in service 1980 with a capacity of 3 MGD. The expansion and upgrade to BNR was completed in 1999, designed to provide operators with maximum operational flexibility for nitrogen and phosphorus removal optimization. The plant can be operated in the following process configurations:

- University of Cape Town (UCT)
- Modified Orange Water and Sewer Authority (OWASA)
- Charlotte, North Carolina (CNC)

The treatment process also includes deep-bed filtration and dewatering filtrate storage and equalization.

Because the plant receives very fresh wastewater with little VFA, acetic acid was originally added to the anaerobic zone. Acetic acid has since been replaced with a sugar waste from a soft drink bottling plant.

Both primary and secondary sludge are thickened and anaerobically digested. The dewatering recycle load is retuned to the plant influent after equalization.

Even though not required for permit compliance the operational strategy aims for secondary phosphate concentration of less then 0.2 mg/L. The target effluent total nitrogen concentration is less then 5 mg/L with an ammonia-nitrogen concentration of less then 0.2 mg/L. Other effluent quality targets are average month BOD and TSS of less then 5 mg/L. After an initial startup and optimization period the EBPR process has delivered very consistent low effluent phosphorus concentration. However, alum addition remains a part of the operation strategy to lower the primary effluent and dewatering recycle phosphate and provide EBPR backup.

Table D-2. McDowell Creek Permit Requirements (Summer)[1].

| Parameter | Units | Monthly | Weekly |
|---|---|---|---|
| cBOD - May-Oct | mg/L | 4.2 | 6.3 |
| cBOD Oct - May | mg/L | 8.3 | 12.4 |
| TSS | mg/L | 15 | 22.5 |
| NH4-N | mg N/L | 1 | 3 |
| Total Nitrogen - May - Oct | lb N/d | 450 | - |
| Total Nitrogen – Oct - May | lb N/d | 500 | |
| Total Phosphorus - May - Oct | lb N/d | 35 | - |
| Total Phosphorus - Oct - May | lb N/d | 450 | |

1. Based on 2003 average flow of 6.6 MGD.

## D.2.2 Facility Process Design

### D.2.2.1 Liquid Stream

Figure D-2 shows the McDowell Creek process layout for the UCT mode. Raw influent is degritted and screened before entering the primary clarifier. Internal recycle flows, including dewatering recycle, are returned to the influent channel upstream of grit removal. The activated sludge BNR process follows primary clarification. Secondary effluent (SE) is filtered and disinfected. The characteristic feature of the process design at McDowell Creek is the sugar waste addition to the primary effluent in combination with a UCT configuration. Originally primary effluent VFA was supplemented using acetic acid. Acetic acid was successful in providing the required substrate for EBPR but was costly. In 2001 sugar waste from a soda bottling plant replaced acetic acid, which proved to be equally effective at a slightly higher dose but much lower cost.

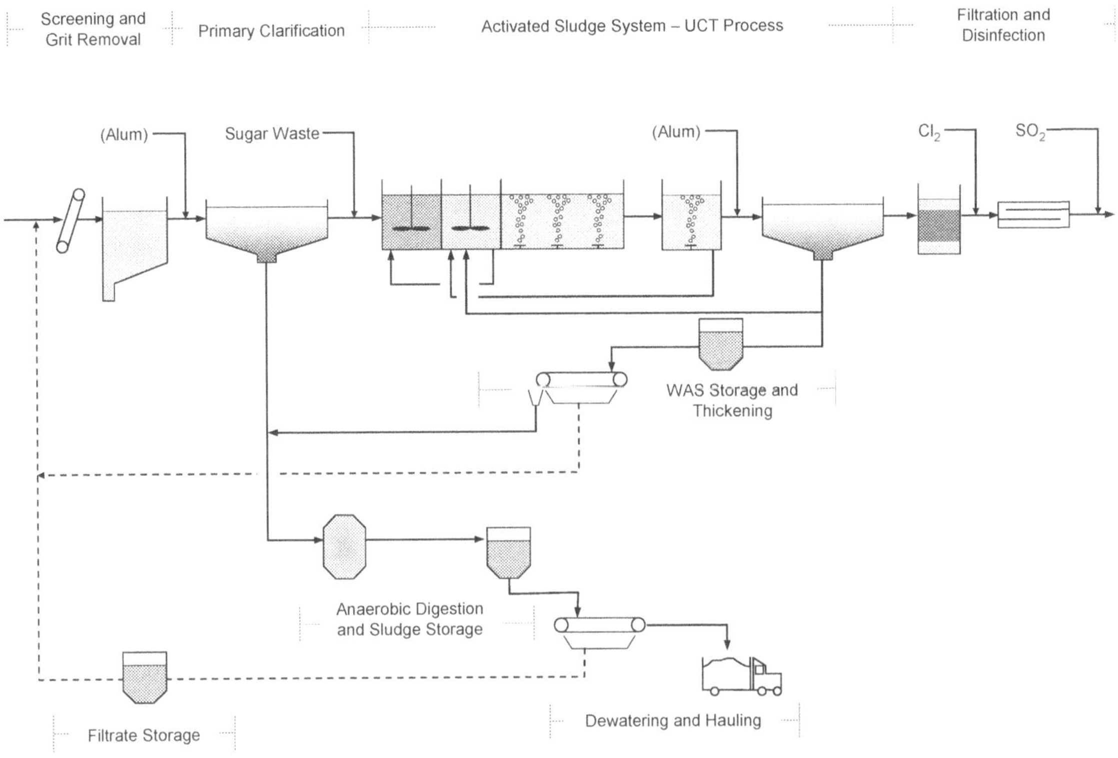


**Figure D-2. Process Schematic for McDowell Creek WWTP Liquid and Solids Treatment.**

The key feature of the UCT process is that RAS and mixed liquor recycle from the aerobic zone enter the process train at the head of the anoxic (ANX) zone. Mixed liquor is recycled from the anoxic zone to the anaerobic (ANR) zone to achieve EBPR. This configuration limits the transfer of nitrates into the anaerobic zone optimizing the VFA availability for phosphorus release. Primary effluent and the recycle from the anoxic zone mix at the head of the anaerobic zone. In the first anoxic cell the mixed liquor recycle from the aeration basin effluent mixes with the anaerobic zone effluent. The process configuration of the modified UCT design is very similar to the VIP process. The only real difference is that the mixed liquor recycle and RAS are returned separately to the anoxic zone in the UCT configuration. The VIP process design calls for the use of a single recycle line where RAS and mixed liquor recycle merger prior to entering the anoxic zone.

The activated sludge system consists of an anaerobic zone that is equally divided into four cells. Downstream of the anaerobic zone two trains receive equal share of the anaerobic zone effluent. Each train has two anoxic cells and one long (plug flow) aerobic (AER) zone. Following each train is a post aeration basin from where the mixed liquor recycle is pumped back to a junction box upstream of both trains.

The combined HRT at design flow is 11.5 hours with an anaerobic retention time of 0.75 hours and anoxic retention time of 2.15 hours. Deep bed filters provide additional solids retention down stream of the secondary clarifier. Effluent BOD and TSS concentration of less than 2 mg/L are consistently accomplished. The filter also offers another alternative for post denitrification and phosphorus precipitation.

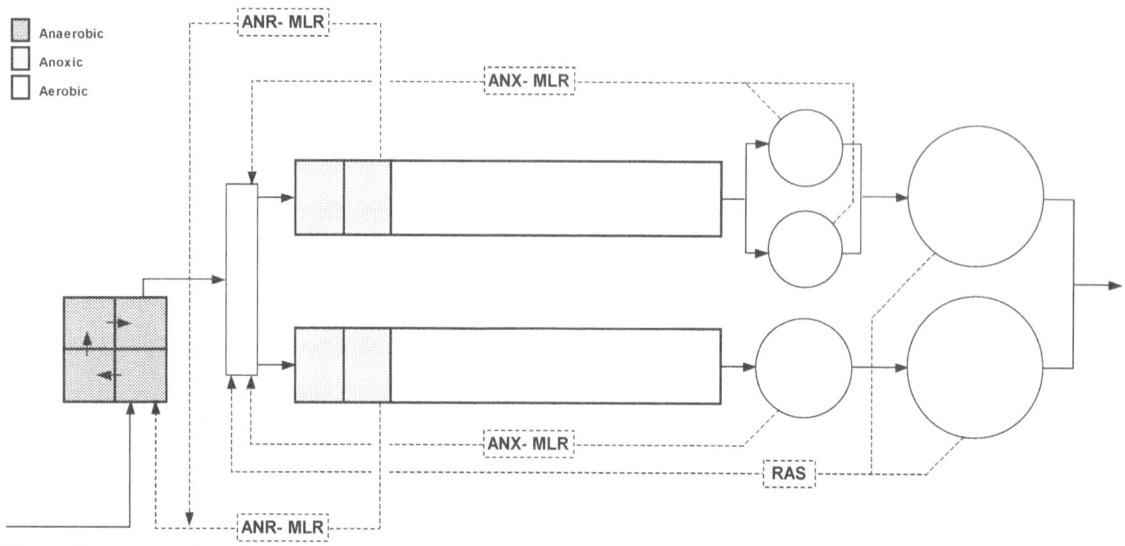


**Figure D-3. Secondary Treatment Layout at Mc Dowell Creek WWTP. (A Similar Third Treatment Train Has Not Been Used As of Completion of This Report).**

**Table D-3. Key Process Parameters (Data from 2001 – 2003).**

| Component | Unit | Design | Average |
|---|---|---|---|
| Anaerobic HRT | hrs | 0.75 | 0.91 |
| Anoxic HRT | hrs | 2.15 | 2.62 |
| Aerobic HRT | hrs | 8.5 | 10.4 |
| Total HRT | hrs | 11.5 | 13.4 |
| RAS Rate | % | 80% | 87% |
| Anaerobic Mixed Liquor Recycle Rate* | % | 65% | 70% |
| Anoxic Mixed Liquor Recycle Rate* | % | 260% | 320% |
| SRT | days | 15 | 6.7 |
| MLSS | mg/L | 1580 | 3400 |
| Solids Recycle Return | | Returned to plant influent | |
| Flow | MGD | 6.0 | 5.04 |

* Based on Design flow and Rate Pump Flows

### D.2.2.2 Solids Stream

Solids treatment consists of thickening, anaerobic digestion, and digested sludge dewatering. Primary sludge is thickened in the primary clarifier to about 3-4% solids before being fed directly to the anaerobic digester. Waste activated sludge is flow equalized in an aerated WAS holing tank and gravity belt thickened to 5-7%. WAS thickening is operated continuously. Currently two anaerobic digesters are in operation at McDowell Creek with an SRT of 25 days.

Digested sludge is dewatered eight hours a day, five days a week. To minimize the adverse impact of the dewatering recycle load on BNR performance filtrate is stored and the return is distributed over 24 hours.

**Table D-4. Solids Processing Summary.**

| Component | Technology | Comments |
|---|---|---|
| Primary Sludge Thickening | Gravity | Thickened in Primary Clarifier |
| Was Thickening | Gravity Belt Thickener | |
| Primary Sludge Stabilization<br>WAS Stabilization | Anaerobic Digestion | SRT = 25<br>TSS = 2%<br>VSS Destruction = 55% |
| Dewatering | Belt Filter Press | |
| Centrate Storage | none | |

### D.2.2.3 Chemical Addition

Since 2000, operators at McDowell Creek have added an industrial sugar waste to supplement influent VFA and soluble BOD. Dosage rates range from 1000 – 60000 gpd. The main driver for the sugar dosage rate is the rate of delivery since the treatment plant has a contract to dispose the sugar waste. According to plant staff most of the time the supply remains the limiting factor.

The analytical results from two sugar samples are shown in Table D-5. The subsequent BOD dosage rates range between 10 and 30 mg/L

Alum is added to the primary clarifier and dewatering filtrate to reduce primary effluent phosphate. It can also be added to the secondary clarifier whenever EBPR supplementation is required.

**Table D-5. Sugar Waste Analysis Results.**

| Component | Unit | Sample |
|---|---|---|
| COD | mg/L | 93,500 |
| SCOD | mg/L | 83,300 |
| BOD | mg/L | 36,700 |

**Table D-6. Chemical Use Summary.**

| Chemical | Unit | 2001 | 2002 | 2003 |
|---|---|---|---|---|
| Lime, 30% Solution | 1000 gal/year | 220 | 378 | 277 |
| Alum | tons | 278 | 218 | 148 |
| Sugar Waste | tons BOD/year | - | - | 93 |

### D.2.2.4 Process Control

The low effluent nutrient requirements require close attention by plant staff. The key operation parameters are aeration basin dissolved oxygen (DO), SRT, primary and secondary effluent $PO_4$-P, and primary effluent alkalinity. The alum dosage to the primary effluent is adjusted to maintain average primary effluent phosphorus concentrations at 6.0 mg/L. Lime addition is controlled to maintain a plant effluent alkalinity of 100 mg/L. the internal recalculation pumps are constant speed and therefore the anaerobic and anoxic recalculation rates cannot be controlled. The RAS rate is available as a variable but is rarely changed by operators.

The aeration basin DO is set to maintain a concentration of 3.0 mg/L. Dewatering filtrate storage and equalization is a critical element for process stability, and nitrogen as well as phosphorus removal performance. Filtrate is stored and returned to distribute the load more equally over a seven-day weekly average and to equalize the daily diurnal load.

## D.3 Process Performance Evaluation

### D.3.1 Approach

The process evaluation was approached from two data activities: 1) a review historical data and 2) the conduct of special sampling to supplement the routinely recorded water quality and process measurements. Several data sets were collected (Table D-7):

- **Routine Data Review.** The reviewed data represent all available water quality and process monitoring data collected between January 2000 and July 2003. This data were analyzed to determine the general process performance, reliability of meeting effluent quality, and evaluation of the interaction between process parameters and performance. This data set is referred to as the **2000-2003 data set**.
- **Special Data Collection.** During both Phase 1 (2002) and Phase 2 (2003), tests were conducted to evaluate the EBPR performance. During Phase 2, two intensive sampling periods provided additional data to supplement the routine data collection. This includes aeration basin phosphorus profiles, mixed liquor samples for microscopic G-bacteria investigation, and phosphorus uptake/release tests. In total, samples were collected in several periods and arranged in the following data sets (note that some data collection procedures changed from Phase 1 to Phase 2):
  - **Phase 1 data set** This data set was collected in Year 1 of the project during Aug and Sep 2002.
  - **Phase 2A data set** This data set was collected in Year 2 of the project during the first intensive sampling period in June 2003.
  - **Phase 2B data set** This data set was collected in Year 2 of the project during the second intensive sampling period in October 2003.
  - **Phase 2F data set** This data set was collected in Year 2 of the project during the site visit in June 2003.

Table D-7. Summary of Field Tests and Sampling.

| Field Testing Segment | Month | Data Set ID | Purpose Conducted |
|---|---|---|---|
| Historical Data Review | January 2000 – August 2003 | 2000 - 2003 | Influent, Effluent Averages<br>Process Parameter Averages |
| Phase 1 – Field Tests | September 2002 (One day) | Phase 1 | Aeration Basin Profiles<br>Uptake and Release Test<br>Operator Interviews |
| Phase 2 – Field Tests | June 2003 | Phase 2A | Uptake and Release Test<br>Operator Interviews<br>Data Collection |
| Phase 2 – Intensive Sampling Period 1 | August 2003 | Phase 2B | Influent Characterization<br>Aeration Basin Profiles<br>Phosphorus Mass Balance<br>Metal Mass Balance |

## D.3.2 Overall Process Characterization

This section of the report summarizes the aeration basin influent characteristics and secondary treatment process performance based on the data collection described in Section 3.1. It also discusses relationships between aeration basin influent loading and process performance.

### D.3.2.1 Aeration Basin Influent Characterization

The McDowell plant service area service area had been under drought conditions for several years until 2003. According to the plant staff the flows prior to 2003 were lower than normal but at the same time the flows in 2003 were higher due to large I&I contributions. Figure D-4 shows both plant influent flow and temperature. The increase in plant flow between 2002 and 2003 was nearly 25%.

From 2000 to 2002, the flow averaged 4.9 MGD annually while in 2003 the average flow jumped to 6.1 MGD. Influent temperature ranged between 14-27°C.

Table D-8 summarizes the average aeration basin influent composition and COD fractions, as well as values for the two intensive testing phases. The primary effluent BOD/P ratio is approximately 30:1. Ratios considered favorable for EBPR are 20 mg BOD/mg P or higher. The VFA to P ratio was measured during the Phase 1 testing but the values have to be questioned. During Phase 2, unexpected budget shortfalls resulted in a termination of the sampling program, resulting in some missing data. However, from the available data could conclude that aeration basin influent characteristics appear adequate to support EBPR.

The influent loads for, TSS, ammonia and phosphorus have not changes significantly for the past three years except for the last few months shown in Figure D-5 and Figure D-6. The decline is probably due to sewer overflows during the months of record precipitation in spring of 2003. Beginning in 2001 the BOD load shows a seasonal peak during the winter months. The reason for this seasonal variation has not been determined.

**Table D-8. Aeration Basin Influent Wastewater Characterization.**

| ABI Wastewater Parameter | Unit | 2001 - 2003 | Phase 1[1] | Phase 2A | Phase 2B |
|---|---|---|---|---|---|
| BOD | mg/L | 190 | 120 | 150 | 150 |
| sBOD | mg/L | - | 52 | - | |
| COD | mg/L | 330 | 350 | 320 | 340 |
| ffCOD | mg/L | - | 160 | - | - |
| VFA | mg/L | - | 42 | - | 47 |
| TP | mg P/L | 6.5 | 6.7 | 3.8 | 4.9 |
| $PO_4$-P | mg P/L | - | 3.4 | - | - |
| pH | | - | - | - | - |
| TKN | mg N/L | 31 | 35 | 27 | 35 |
| $NH_4$-N | mg N/L | 24 | 26 | 20 | 23 |
| TSS | mg/L | 86 | 68 | 92 | 87 |
| Flow | MGD | 5.0 | 4.7 | 8.9 | 5.5 |
| Temperature | °C | 20 | 26 | - | 28 |
| VFA : P Ratio | - | - | 9.1 | - | 8.5 |
| ffCOD : P Ratio | - | - | 35 | - | - |
| COD : P Ratio | - | 50 | 75 | 84 | 70 |
| BOD:P Ratio | - | 27 | 26 | 39 | 31 |
| ffCOD : COD Ratio | - | - | 0.46 | - | - |
| VFA : ffCOD Ratio | - | - | 0.26 | - | - |

[1] Single composite sample.

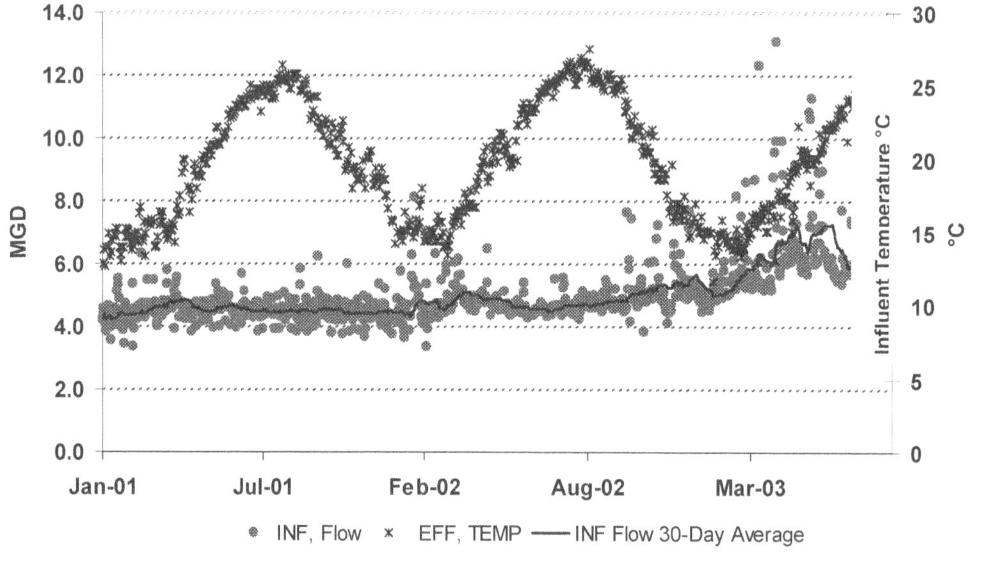


**Figure D-4. Influent Flow, Temperature, and 30-day average Flow Between 2000 - 2003.**

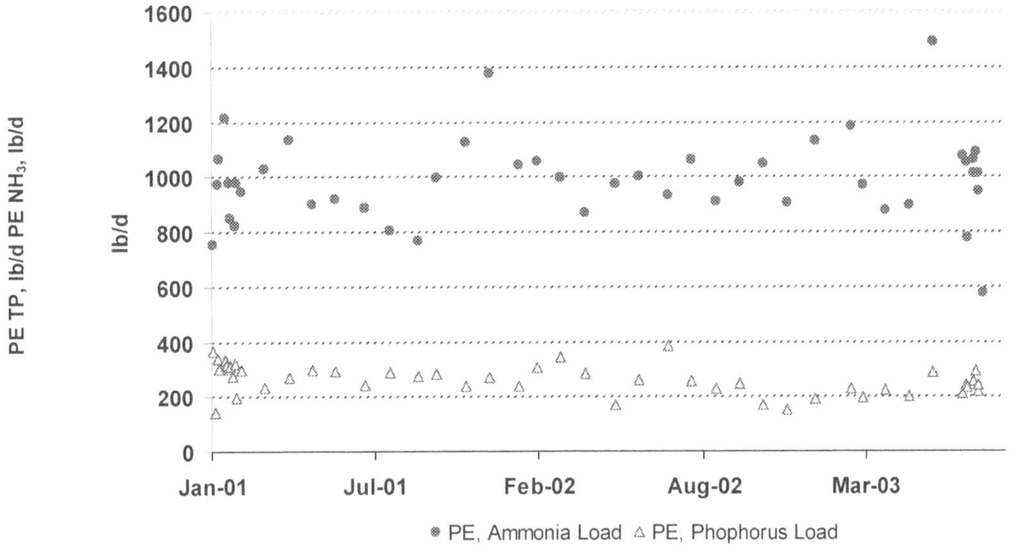


**Figure D-5. Primary Effluent Phosphorus and Ammonia Load.**

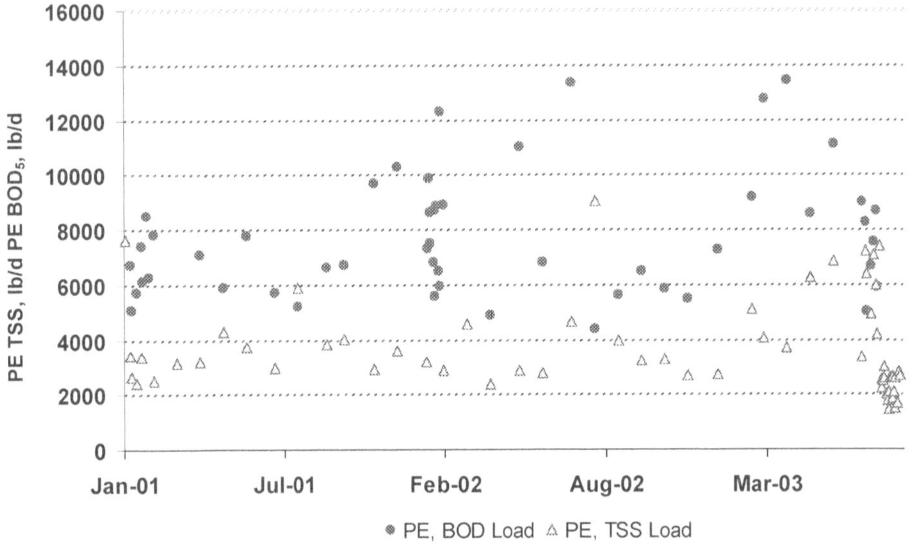


**Figure D-6. Primary Effluent BOD and Suspended Solids Load.**

Table D-9 shows minimum, maximum, and average values for key aeration basin influent (ABI) parameters.

Table D-9. Selected Water Quality Measurement and Process Parameters from 2001 - 2003.

| Parameter | Unit | Min | Ave | Max |
|---|---|---|---|---|
| Influent Flow | MGD | 3.4 | 5.0 | 13.1 |
| ABI Temperature | °C | 11.7 | 20.0 | 27.5 |
| ABI BOD | mg/L | 82 | 186 | 344 |
| TKN | mg N/L | 15.2 | 30.64 | 45.2 |
| ABI $NH_4$-N | mg N/L | 13.0 | 24.0 | 35.0 |
| ABI TP | mg P/L | 3.6 | 6.5 | 9.9 |
| ABI TSS | mg/L | 33 | 86 | 248 |

### D3.2.2 EBPR Performance

Table D-10 summarizes the effluent water quality and process parameters based on the 2000 - 2003 data set. The plant has a year-round permit and produced average effluent ammonia concentration of 0.25 mg/L and effluent nitrate of 4.1 mg/L. The plant operated since 2001 without any major process upset. Even the unusually wet weather during 2003 that generated record flows to the plant did not impact the plant nitrogen removal performance. The low average and maximum SVI further indicate very stable activated sludge process.

Table D-10. Selected Water Quality Measurement and Process Parameters from 2001 – 2003.

| Parameter | Unit | Min | Ave | Max |
|---|---|---|---|---|
| SE $NH_4$-N | mg N/L | 0.2 | 0.25 | 0.9 |
| SE $NO_3$-N* | mg N/L | 1.2 | 4.1 | 17.8 |
| SE TN | mg N/L | - | - | - |
| SE TP* | mg P/L | 0.1 | 0.15 | 2.7 |
| SE $PO_4$-P | mg P/L | - | - | - |
| SE BOD | mg/L | 2.0 | 3.4 | 9.6 |
| SE TSS | mg/L | 1.0 | 3.1 | 11.0 |
| SE Total Alk** | mg/L | 60 | 86 | 110 |
| AB TP Load | Lb/ 1000 lb VSS | 2.63 | 5.1 | 7.11 |
| SRT | Days | 4.0 | 6.72 | 9.0 |
| ANR HRT | Hrs | 0.3 | 0.9 | 1.3 |
| ANX HRT | Hrs | 1.0 | 2.6 | 3.8 |
| AER HRT | Hrs | 3.4 | 10.4 | 15.1 |
| MLSS | mg/L | 1640 | 3600 | 4860 |
| MLVSS | mg/L | 1200 | 2520 | 3650 |
| MLSS/MLVSS | | 48.4 | 69.7 | 90.4 |
| SVI*** | ML/g | 47.7 | 79.4 | 132 |
| TP – % Removal | % | 98.77 | 99.93 | 99.99 |
| TN - % Removal | % | - | 88.0 | - |

* Final Effluent
** 2003 Data only
* 2003 Data only

With no aeration basin effluent phosphate data available it is difficult to assess the true EBPR performance at McDowell Creek. No relationship between plant load and performance could therefore be determined, other than that the plant performed well under loading conditions.

The plant as a whole delivers effluent phosphorus concentrations (Figure D-7) of less than 0.5 mg/L 95% of the time. From Figure D-8 it becomes apparent that phosphorus removal has been optimized over the last three years including some of the process improvements described in the previous section contributed. There is no visible correlation between alum use and effluent TP concentration (Figure D-9) although the average use during the summer is lower than during the winter as one would expect.

The primary effluent wastewater composition (excluding the added sugar) would suggest good EBPR performance. However, the denitrification goal and its subsequent soluble BOD demand provide significant competition for biological phosphorus uptake. The sugar waste addition to the anaerobic zone, however seems to shield phosphorus release reaction from much of the competing denitrification.

Given the minimum water temperature of 15°C one would expect solid nitrification performance. However, most plants with similar environmental conditions show occasional upsets, which are completely absent at McDowell Creek (Figure D-10). Perhaps the lack of problematic industrial discharge protects the plant from mixed liquor poisoning often suspected at other facilities to be responsible for sudden and inexplicable loss of nitrification.

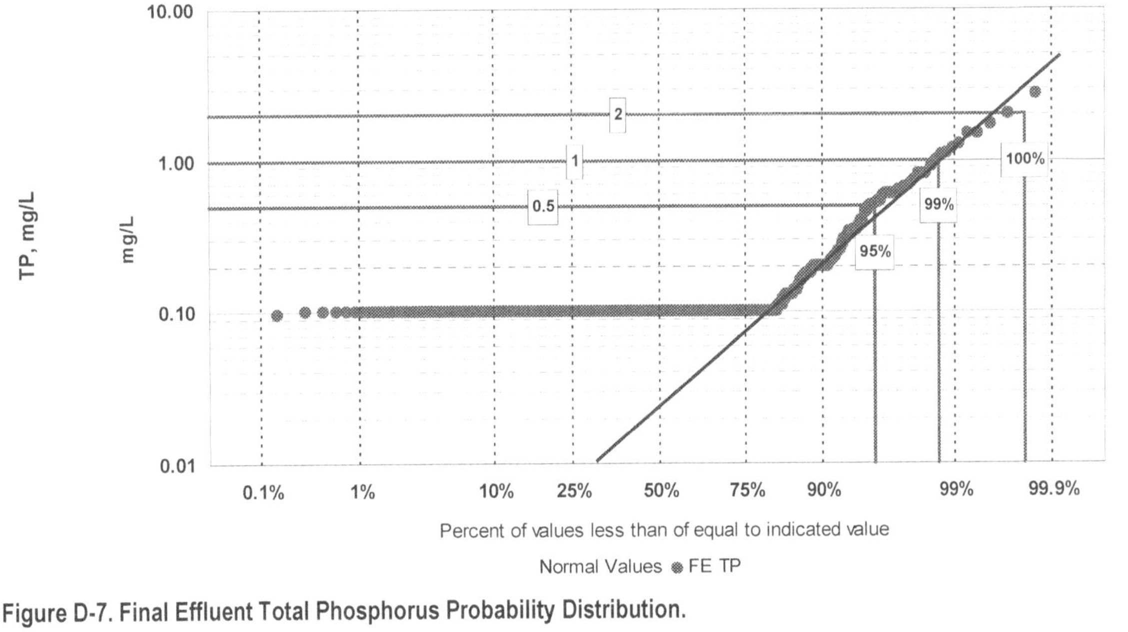


**Figure D-7. Final Effluent Total Phosphorus Probability Distribution.**

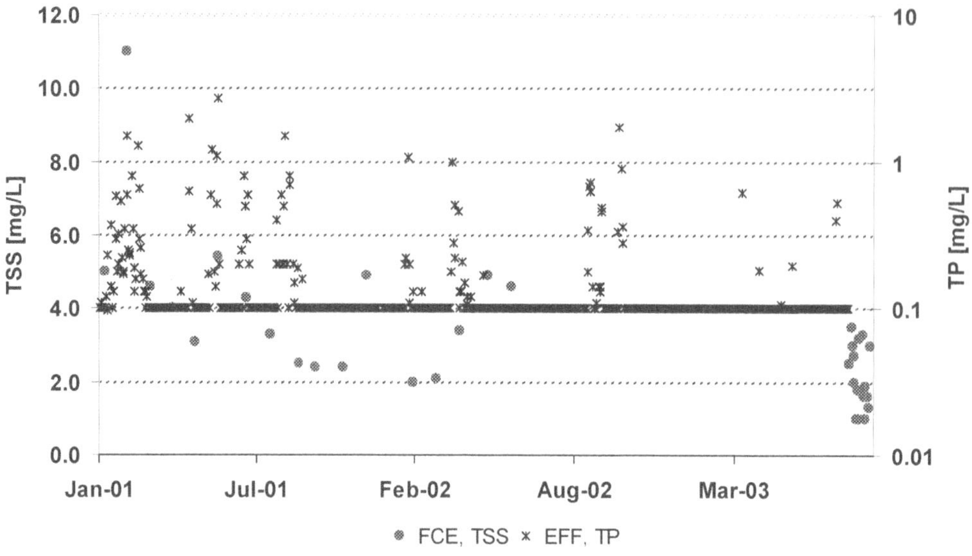


**Figure D-8. Effluent Total Phosphorus (log scale) and Total Suspended Solids.**

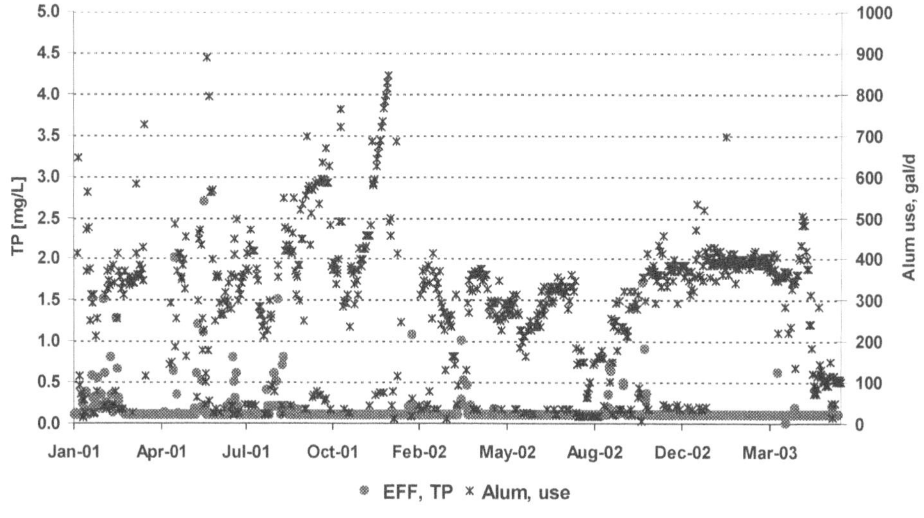


**Figure D-9. Effluent Total Phosphorus Concentration Since 2001 Compared to the Total Daily Alum Use.**

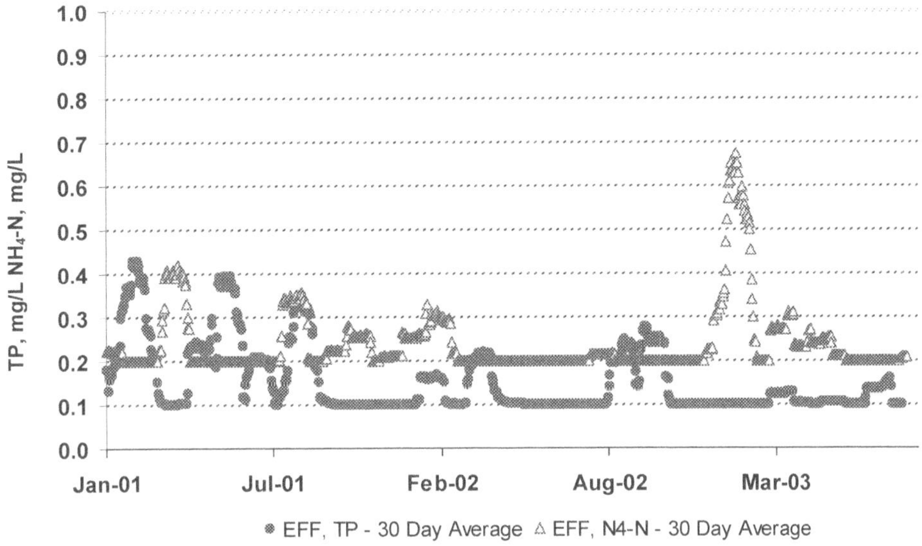


**Figure D-10. 30-day Average for Effluent Total Phosphorus and Ammonia.**

### *D.3.2.2.1 Phase 1 Testing*

Since the introduction of BNR in 1999 the performance and reliability of the plant has improved each year. During Phase 1, June 2002 through September 2002 the plant operated without upsets or any significant operational problems, producing effluent quality well below the permit requirements. Due to the additional BOD addition in form of sugar waste, high mixed liquor recycle rates and long anoxic retention time the effluent total nitrogen remained below 5 mg/L. Effluent phosphorus was at or below the detection limit except during September.

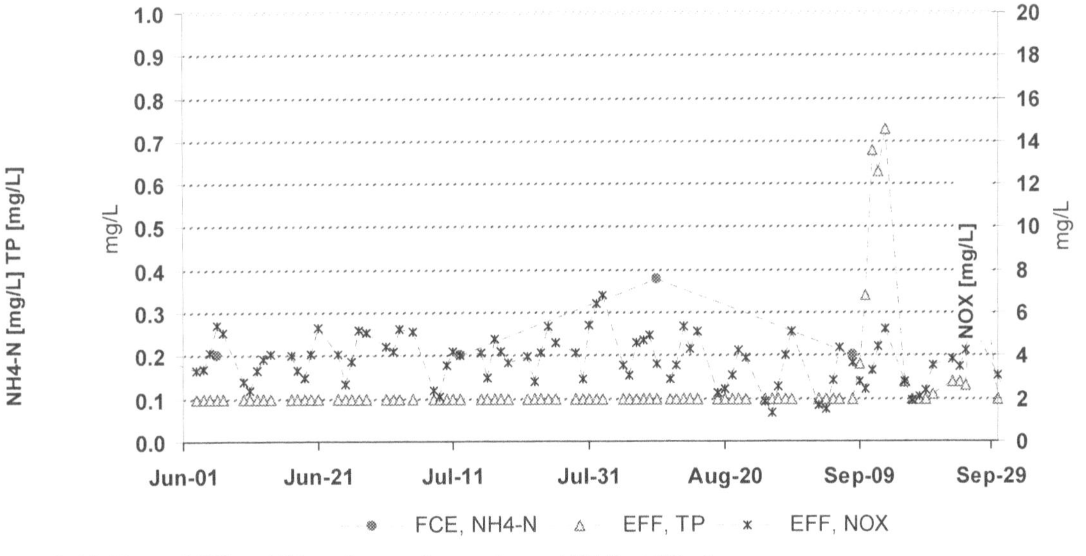


**Figure D-11. Phase 1 Effluent Phosphorus, Ammonia, and Nitrite + Nitrate.**

#### *D.3.2.2.2 Phase 2 Testing*

During the year of 2003 much of North Carolina experienced record rainfalls. Average plant flows at McDowell Creek increased from 4.8 MGD in 2002 to 6.1 MGD in 2003. The maximum month flow reached 7.3 MGD with a maximum day of 13 MGD (Figure D-12). At 7.3 MGD the plant operated essentially 21% over its rated capacity. The plant handled the peak flows well. While effluent concentration of ammonia and phosphorus were observed at times higher (Figure D-13), the process held up extremely well indicating that its true capacity may be considerably higher its rated capacity. During August 2003 the plant seemed to experience some degree of difficulty with phosphorus removal and denitrification.

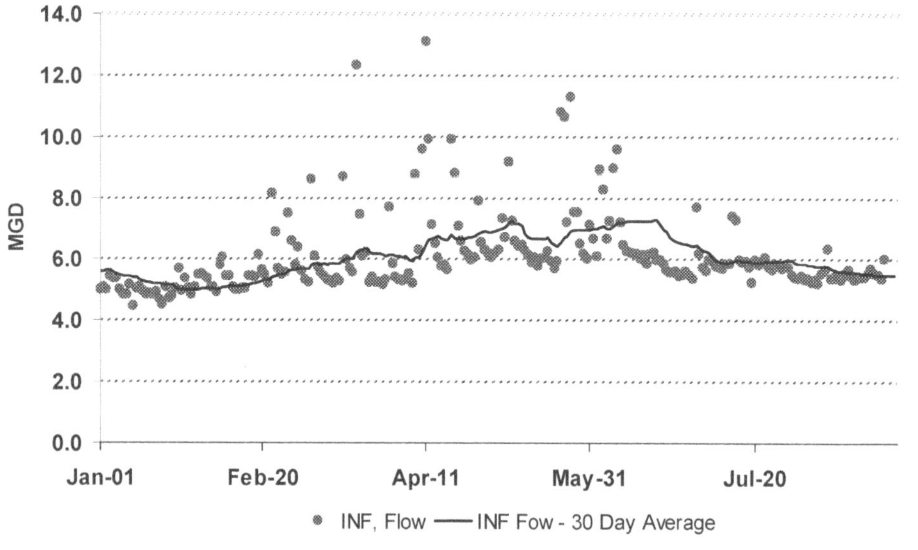


**Figure D-12. Influent flow from January through August 2003.**

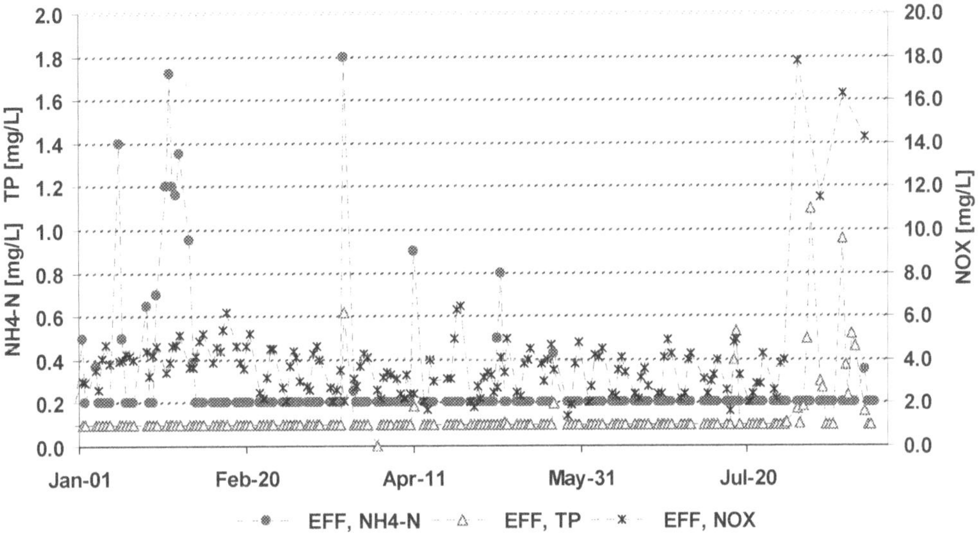


**Figure D-13. Phase 2 Secondary Effluent Phosphorus, TKN, and Suspended Solids.**

### D3.2.3 Relationship between Performance and Process Loading

With no aeration basin effluent phosphate data available it is difficult to assess the true EBPR performance at McDowell Creek. The plant as a whole delivers effluent phosphorus concentrations (Figure D-7) of less than 0.5 mg/L 95% of the time. From Figure D-8 it becomes apparent that phosphorus removal has been optimized over the last three years including some of the process improvements described in the previous section. There is no visible correlation between alum use and effluent TP concentration (Figure D-9) although the average use during the summer is lower than during the winter as one would expect.

The primary effluent wastewater composition (excluding the added sugar) would suggest good EBPR performance. However, the denitrification goal and its subsequent soluble BOD demand provide significant competition for biological phosphorus uptake. The sugar waste addition to the anaerobic zone, however shields the phosphorus release reaction from much of the competing denitrification.

Given the minimum water temperature of 15°C one would expect solid nitrification performance. However, most plants with similar environmental conditions show occasional upsets, which are completely absent at McDowell Creek (Figure D-10). Perhaps the lack of problematic industrial discharge protects the plant from mixed liquor poisoning often suspected at other facilities to be responsible for sudden and inexplicable loss of nitrification.

Primary clarification removes an average of 40% of the influent BOD (Figure D-14) while the phosphorus load shows almost no change (Figure D-15). It appears that the recycle phosphorus from digested sludge dewatering is compensated by alum addition to the filtrate and primary clarifier. The primary effluent BOD/P ratio at McDowell Creek of 30.9 mg BOD/ mg P while favorable for EBPR (Figure D-16) does become limiting for a treatment flow with low effluent phosphorus and low effluent total nitrogen. This ratio further improves after the sugar waste addition.

Between influent and primary effluent ammonia increases 11% (Figure D-17). This indicates a very moderate recycle ammonia load from dewatering operation.

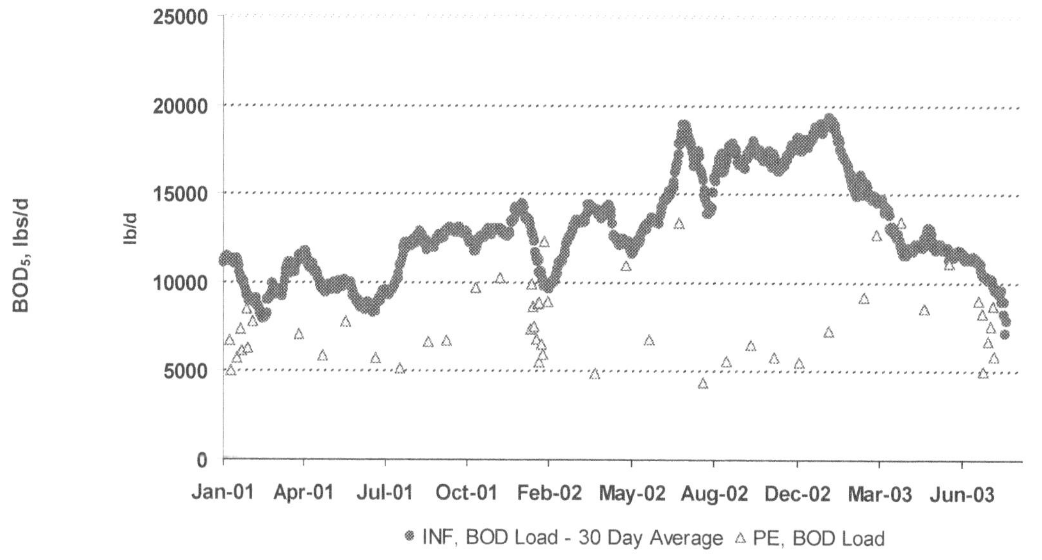


**Figure D-14. 30-day Averages for Raw Influent and Primary Effluent BOD between 2001 and 2003.**

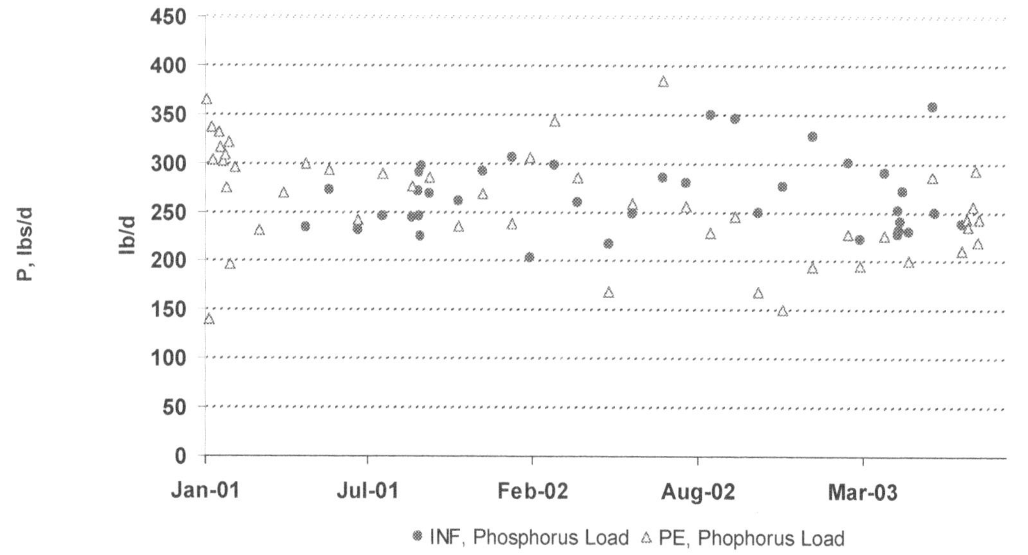


**Figure D-15. Influent and Primary Effluent Total Phosphorus Loads from 2001 - 2003.**

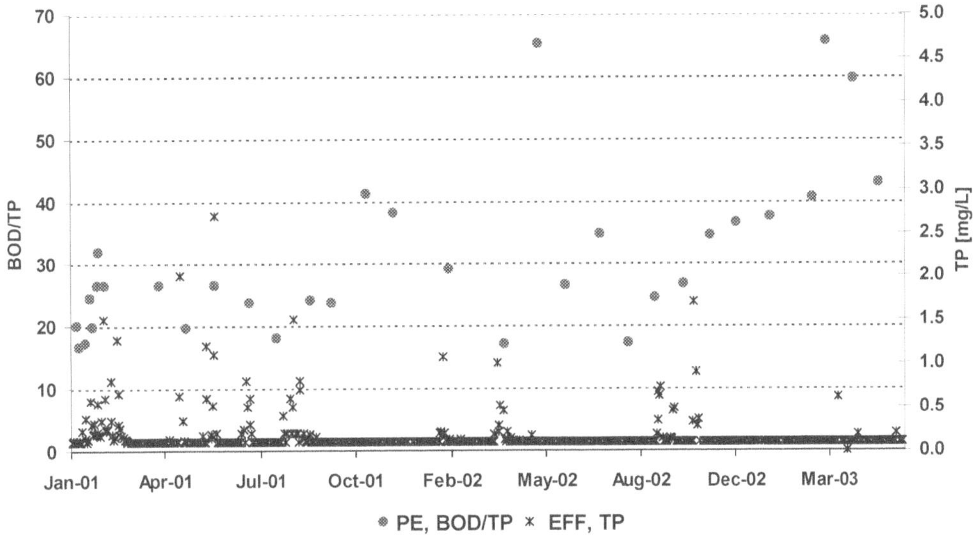


Figure D-16. 30-day Averages of Primary Effluent BOD/TP Ratio and Effluent Total Phosphorus.

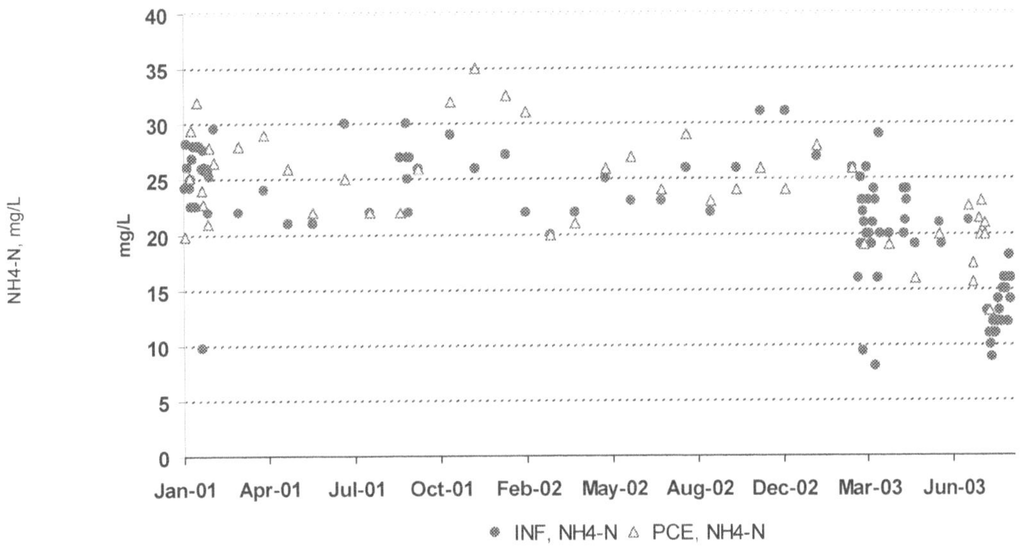


Figure D-17. Raw Influent and Primary Effluent Ammonia from between 2000 and 2003.

## D.3.3 Additional EBPR Process Evaluations

### D.3.3.1 Aeration Basin Phosphorus Profiles

The aeration basin phosphate profile was measured during each sampling phase. Charts of the profiles (Figure D-19 and Figure D-21) show four lines:

- The individual daily profiles (thin "shaded grey" lines in background with diamond points)

- Average phosphate profiles calculated from the daily profiles (thick line with circular points)
- Flow corrected average phosphate profile (thick line with triangular points). The flow corrected concentration is calculated as the theoretical phosphorus concentration based on the influent flow. This allows for a direct comparison of the concentration values without the influence of dilution from recycle flows.
- Phosphorus release rates (columns). The release rate (lb/d P / 1000 lb VSS) are calculated as phosphorus released per day compared to the aeration basin volatile solids content.

In order to understand the phosphate profile, the flow scheme must first be investigated. The flows are relatively complicated for the McDowell Creek WWTP mode of operation since three recycle streams must be accounted for: RAS, ANR, and ANX recycle. Figure D-20 and Figure D-22 illustrate the internal recycle flows for the three periods when the aeration basin profiles where measured.

### *D.3.3.1.1 Phase 1 Profiles*

Figure D-19 shows the phosphate profiles measured during the Phase 1 field test. The two profiles were taken on the same day, one mid morning and one during the early afternoon. Because of the high dilution by recirculation flows (Figurc D-20) one would not expect big difference in profiles measured only a few hours apart.

The flow corrected profile (normalized to aeration basin influent flow) shows that phosphorus release continues in the anoxic zone. The additional release however, did not appear to negatively impact phosphorus uptake. The phosphorus release rate amounts to 26.6 lb P/1000 lb VSS/d (Table D-11). This includes the release in the anoxic zone. One good indicator for the PAO population is the initial release in the first anaerobic cell where VFA is more readily available. With a retention time of 0.4 hours and a release of 6.8 lb P/ 1000 lb VSS/d the release rate is 14.2 lb P/ 1000 lb VSS/hr, which is at the low end comparing with the other investigated facilities

The internal recycle rates at McDowel are very high. In average the combined flow downstream of the anaerobic zone is almost six times the influent flow. This results in relatively flat profiles. However the flow corrected profiles and net phosphorus release provide more clarity with respect to the release and uptake activity.

While nitrates profiles were not measured, it is unlikely that the anoxic zone has enough nitrate to sustain anoxic conditions. That is effluent nitrate is already very low at only 3 – 5 mg/L and the anoxic retention time is nearly three hours.

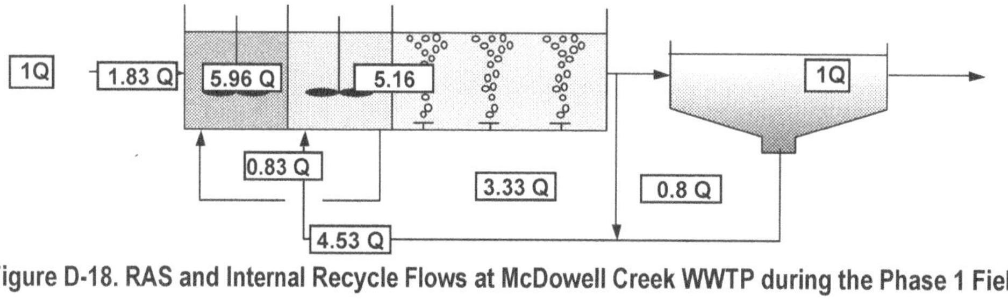


**Figure D-18. RAS and Internal Recycle Flows at McDowell Creek WWTP during the Phase 1 Field Testing.**

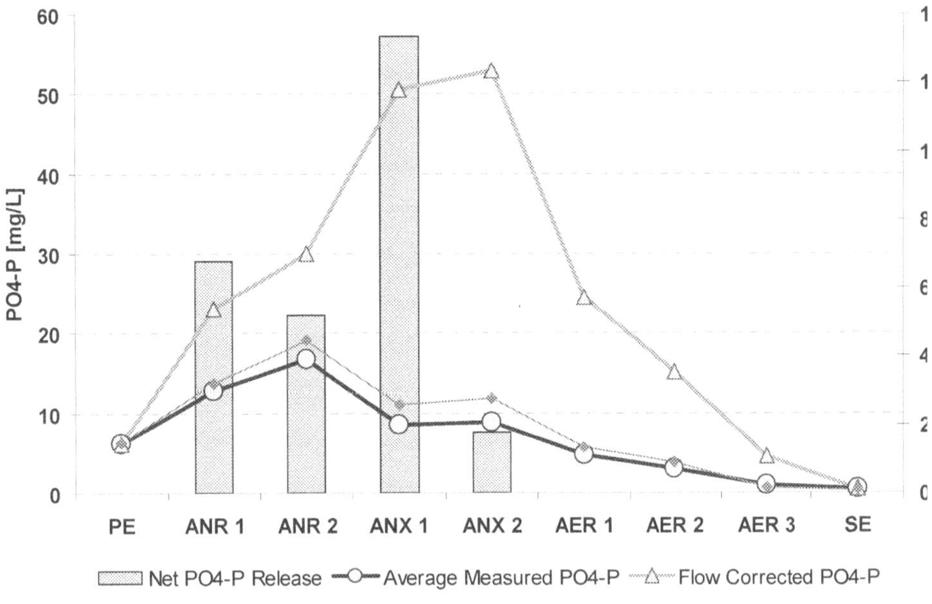


**Figure D-19. Phosphate Profiles Measured During Phase 1 Field Testing.**

### *D.3.3.1.2 Phase 2 Profiles*

The profile measured during Phase 2 shows a lower phosphorus release compared to Phase 1. The plant did experience some difficulty in phosphorus removal at the time but did not show clear indication for a process upset. The flows during Phase 2 were 15% higher. The total phosphorus release was only 15.5 lb P/ 1000 lb VSS/d with an initial release rate of 13.7 lb/1000 lb VSS/hr in the first anaerobic zone.

During Phase 2 the process showed uniquely low phosphorus content for an EBPR sludge. The measured MLSS total phosphors content was measured at 2.3% vs 4.3% during Phase 1. Taking into account the normal cell phosphorus the amount of stored phosphorus was very small during Phase 2. Whether this was caused by a change in phosphorus or VFA supply at the time, or a change in PAO population cannot be determined based on available information. However, the initial release (Table D-11) calculated from the both average profiles is almost identical, which could be an indication that the PAO population was indeed similar. The change in the profile does not occur until downstream of the first anaerobic cell. This could be an indicator that the release was VFA limited and explain the reduction in phosphorus uptake, thus cell phosphorus content.

In both cases there was sufficient BOD in the primary effluent to sustain EBPR. However, the supplementary BOD added to the process is added in the form of sugar, which has to be converted to VFA in the anaerobic zone. Anaerobic retention time as well as how fresh the sugar waste was at the time of use would determine how much of the sugar would actually be converted into VFA before leaving the anaerobic zone.

Another noticeable difference between the two phases was that during Phase 1 the anoxic phosphorus uptake was almost 10 times higher than during Phase 2.

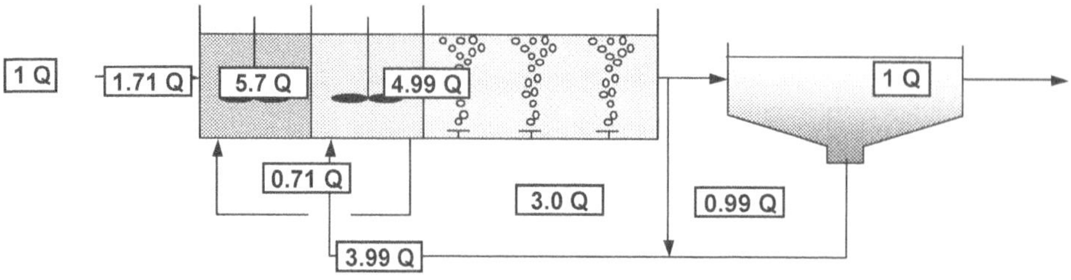


**Figure D-20. RAS and Internal Recycle Flows at McDowell Creek WWTP During Phase 2 Testing.**

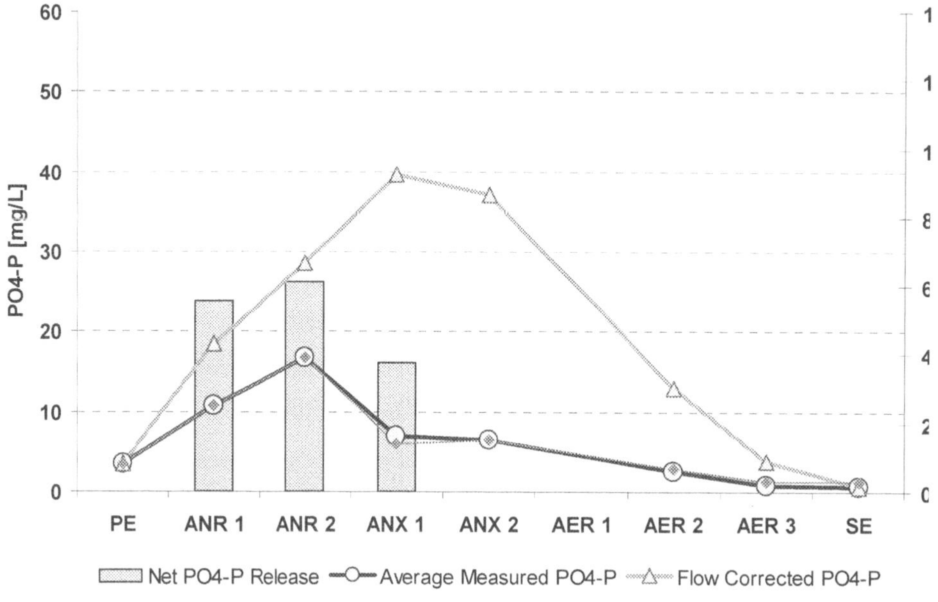


**Figure D-21. Phosphate Profiles During Intensive Sampling Period.**

**Table D-11. Process Control and Water Quality Measurement Results During Intensive Testing (two-week average).**

| Raw Wastewater Parameter | Unit | Phase 1 | Phase 2B |
|---|---|---|---|
| Total Phosphorus | mg/L | 35.0 | 25.5 |
| Anoxic Phosphorus Uptake | mg/L | 14.2 | 1.6 |
| Phosphorus Release Rate | lb P per 1000 lb VSS per day | 26.6 | 15.5 |
| Phosphorus Uptake Rate | lb P per 1000 lb VSS per day | 31.4 | 20.9 |
| Total Aeration Basin Phosphorus Inventory | Lb P | 2880 | 1028 |
| Release of P Content | Lb/d | 1520 | 1155 |
| Initial Phosphorus Release Rate | lb/lb VSS/hr | 14.2 | 13.7 |

### D3.3.2 Uptake and Release Test

The phosphorus uptake and release tests (URT) provide a quantification of the PAO population in the basin. By adding VFA to the influent, the phosphorus release is limited by the PAO population in the basin and amount of stored phosphorus. Theoretically, the phosphorus release under anaerobic conditions is 0.5 mg P/mg VFA as COD. Deviation (reduction) from this value could indicate competition from other bacteria such as GAOs.

The URT test can be used for the following:

- Determine the amount of PAO activity from phosphorus release rate. The release rate (mg P/g VSS/h) is indicative of the PAO activity.
- Determine the potential for competition in the anaerobic zone for VFA. The phosphorus release ratio (mg P released/mg VFA removed) indicates the PAO activity. A phosphorus release ratio of 0.5 mg P released/mg VFA removed as COD indicates PAO activity. A lower ratio is indicative of competition from other VFA consuming organisms such as GAOs.
- Determine key kinetic parameters for process simulation model calibration such as$\Delta$P/acetic acid and Magnesium and potassium uptake and release rates.
- Provides information regarding the phosphorus uptake kinetics, which represents the time required in the aerobic zone to remove the phosphorus.

Table D-12 summarizes the results from URTs conducted with mixed liquor from McDowell Creek WWTP. Figure D-22 and Figure D-23 show the phosphate and metal curves measured during the tests. The first test was conducted during the site visit in September 2002; the second test was conducted in August 2003. The difference in URT results are consistent with the observed process performance.

The metal uptake release ratios measured during both tests were in the range observed at other facilities. Interestingly while the uptake and release ratios (ratio between phosphorus and magnesium) are almost identical for each test but differ almost 50% between both tests. The phosphorus release however, was less then half for the Phase 2 sample. The profiles showed identical initial phosphorus release, indicating similar PAO populations. The lower $\Delta$P/acetic acid ratio suggests increased competition by other VFA consuming organisms such as GAO or denitrifiers. This aspect could have been overpowered in the field by a higher supply of VFA that provides sufficient substrate for PAOs even after other organisms consumed a fraction of the available VFA. The high primary effluent BOD/P ratio plus the additional BOD addition at McDowell would support this theory.

**Table D-12. Uptake and Release Tests Results.**

| Test Parameter | Unit | Phase 2A | Phase 2B |
|---|---|---|---|
| Phosphorus Release Rate | mg P /g VSS /hr | 16.6 | 7.1 |
| Phosphorus Uptake Rate | mg P /g VSS /hr | 5.5 | 1.7 |
| Release Rate/Uptake Rate | -- | 3.0 | 4.2 |
| ΔP/Acetic acid | mg $PO_4$-P/ mg COD | 0.34 | 0.2 |
| ΔP/ΔMg - Release | mg $PO_4$-P / mg Mg | 6.2 | 4.8 |
| ΔP/ΔMg - Uptake | mg $PO_4$-P / mg Mg | 6.0 | 4.1 |
| ΔP/ΔK - Release | mg $PO_4$-P / mg K | - | 2.7 |
| ΔP/ΔK - Uptake | mg $PO_4$-P / mg K | - | 2.4 |
| MLVSS | mg/L | 2400 | 2000 |

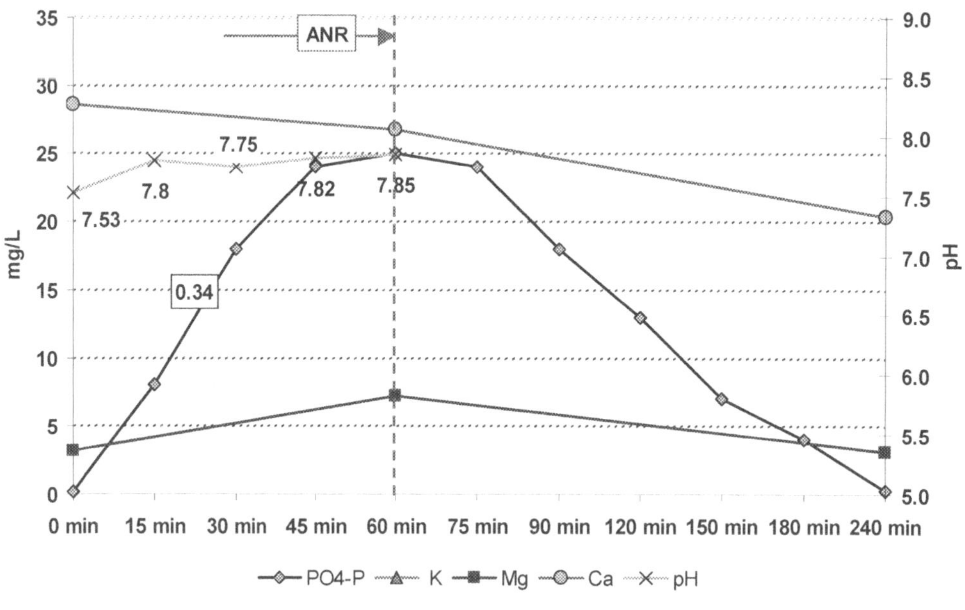


**Figure D-22. Phosphorus Uptake and Release Test Results from Phase 1.**

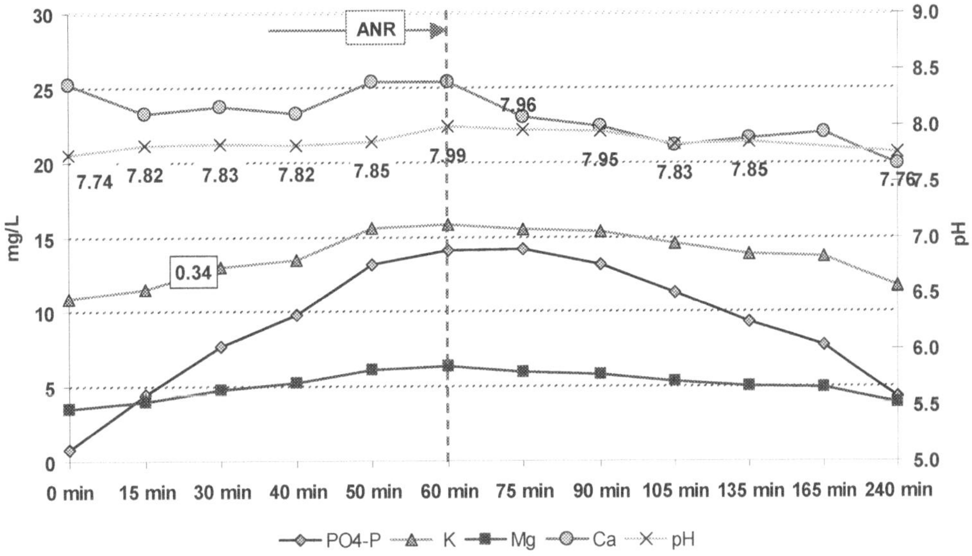


Figure D-23. Phosphorus Uptake and Release Test Results from Phase 2B.

### D.3.3.3 Phosphorus and Metal Mass Balance

The purpose of the phosphorus mass balance is to track the average mass flow of phosphorus throughout the system. Values presented in this section are either 2002 annual averages or grab sample measurements from the Phase 1 testing.

For any EBPR and BNR facility with anaerobic digestion the recycle load has a significant impact on secondary process stability and process capacity. At McDowell Creek the dewatering filtrate is equalized after a portion of the phosphate was precipitated by alum addition. The latter reduces the recycle load such that the net increase between raw influent and primary clarifier only amounts to 8% while the recycle phosphorus load is actually much higher (23%). But the majority of the recycle phosphorus is chemically bound and settled out in the primary clarifier.

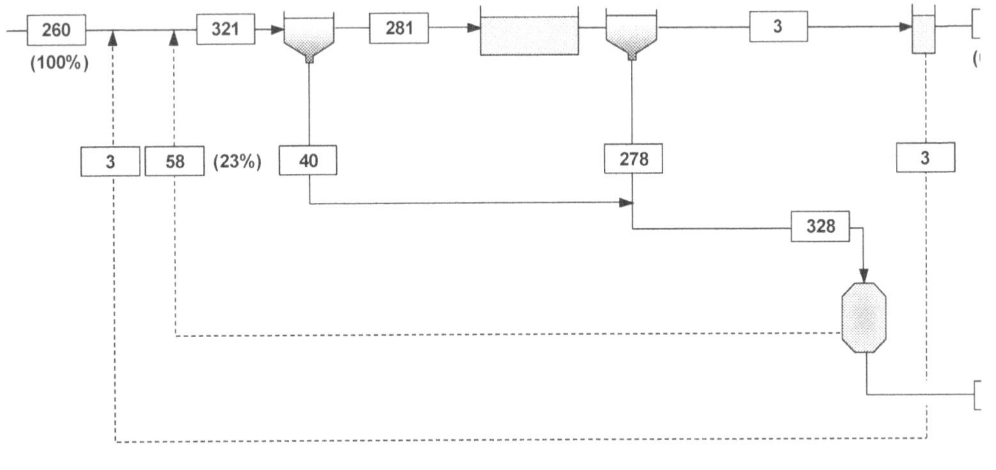


Figure D-24. Phosphorus Mass Balance for McDowell Creek WWTP (2002 annual average).

Table D-13. Results of Metals Testing.

| Analysis | Aeration Basin Influent mg/L | Anaerobic Zone mg/L | Secondary Effluent mg/L |
|---|---|---|---|
| Mg | 3.6 | 6.6 | 46 |
| Ca | 17 | 32 | 110 |
| K | 16 | 19 | 82 |

### D.3.4 Microscopic Examination of Digested Sludge

At facilities that operate anaerobic digestions and especially where waste activated sludge is processed by means of anaerobic digestions, struvite (magnesium ammonium phosphate) can be found in digested sludge.

The pictures in Figure D-25 do not show any crystals, which by itself is not conclusive evidence of crystals absence, but could be explained with the primary alum addition. Aluminum phosphate has a lower solubility than magnesium ammonium phosphate, thus struvite would not begin to form until all residual (soluble) alum has been consumed.

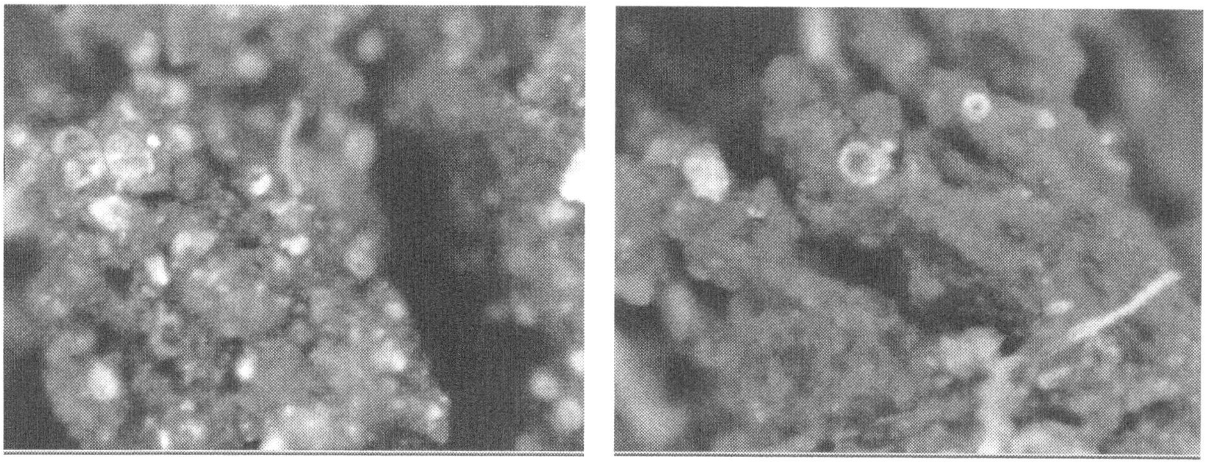

Figure D-25. Digital Microscope Pictures of Dewatered Sludge at 200x Magnification.

### D.3.5 PAO and GAO Investigations

A mixed liquor sample from the McDowell plant were examined to observe the presence of polyphosphate granules (indicative of EBPR), and to look for the presence of GAOs, which typically consist of large cocci (2-3 μm) arranged in tetrad clusters within the floc. The sample was analyzed on September 3, 2003.

Both microscopic study and molecular techniques were used to examine the presence of GAOs in the mixed liquor from the plant. Light microscopy observation was used to observe the tetrad-forming GAOs, and molecular probing using specific oligonucleotide probes targeting GAOs was completed to identify both tetrad-forming and non-tetrad-forming GAOs. In addition Neisser and DAPI staining were applied to observe the presence and relative abundance of PAOs.

Figure D-26 and Figure D-27 show the observation of PAOs and GAOs using a Neisser stain. Figure D-28 shows the results of DAPI stain. The GAOs have Neisser positive stained cell walls, but are not positive for cell inclusions as are the PAOs. The Neisser positive stained PAOs are dark blue to purple in color, and usually group in clusters within the flocs. The DAPI-stained PAOs are yellow in color, which is characteristic of polyphosphate granules observed under the fluorescence microscope. Other non-polyphosphate-accumulating bacteria including GAOs show up in blue with the DAPI stain.

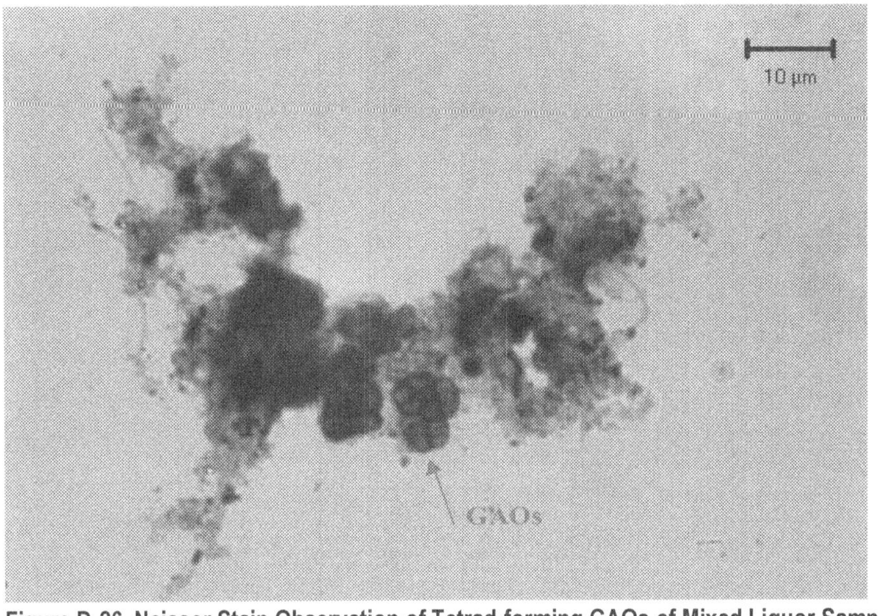


**Figure D-26. Neisser Stain Observation of Tetrad-forming GAOs of Mixed Liquor Sample from McDowell WWTP. Sample was taken on 9-9-03.**

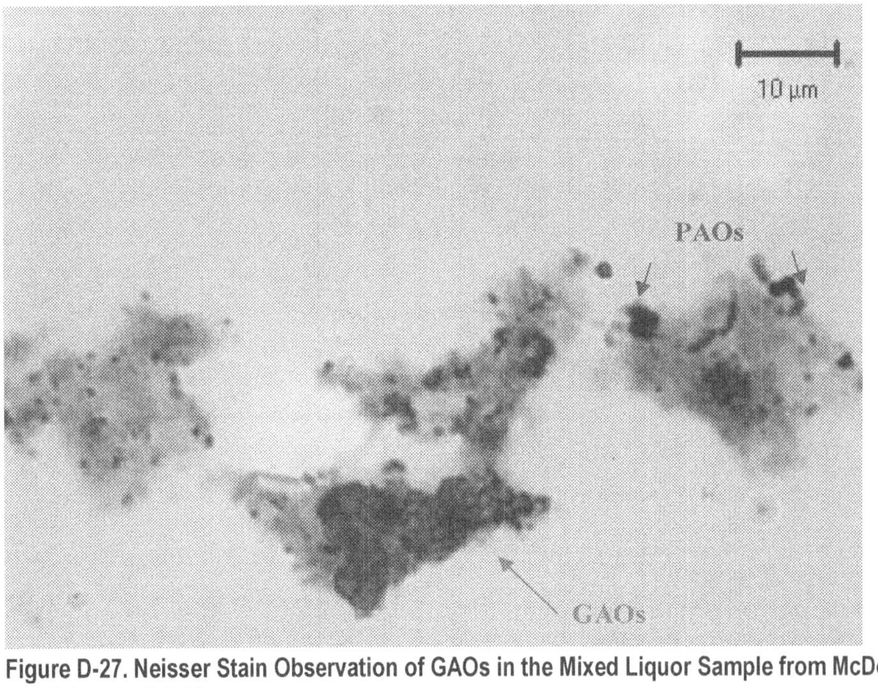


**Figure D-27. Neisser Stain Observation of GAOs in the Mixed Liquor Sample from McDowell WWTP. Sample was taken on 9-9-03.**

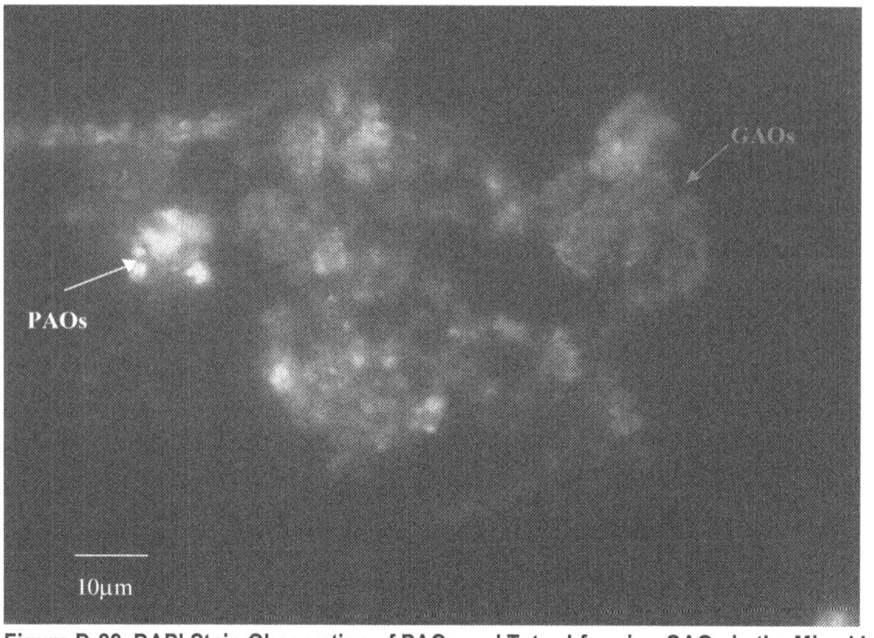


**Figure D-28. DAPI Stain Observation of PAOs and Tetrad-forming GAOs in the Mixed Liquor Sample from McDowell WWTP. Sample was taken on 9-9-03.**

summarizes the DAPI and Neisser stain observation of PAOs and GAOs for the mixed liquor sample from McDowell. Tetrad-forming GAOs were predominant in the mixed liquor sample and they group together in flocs. Some PAOs were observed as well but they were not as abundant as GAOs.

Examination of the mixed liquor samples via molecular methods (fluorescence in situ hybridization (FISH))using 16S rDNA-based probes targeting *Accumulibacter* (represent PAOs) (Crocetti et al. 00) and some non tetrad-forming GAOs-*Competibacter* (GAOQ989 andGB-G2, Crocetti et al. 00 and Kong et al. 00) showed that only 1-10% of the Eubacteria populations was *Accumulibacter*-GAOs. No *Competibacter*-GAOs was observed. It should be noted that FISH methods only targeted certain PAO and GAO species for which the probes are available. DAPI and Neisser stain were targeting at all cells that are capable of accumulating poly-P, therefore should reveal more PAO populations than FISH method.

**Table D-14. Summary of PAOs and GAOs Observation for McDowell WWTP.**

| Date of Sample | GAO observation (Tetrad-forming) | GAO observation (*Competibacter*) | PAO observation | Comments |
|---|---|---|---|---|
| McDowell 9-9-03 | ***Abundant***, 8 out of 10 fields observed GAOs, grouped together in focs, some fields almost had all GAOs. | Not observed. | Not many PAOs flocs, small, poly-P granules | GAOs seemed to be dominant, more abundant than PAOs. |

The phosphorus removal performance at the plant was evaluated for the time period when the sample was taken for the GAOs study. The results of phosphorus removal performance evaluation data are shown in Table D-15.

Table D-15. Comparison of GAOs Abundance and P removal Performance at McDowell WWTP.

| Date of Sample | GAOs abundance | P Release Rate gP/gVSS.hr | P Uptake Rate gP/gVSS.hr | P Release Ratio gΔ P/gΔAcetic Acid | Release Rate/ Uptake Rate |
|---|---|---|---|---|---|
| 9-9-03 | Predominant | 5.5 | 1.4 | 0.21 | 4 |

Evaluation of phosphorus removal performance at the plant during June to September 2003 indicated decreasing phosphorus removal efficiency as shown by the declining specific phosphorus release and uptake rate and the amount phosphorus removed per unit amount of acetate taken up. The results of the GAO study indicated that some factor favored the growth of tetrad-forming GAOs during the July through September period, and they were becoming more abundant in the microbial community. The deteriorating phosphorus removal performance during that that time was consistent with the understanding that abundant GAOs would compete with PAOs for substrate and hamper the phosphorus removal efficiency. The population level of non-tetrad-forming GAOs (*Competibacter*) was not observed in the plant sample.

## D.4 Conclusions

After an initial startup and optimization period the EBPR process at McDowell Creek has delivered very consistent low effluent phosphorus concentrations. Primary effluent BOD:P ratios were measured at 30:1 before the addition of sugar waste that further increases VFAs available in the anaerobic zone. However, alum addition remains a part of the operation strategy to lower the primary effluent and dewatering recycle phosphate and provide EBPR backup.

Because phosphorus removal is not a regulatory requirement, secondary effluent phosphorus concentrations are not routinely measured. Final effluent phosphorus concentrations, however, are under 0.5 mg/L 95% of the time, indicating very stable performance. During the intensive sampling periods, secondary effluent total phosphorus was typically under 0.2 mg/L.

During the year of 2003 much of North Carolina experienced record rainfalls. The maximum month flow reached 7.3 MGD with a maximum day of 13 MGD. At 7.3 MGD, the plant operated essentially 21% over its rated capacity. During the high flows, effluent total phosphorus generally remained under 0.2 mg/L. This performance is indicative of the stable phosphorus removal observed at the McDowell Creek facility.

APPENDIX E

# Nansemond Enhanced Biological Phosphorus Removal (EBPR) Evaluation Report

## E.1 Summary of Findings

The Nansemond WWTP, located 10 miles northwest of Norfolk VA, is operated by the Hampton Roads Sanitation District (HRSD). The plant was designed to operate in Virginia Initiative Plant (VIP) mode, a BNR process design developed at HRSD. Currently flows average 20 MGD, which is 67 percent of the 30 MGD design capacity. The plant adds ferric chloride on an as-needed basis for additional phosphorus removal.

Wastewater composition is considered unfavorable for EBPR based on conventional criteria. The $BOD_5$/TP ratio average of 14 mg $BOD_5$/mg P for the past three years has dropped below 10 mg $BOD_5$/mg P during the last eight months; plants with good EBPR performance typically have ratios over 20 mg $BOD_5$/mg P. In addition, the measured VFA to P ratio ranged between 0.9 and 2.6 mg VFA/mg P; a ratio of 6 – 8 mg VFA/mg P is considered adequate for EBPR. Even so, operating data indicate that the plant is achieving EBPR at times. This could indicate the degradation of readily biodegradable organics to volatile fatty acids in the anaerobic zones.

Primary effluent total phosphorus concentrations are very high, often more than double that typical for municipal wastewater. Primary effluent average total phosphorus concentration is 14.2 mg P/L, 81% of which is orthophosphate. Return flows from solids processing add 60% to the raw influent total phosphorus load. This is reflected in the elevated primary effluent phosphorus concentration.

The Nansemond plant operates in EBPR mode year round and targets a 2 mg/L effluent total phosphorus concentration. Operating data since 2003 indicate that the target is met 85% of the time. Ferric chloride is added to the secondary clarifier only when EBPR consistently fails to meet treatment goals. While EBPR appears to be present for the majority of the time, the plant periodically loses biological phosphorus uptake. Data show that the loss of EBPR occurs when influent $BOD_5$ concentration decreases.

Because of occasional undocumented chemical addition at the plant it is difficult to determine the exact contribution of EBPR to the overall phosphorus removal. However, metal uptake (Mg, Ca, K) information is evidence for significant phosphorus transfer (about 50%) to digesters through luxury uptake.

The following table summarizes the key findings from this study.

| | |
|---|---|
| Reliability | Plant achieves effluent Total Phosphorus (TP) target of 2.0 mg/L with 85% reliability, but effluent TP concentrations are highly variable. Aeration basin influent phosphorus is high (over 14 mg P/L), and includes 60% P contributed by the recycle stream. |
| Operational Criteria | Solids retention time (SRT) is maintained at 6-12 days. Anaerobic hydraulic retention time (HRT) is maintained at 1-2 hours, anoxic HRT typically 2 hours, and aerobic HRT under 4 hours. |
| Influent Organic Characterization | Aeration basin influent VFA:P ratio ranges from 0.9 to 2.6, but periods of poor plant performance appear more closely linked to *change* in influent organic concentration than the value of the ratio. Influent BOD:P ratio averages 14 mg BOD/mg P. |
| P Removal Capacity | During the uptake and release test (URT), phosphorus release and uptake were higher than measured in the field release in the anaerobic zone. This implies that P removal capacity is higher than currently experienced in the field. |
| Measures Showing PAO Growth | Anaerobic P release and low effluent P during intensive sampling periods (with no chemical addition) demonstrate that phosphorus accumulating organism (PAO) growth is occurring. Microscopic evaluation also identified PAOs. Uptake and release tests confirmed that PAOs were present. |
| Factors Hindering PAO Growth | Low VFA concentration in aeration basin influent hinders PAO growth, possibly due to discharge practices of a large major industry. Also, uptake and release test (URT) showed $\Delta$P/acetic acid ratio of 0.2 mg P/mg acetic acid, lower than the theoretical value of 0.5. This indicates that not all VFAs are being consumed by PAOs. |
| P and Metals Mass Balance | Phosphorus, calcium, magnesium, and potassium all decrease between the primary and secondary effluents. When expressed in equivalent fractions, calcium is the dominant cation removed across the process. Digested biosolids samples suggest that a significant fraction of the phosphorus is potentially tied up in the form of an iron or calcium phosphate precipitate. |
| Secondary $P_{sol}$ Release | Secondary $P_{sol}$ release was not observed. $P_{sol}$ uptake began in the last anoxic cell during phosphate profile testing. |

## E.2 Introduction

The analysis of this facility was completed as part of a project sponsored by the Water Environment Research Foundation (WERF). The project objective is to develop information for evaluating the ability of municipal wastewater treatment plants to efficiently and cost effectively remove phosphorus to low levels through processes that involve EBPR. This was completed in large part through detailed studies at several treatment facilities that employ EBPR. The goals of the study were to:

- Determine process reliability.
- Determine the biological phosphorus removal efficiency.
- Determine the biological phosphorus removal capacity.
- Determine how chemical addition affects Bio-P removal.

Definitions of terms used in this analysis are described in Appendix P.

### E.2.1 Facility Introduction

Figure E-1. Nansemond WWTP, Hampton Roads Sanitation District. Suffolk, VA.

The Nansemond Plant is located in Suffolk, VA, and is operated by the Hampton Roads Sanitation District (HRSD). Nansemond has a 30 mgd design treatment capacity with a current annual average flow of approximately 20.1 mgd. The Nansemond plant was designed for biological nutrient removal, including EBPR, and has been in operation since 1998. Effluent is discharged to the mouth of the James River. The activated sludge process was designed for and is operated using the VIP process configuration. Both primary and secondary sludges are thickened and anaerobically digested. The dewatering recycle load is returned to the plant influent stream.

The plant's operational strategy aims for 1 mg/L secondary effluent NH4-N and 9 mg/L secondary effluent (SE) total nitrogen (65 percent N-removal) even though nitrogen removal is not required for permit compliance. Nansemond has a year-round phosphorus limit of 2 mg/L (average month). Other effluent requirements are an average month $BOD_5$/TSS of 30 mg/L. While the facility operates EBPR successfully the majority of the time, chemical addition to supplement EBPR and in cases of contingency are required.

The facility also experiences periodic process upsets which have not been definitively linked to monitored process and water quality parameters, and are suspected to the be result of mixed liquor poisoning or shock loading within the system. The Nansemond plant serves several major food processing facilities that discharge into the sewer system. Among these industrial customers, it is known that a meat processing plant pretreats it waste in lagoons and discharges lagoon effluent into the sewer. The lagoons are operated to achieve nitrification (i.e., BOD and ammonia removal), but are not required to remove phosphorus or nitrates.

### E.2.2 Facility Process Design

#### E.2.2.1 Liquid Stream

The biological nutrient removal (BNR) process follows screening, grit removal, and primary clarification in the plant system. The configuration matches the VIP design developed by HRSD. Figure E-2 shows the Nansemond WWTP process design. Table E-1 summarizes key design parameters compared to 2002 average values.

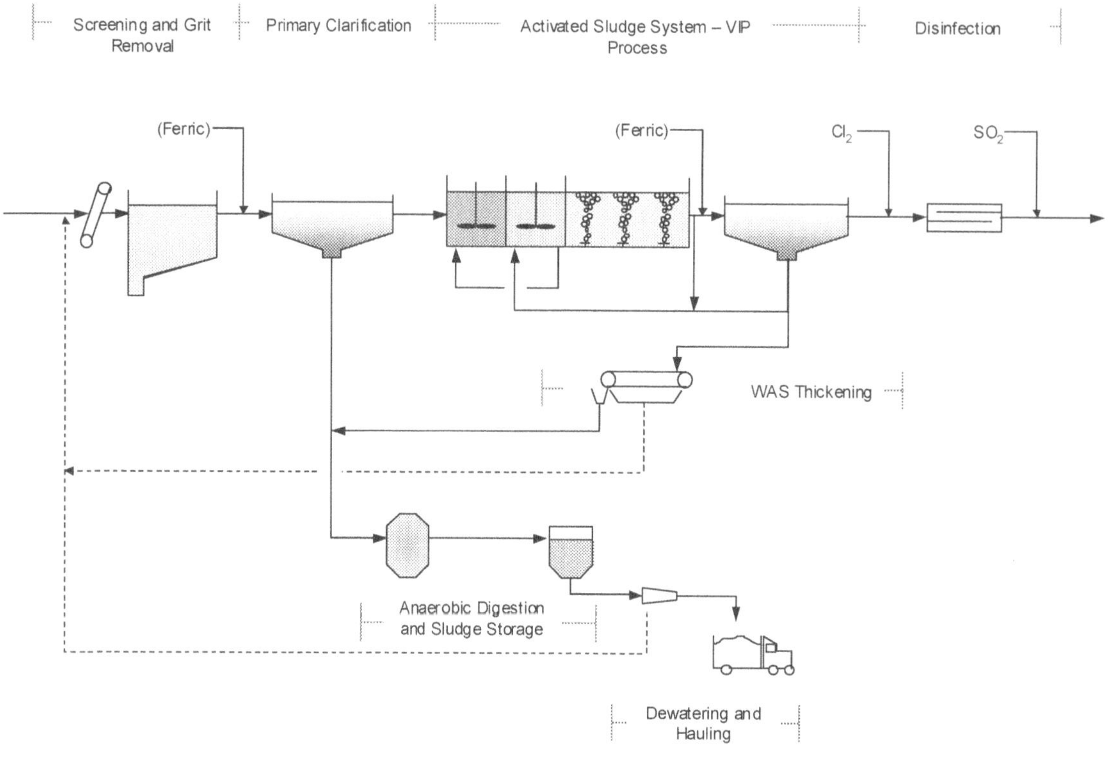


**Figure E-2. Process Schematic for Nansemond WWTP Liquid and Solids Treatment.**

The key feature of the VIP process is that return activated sludge (RAS) is combined with mixed liquor recycled from the aerobic zone. RAS enters the process train at the head of the anoxic zone and mixed liquor is recycled from the anoxic zone to the anaerobic zone to achieve EBPR. This configuration limits the transfer of nitrates into the anaerobic zone, optimizing volatile fatty acid (VFA) availability for phosphorus release. Primary effluent and the recycle stream from the anoxic zone mix at the head of the anaerobic zone. In the first anoxic cell, mixed liquor from the aeration basin effluent mixes with the anaerobic zone effluent. Operators also have the option to send the RAS/mixed liquor recycle stream to one of the two downstream anoxic cells if longer anaerobic retention is desired.

At Nansemond there are seven unaerated trains, each with two anaerobic cells (ANR) and four anoxic cells (ANX) followed by five aerobic trains (AER). Each aerobic train has four zones (Figure E-3).

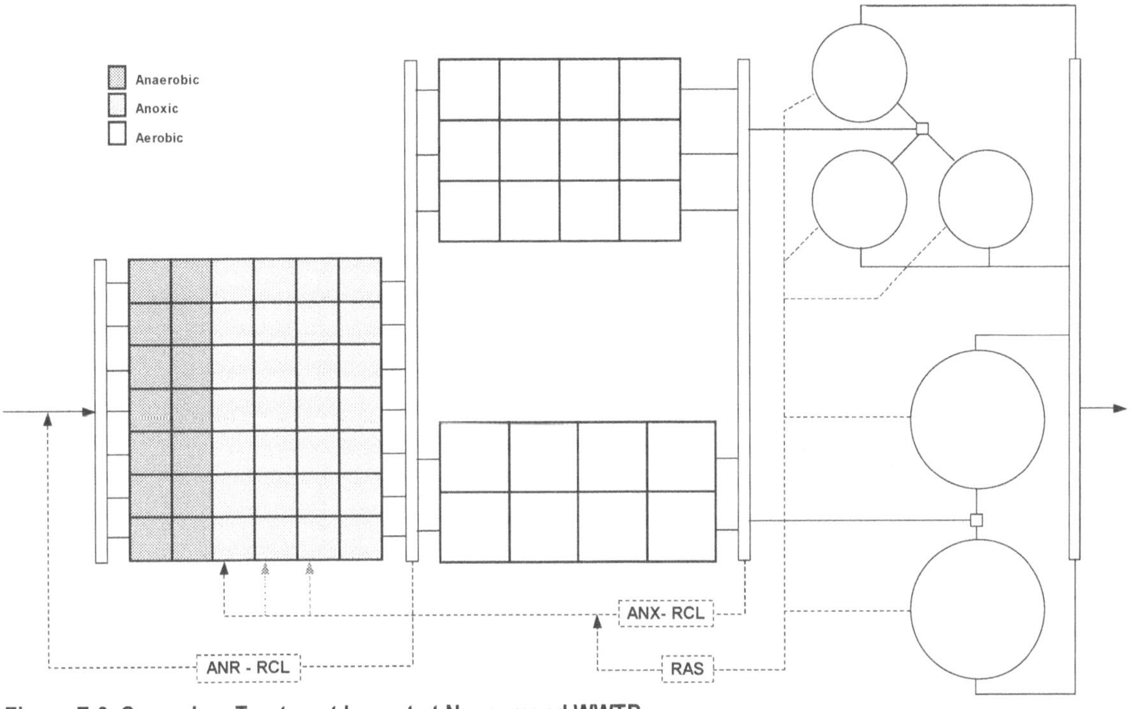


**Figure E-3. Secondary Treatment Layout at Nansemond WWTP.**

**Table E-1. Key Process Parameters.**

| Component | Unit | Design | Average[1] |
|---|---|---|---|
| ANR HRT | hrs | 1.0 | 0.83 |
| ANX HRT | hrs | 4.2 | 2.9 |
| AER HRT | hrs | 5.6 | 3.8 |
| Total HRT | hrs | 11 | 7.5 |
| RAS Rate | % | 75 | 84 |
| ANR ML Recycle Rate | % | NA | 91 |
| ANX ML Recycle Rate | % | NA | 99 |
| SRT | days | NA | 8.5 |
| MLSS | mg/L | NA | 3520 |
| Solids Recycle Return | | to plant influent | |
| Flow | MGD | 30 | 18.1 |

[1] Based on 2002 data.
[2] Retention time is calculated based on aeration basin influent flow and does not include the recycle volume.

### E.2.2.2 Solids Stream

Solids treatment processes consist of thickening, anaerobic digestion, and digested biosolids dewatering. A gravity belt thickener (GBT) increases waste activated sludge (WAS) TSS concentration to 5-7% while primary sludge is thickened in the primary clarifier to about 3-5% solids before conveyance to the digester. Primary sludge and WAS are mixed prior to digester feed. The two digesters are operated as one and are fed intermittently, with the feed alternated between the two digesters every two hours.

The plant operates two mixed mesophilic digesters at a target MCRT of 17-20 days to accomplish on average 38% volatile solids destruction. Digester operation has been plagued by periodic foaming problems. Controlling activated sludge foaming to minimize the influx of filamentous organism has not translated into a reduction of digester foaming. Operators experimented with chlorinating the gravity belt thickened biosolids before entering the digester. Improved control of digester feeding has proven to reduce the foaming problem but not entirely eradicated it.

The digesters are being upgraded with a gutter system to catch the foam, using nonpotable water to carry the foam to the plant drain system. While this minimizes the foaming overflow and odor problems it does not minimize the foaming problem itself. Discussions to capture the foam in a separate tank and chlorinate to prevent the recycling of filaments back the secondary treatment trains are underway.

Return centrate is sent to the headworks. Dewatering centrifuges operate approximately 8-10 hours/d, 7 d/wk during graveyard shift. Centrifuge operation starts around 11 PM and continues until completed. Digested biosolids are incinerated off site. The agency has permits for composting, land application, and landfill disposal, all of which are only used in emergencies.

**Table E-2. Solids Processing Summary**

| Component | Technology | Comments |
|---|---|---|
| Primary Sludge Thickening | Gravity | PSL is thickened in the primary clarifiers to 3– 5 percent |
| WAS Thickening | GBT | 5– 7 percent |
| Primary Sludge and WAS Stabilization | Anaerobic Digestion | SRT = 17 – 20 days<br>MLSS = 2.5 – 3 percent<br>VSS Destruction = 38% |
| Dewatering | Centrifuge | Operated during night shift between 11 PM and 7 AM |
| Filtrate/Centrate Storage | none | |
| Biosolids Disposal | Incineration | Option to land apply, compost and landfill disposal |

### E.2.2.3 Chemical Addition

At Nansemond WWTP operators have the option of adding ferric chloride to the primary clarifier, secondary clarifier, and to the digested biosolids upstream of the dewatering centrifuges. Ferric doses are set based on operator experience. Feed is flow paced. Ferric is typically added to the secondary clarifier influent. When EBPR produces adequate phosphorus concentration in effluent, ferric is only added to the dewatering feed in order to reduce the recycle phosphorus load and aid dewatering process.

In the past, primary ferric addition reduced BOD but it negatively impacted EBPR, so this mode has been abandoned. At times ferrous chloride is added to the aeration basin.

Sodium hydroxide is used for pH control and is added to the primary clarifier effluent or to the contact tank effluent as needed. It was not used during the intensive sampling.

Anionic polymer is used in combination with ferric chloride to address settling problems in the secondary clarifiers. Cationic polymer is used for dewatering and thickening.

### E.2.2.4 Process Control

The operational strategy is to maintain a solids retention time (SRT) between six and 12 days. SRT is calculated based on total activated sludge basins, excluding clarifiers. Dissolved oxygen (DO) is controlled by one probe in the aeration basin to maintain at least 2 mg/L DO in the last one-third of the aeration tank. The number of aerated and non-aerated trains in service is determined based on flow. The operational goal is to stay within minimum and maximum design hydraulic retention time (HRT). Operators also have the option of changing the number of anoxic and anaerobic cells per train through a swing zone that can be operated as either anoxic or anaerobic, depending on where RAS/mixed liquor recycle is returned.

Currently the plant is testing depth probes to automate the secondary clarifier blanket maintenance. The plant does not operate any online analytical equipment except for DO probes and pH meters.

## E.3 Process Performance Evaluation

### E.3.1 Approach

The process evaluation entailed two data collection activities: a review of historical data and the conduct of special sampling to supplement routinely recorded water quality and process measurements. The following data were collected (see Table E-3).

**Routine Data Review**

The reviewed data represent all available water quality and process monitoring information collected from January 1, 2000 through July 31, 2003. Data were analyzed to determine general process performance, reliability of meeting effluent quality goals, and evaluation the interaction between process operating parameters and performance. This routine data is referred to as the **2000-2003 data set**.

**Special Data Collection**

Project team members conducted field tests and interviewed Nansemond plant staff during each phase of the project (Table E-4). During Phase 1 the focus of site visits was on supplementing detailed survey information and data. Aeration basin phosphate and ammonia profiles were measured and preliminary phosphorus uptake and release tests were conducted. Results of the Phase 1 uptake and release tests are not be presented in this report because test methodology was modified for Phase 2 to permit better interpretation of the results. Wastewater characteristic samples were collected for one day during Phase 1. During Phase 2, data collection for each period extended over two weeks and generally included collection of six samples for each water quality parameter.

During the intensive sampling periods the plant staff was asked to take additional samples that were either not taken or only taken infrequently as part of the routine sampling plan. The sampling plan included influent VFA, $BOD_5$, COD, readily biodegradable COD as measured by the flocculated filtered COD test (FFCOD), and soluble $BOD_5$ ($sBOD_5$) as well as phosphorus and nitrogen fractions. Influent, effluent, and digested biosolids phosphorus and metals samples were included in the sampling plan to allow performance of phosphorus and metals mass balances.

The results of special data collection efforts at Nansemond are included in the following data sets:

- **Phase 1 data set** This data set was collected during September 2002.
- **Phase 2A data set** This data set was collected during the first intensive sampling period in June 2003.
- **Phase 2B data set** This data set was collected during the second intensive sampling period in July 2003.
- **Phase 2F data set** This data set was collected during the field visit in June 2003.
-

**Table E-3. Summary of Field Tests and Sampling.**

| Field Testing Segment | Month | Data Set ID | Purpose Conducted |
|---|---|---|---|
| Historical Data Review | January 2000 – August 2003 | 2000 - 2003 | Influent, Effluent Averages<br>Process Parameter Averages |
| Phase 1 – Field Tests | September 2002 (One day) | Phase 1 | Aeration Basin Profiles<br>Uptake and Release Test<br>Operator Interviews |
| Phase 2 – Field Tests | June 2003 | Phase 2F | Uptake and Release Test<br>Operator Interviews<br>Data Collection |
| Phase 2 – Intensive Sampling Period 1 | June 2003<br>Two weeks | Phase 2A | Influent Characterization<br>Aeration Basin Profiles<br>Phosphorus Mass Balance<br>Metal Mass Balance |
| Phase 2 – Intensive Sampling Period 2 | July 2003<br>Two weeks | Phase 2B | Influent Characterization<br>Aeration Basin Profiles<br>Phosphorus Mass Balance<br>Metal Mass Balance |

## E.3.2 Overall Process Characterization and Performance

This section of the report summarizes the aeration basin influent characteristics and secondary treatment process performance based on routine data gathering and additional data collection described in Section 3.1. It also discusses relationships between primary effluent loading and process performance.

### E.3.2.1 Aeration Basin Influent Characterization

Influent flows at Nansemond have returned to 2000 levels after a steady decline between 2000 and 2001. Average influent flow for the past three years was 20.1 mgd. During Phase 1 testing, average flow was slightly lower at 18.8 mgd (Figure E-4). Diurnal flow has a typical pattern for this size service area, with peaking factors for minimum flow of 0.54 and maximum flow of 1.35 during dry weather conditions. Wastewater temperatures range between 12°C and 30°C.

The aeration basin influent (ABI) at Nansemond has some unique characteristics, which in part originate within the facility. Primary effluent ammonia load (Figure E-5) has increased 12% in the last three years, starting in April 2002 and increasing through September 2002. Primary effluent total phosphorus has not changed since 2000 and the same is true for primary effluent phosphate.

The most dramatic change in aeration basin influent involves primary effluent BOD and soluble BOD (Figure E-6) load. BOD decreased 33 percent (over 12,000 lb/d) from 2002 to 2003. Similar to ammonia, change in loads has occurred since September 2002, but the decrease in BOD was much more pronounced than the increase in ammonia. Primary effluent soluble BOD follows this trend, but not to the same extent of total BOD. Industrial dischargers (i.e. meat processing plants) contribute an unknown amount of COD to the wastewater. This industrial discharge is believed to cause sporadic temporary influent BOD load changes noted in the data. All measured influent parameters, including flow, experienced significant but temporary reductions in August 2001. The reason for this decline is not known.

Table E-5 summarizes aeration basins influent parameters for all testing phases. The average primary influent BOD/TP ratio is 14 mg BOD/mg P over the three year period, but due to recent primary effluent BOD reductions, the ratio has dropped to 9.8 mg BOD/mg P. Ratios considered favorable for EBPR are 20 mg BOD/mg P or higher. The VFA to P ratio was measured at 1.18 mg VFA /mg P. A ratio of 6 – 8 mg VFA /mg P would be considered favorable for EBPR. The raw influent total phosphorus is on the high end of municipal wastewater phosphorus concentrations, and may be caused by industrial discharges.

**Table E-4. Aeration Basin Influent Wastewater Characterization.**

| ABI Wastewater Parameter | Unit | 2000 – 2003[1] | Phase 1[2] | Phase 2A | Phase 2B |
|---|---|---|---|---|---|
| $BOD_5$ | mg/L | 200 | 350 | 81 | 150 |
| $SBOD_5$ | mg/L | 88 | 39 | 27 | 31 |
| COD | mg/L | - | 570 | 290 | 410 |
| ffCOD | mg/L | - | 70 | 76 | 120 |
| VFA | mg/L | 18 | 20 | 14 | 48 |
| TP | mg/L | 14 | 20 | 12 | 12 |
| $PO_4$-P | mg P/L | 12 | 12 | 8.4 | 9.0 |
| Total Alkalinity (as $CaCO_3$) | mg/L | 310 | - | 270 | 300 |
| pH | | - | 7.0 | 7.0 | 7.1 |
| TKN | mg N/L | - | 62 | 42 | 49 |
| $NH_4$-N | mg N/L | 35 | 40 | 38 | 39 |
| TSS | mg/L | 150 | 470 | 79 | 140 |
| Flow | MGD | 20 | 20 | 23 | 24 |
| Temperature | °C | 21 | 26 | 22 | 26 |
| VFA : P Ratio | - | 1.3 | 0.94 | 1.2 | 2.6 |
| ffCOD : P Ratio | - | - | 3.4 | 6.4 | 10 |
| COD : P Ratio | - | - | 27 | 24 | 35 |
| BOD: P Ratio | | 14 | 17 | 6.8 | 13 |
| ffCOD : COD Ratio | - | - | 0.12 | 0.26 | 0.30 |
| VFA : ffCOD Ratio | - | - | 0.28 | 0.18 | 0.25 |
| Period | - | 2000 – 2003 | 9/5/02 | 6/1/03 – 6/14/03 | 7/13/03 – 7/27/03 |

[1]VFA data were only collected since January 2003.

[2] Single grab sample.

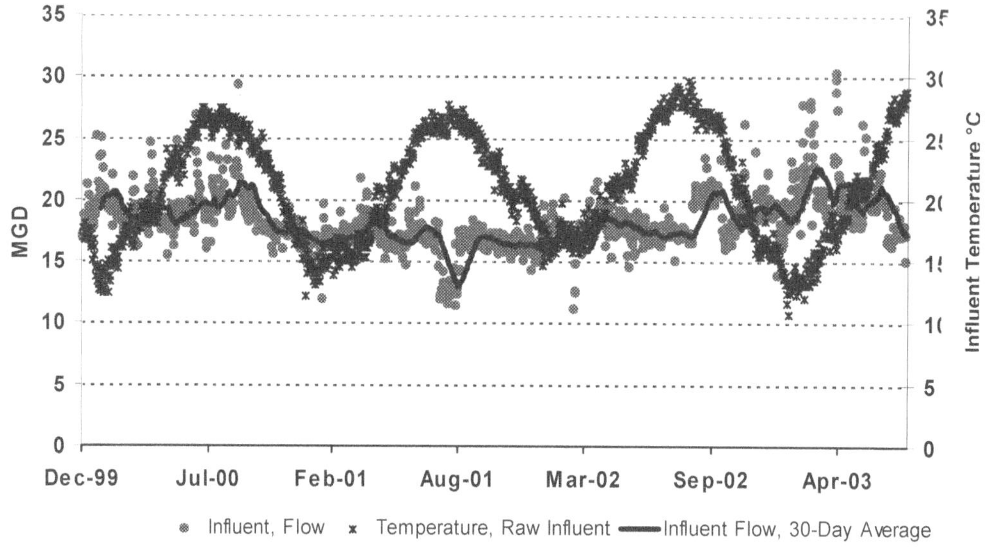


**Figure E-4. Influent Flow, Influent Temperature, and 30-Day average Influent Flow Between 2000 – 2003.**

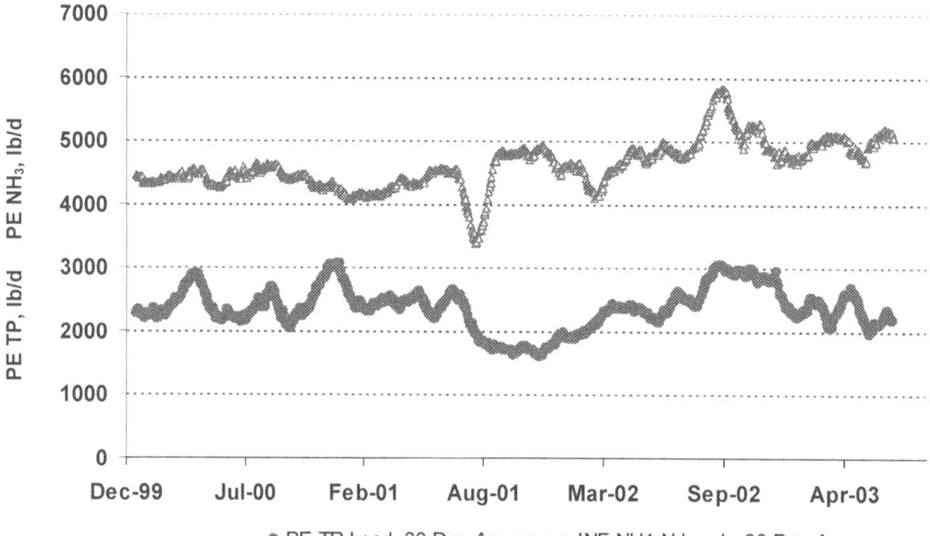


**Figure E-5. Primary Effluent Phosphorus and Ammonia Load.**


WERF

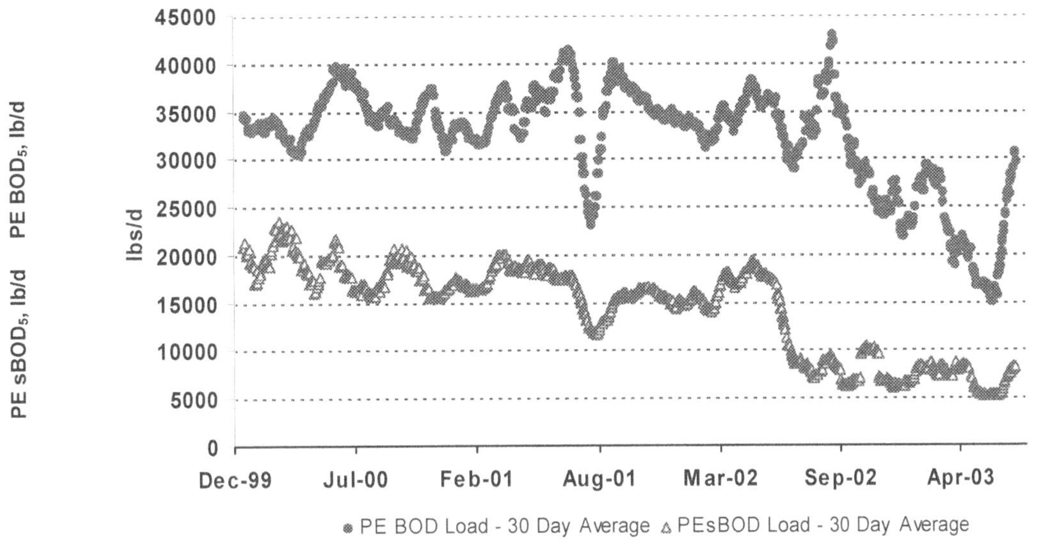


**Figure E-6. Primary Effluent $BOD_5$ and soluble $BOD_5$ Load.**

The aeration basin influent shows large fluctuations in flow and concentration. Concentrations of aeration basin influent BOD, TSS, and nutrients can vary by –87% to +300% (Table E-5). Even the 30-day average loads show large fluctuations, indicating that daily fluctuations can be significant.

**Table E-5. Selected Aeration Basin Influent (ABI) Parameters from 2000 – 2003.**

| Parameter | Unit | Min | Ave | Max |
|---|---|---|---|---|
| ABI Flow | MGD | 13 | 20 | 33 |
| ABI Temperature | °C | 11 | 21 | 30 |
| ABI BOD | mg/L | 58 | 200 | 410 |
| ABI sBOD | mg/L | 17 | 88 | 240 |
| ABI TSS | mg/L | 21 | 150 | 750 |
| ABI $NH_4$-N | mg/L | 19 | 35 | 50 |
| ABI TP | mg/L | 7.0 | 14 | 23 |
| ABI $PO_4$-P | mg/L | 1.5 | 12 | 20 |
| ABI VFA[1] | mg/L | 6.0 | 18 | 48 |

## E.3.2.2 EBPR Performance

Table E-6 summarizes effluent water quality measurements and process parameters for the four data sets. Of particular interest are the anoxic and aerobic hydraulic retention times. The average anoxic HRT is nearly three hours. In contrast, aerobic HRT is relatively short at 3.8 hours. Nitrate concentration in the first anoxic zone was measured at 0.5 mg/L during Phase 2 testing at a time when the plant was nitrifying. Low nitrate concentrations can be attributed to high dilution rates through internal recycle and to the fact that the first anoxic zone already has a hydraulic retention time of 45 minutes (25% of the total anoxic time in Table E-6). The net result

[1] VFA data were only collected since January 2003.

of these concentrations is that the anoxic zone can become completely depleted of nitrate, creating an extended anaerobic zone at the end of the anoxic zone. No correlation or cause-and-effect relationship between effluent quality and hydraulic retention has been found.

The plant also carries mixed liquor with a low volatile solids percentage of 68%. This could be attributed to industrial discharge in the Nansemond service area. The plant rarely experiences problems with filamentous growth and consistently shows low SVIs, on average 82 mL/g. The latter is reflected in the average effluent TSS of 7.7 mg/L (Table E-6).

EBPR is ultimately achieved at Nansemond, but its magnitude is difficult to ascertain. Ferric chloride is added to the primary clarifiers on a routine basis to aid in solids processing, and to secondary clarifiers on an as needed basis for enhanced phosphorus removal. Because plant data does not distinguish between these uses, it is difficult to determine EBPR contribution to total phosphorus removal at the plant. The presence of EBPR however has been proven through uptake and release tests, aeration basin profiles, and microscopic evaluation of mixed liquor (see discussions in later sections of this report).

**Table E-6. Process Control and Water Quality Measurement Results During Intensive Testing (two-week average).**

| Raw Wastewater Parameter | Unit | 2000 – 2003 | Phase 1 | Phase 2A | Phase 2B |
|---|---|---|---|---|---|
| ABE $PO_4$-P | mg P/L | - | - | 0.2 | - |
| SE $PO_4$-P | mg P/L | - | 0.74 | 0.2 | 0.2 |
| SE TP | mg P/L | 1.2 | 1.0 | 0.5 | 0.6 |
| SE $NH_4$-N | mg N/L | 12 | 1.4 | 12 | 6.5 |
| SE $NO_3$-N | mg N/L | 6.4 | 10 | 6.4 | 12 |
| SE TN | mg N/L | 19 | 12 | 19 | 18 |
| SE BOD | mg/L | 6.1 | 3.5 | 7.9 | 6.0 |
| SE TSS | mg/L | 7.7 | 4.8 | 8.0 | 6.3 |
| SE Total Alk (as $CaCO_3$) | mg/L | 170 | - | - | 140 |
| Secondary P-Removal | % | 87 | 88 | 98 | 95 |
| Secondary N-Removal | % | 57 | 74 | 45 | 74 |
| RAS $PO_4$-P | mg/L | - | - | 1.4 | 8.3 |
| RAS Nitrate | mg/L | - | - | 1.3 | 1.3 |
| RAS DO | mg/L | - | - | 0.25 | 0.4 |
| SRT | Days | 8.3 | 6.5 | 6.7 | 8.7 |
| RAS Rate | % Q | 94 | 72 | 79 | 56 |
| ML Recycle to ANR | % Q | 110 | 91 | 86 | 89 |
| ML Recycle to ANX | % Q | 100 | 92 | 95 | 83 |
| Plant Flow (Q) | MGD | 18 | 21 | 20 | 24 |
| ANR HRT | Hours | 0.8 | 0.6 | 1.3 | 0.4 |
| ANX HRT (4 tanks) | Hours | 2.9 | 2.3 | 2.1 | 1.7 |
| AER HRT | Hours | 3.8 | 4.6 | 2.9 | 4.0 |
| MLSS | mg/L | 3520 | 3450 | 3990 | 4310 |
| MLVSS | mg/L | 2420 | 2450 | 2710 | 3060 |
| MLSS/MLVSS | % | 68 | 71 | 68 | 71 |
| SVI | ML/g | 82 | - | - | - |

Note: No chemical addition to secondary process during Phase 2A and 2B testing.

#### *E.3.2.2.1 Phase 1 Testing*

During **Phase 1** field tests, the Nansemond plant performed well with respect to both nitrogen and phosphorus removal. Effluent ammonia concentrations were less than 2 mg/L, total nitrogen less then 12 mg/L, and total phosphorus was less then 1.5 mg/L Figure E-7 shows effluent total phosphorus concentration prior to and during Phase 1 testing. Aeration basin influent BOD data are also shown. BOD levels do not show any appreciable variation in this period, but EBPR appears to go through an upset before settling down at approximately 1 mg/L. Centrifuge dewatering operated until about 10 a.m. In the past ferric chloride was added to the process at the secondary clarifier influent, but was discontinued about three weeks prior to the Phase 1 field test. Ferric was added to the primary clarifiers during field testing.

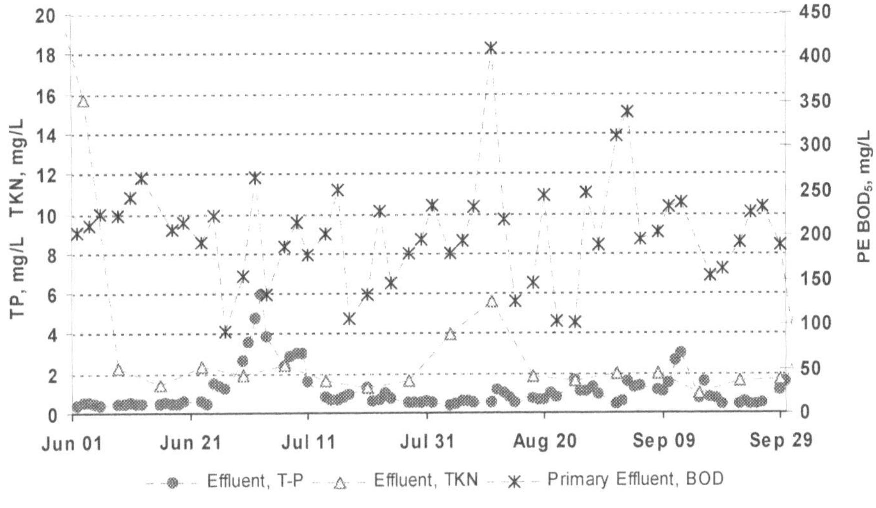


**Figure E-7. Effluent TP and TKN, and Primary Effluent BOD Prior and During Phase 1 Testing (2002).**

In 2002 the plant's only operational change was to reduce the number of anaerobic/anoxic trains from five to four in order to adjust basin volumes for lower flows. Prior to this change, low influent flow resulted in higher detention times in the anaerobic zones (between two and three hours). While no direct negative impact was observed due to increased retention times, operational strategy intends to keep retention times within plus or minus fifty percent of the design value (see Table E-1).

#### *E.3.2.2.2 Phase 2 Testing*

Figure E-8 shows primary effluent $BOD_5$ and effluent total phosphorus levels prior to and during Phase 2 testing. During Phase 2A testing (June 1$^{st}$, 2003) the plant was recovering from a recent process upset; nitrification was only occurring marginally. Ferric chloride was added to the primary clarifiers (50 lb/d, or 0.29 mg/L) only. Despite poor influent nutrient ratios, the plant delivered good EBPR results.

With an aeration basin influent VFA:P ratio of only 1.2, the phosphorus removal performance is impressive despite conventional assumptions of minimum VFA or BOD requirements. It is important to keep in mind nitrification was essentially absent in Phase 2A.

The phosphate profile presented in the next section shows that the absence of nitrate extends the anaerobic phase, allowing phosphorus release to plateau with phosphorus uptake occurring only in the aerobic zones.

In the period between Phase 2A and 2B testing, the plant started to nitrify. Figure E-8 shows that effluent TKN levels dropped in the beginning of July. This transition was accompanied by loss of EBPR. However the EBPR loss was likely unrelated to the changes in nitrification performance but rather to a shift in influent BOD supply, which is discussed in more detail in the following section. No ferric chloride was added during this time.

One month later during Phase 2B testing (in the last two weeks of July), the plant was performing well again. Anaerobic and anoxic hydraulic retention times were significantly reduced while additional aerobic retention time was increased. With 83% mixed liquor and 56% RAS recycle in the system, the plant removed 73% total nitrogen—slightly better than the predicted 58% at 139% total recycle. There was a slight reduction in the rate of phosphorus removal compared with test results during Phase 2A. Phosphate profiles also show that uptake starts in the anoxic zones (see discussion in next section). One noteworthy observation during Phase 2B testing was the large increase in RAS phosphate (1.4 mg/L to 8.3 mg/L), which can be attributed to reduced RAS rates with some secondary release.

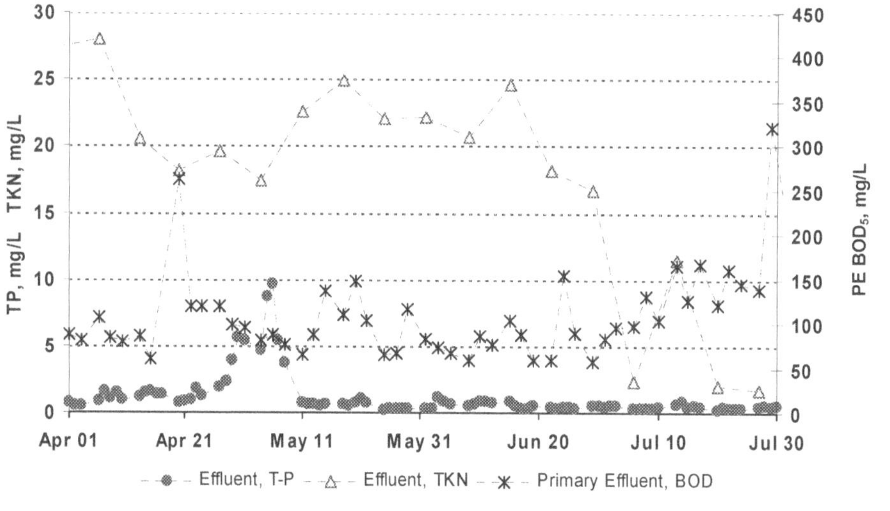


**Figure E-8. Effluent TP and TKN, and Primary Effluent BOD During Phase 2 Testing (2003).**

### E3.2.3 Relationship Between Performance and Process Loading

Traditional analysis of the primary effluent wastewater characteristics would predict marginal EBPR performance. Figure E-9 shows effluent total phosphorus and TSS concentrations between 2000 and 2003. As described earlier, the multiple chemical addition points and objectives make it difficult to assess how much phosphorus removal is due to EBPR and how much due to chemical precipitation. During Phases 1 and 2 intensive sampling periods, no chemical additives were used in the secondary process.

Figure E-10 shows effluent phosphorus probability distribution for all measured days between 2000 and 2003. The average effluent is 0.79 mg/L total phosphorus and the spread in effluent concentrations is quite high. The effluent total phosphorus target of 2 mg/L is met 85% of the time at the plant; 1 mg/L or less was measured 64% of the time, and 24% of all effluent samples were under 0.5 mg/L. When comparing Nansemond effluent phosphorus with other facilities, it is important to remember effluent target concentration is 2 mg/L, and that the average primary effluent total phosphorus concentration is 14.2 mg P/L—unusually high for a municipal treatment plant.

Primary effluent VFA concentration is only 15-20 mg/L, which is very low to support EBPR. In addition, primary effluent phosphorus concentration is very high (7-22.6 mg/L total phosphorus), which makes it very difficult to sustain EBPR over time in this wastewater. VFA concentration decreases significantly between primary influent and primary effluent (Figure E-11). Reasons for the low VFA can only be speculated upon at this point. One potential explanation could be the discharge of active biomass from an industrial pretreatment lagoon that supports biological uptake in sewers, or the discharge of nitrate from this pretreatment facility, which would cause denitrification in the sewers with VFA uptake.

Primary clarifier $BOD_5$ removal has improved significantly since September, 2002, reducing primary effluent BOD by about 40% (Figure E-12). The cause for this improvement is not known. Speculation points to changes in the nature of the industrial discharge in the system.

The plant has been able to deliver good effluent total phosphorus, presumably in part due to EBPR despite such low $BOD_5$/TP ratios. (See discussion of intensive testing periods to follow). EBPR performance at Nansemond has been inconsistent and shows periodic effluent phosphorus spikes along with complete loss of EBPR at times (Figure E-9). Operators suspect the plant is VFA deficient and the occasional complete loss of EBPR is due to mixed liquor poisoning. Currently HRSD is conducting an investigation to determine the type and origin of such potential toxic industrial discharge. Operators have also reported high degrees of biological activity in the sewer system in the past, resulting in air locking of forcemains and requiring installation of gas vents.

The *variability* of primary effluent $BOD_5$ load seems to trigger EBPR upsets periodically, or they consistently appear to coincide with other primary effluent composition changes. Figure E-15 and Figure E-16 show the relationships between effluent phosphorus and primary effluent $BOD_5$. Data clearly show that at almost every occasion when primary effluent $BOD_5$ or the primary effluent $BOD_5$/TP ratio decreases, effluent phosphorus increases. The data also suggest that the decrease in $BOD_5$ load triggers loss in EBPR performance rather than loss of the particular $BOD_5$ load.

Absolute values of the $BOD_5$/P ratio or $BOD_5$ loads do not correlate with phosphorus removal. For example, Figure E-16 shows that in July 2000, effluent total phosphorus was below 1 mg/L when the $BOD_5$/P ratio was around 15 mg $BOD_5$/mg P. However, in March 2003, effluent total phosphorus is again at 1 mg/L while $BOD_5$/P ratio is below 10 mg $BOD_5$/mg P. Even at the highest $BOD_5$/P ratio in August-September 2001, the effluent total phosphorus remains highly variable. The figure shows, however, that *changes* in the BOD/TP ratio result in a change in total phosphorus.

In addition to high raw influent total phosphorus averaging 9.5 mg/L, recycled phosphorus from solids processing increases phosphorus levels by more than 50%, ultimately producing 14.1 mg/L primary effluent total phosphorus (Figure E-14). Dewatering recycled ammonia (Figure E-13) contributed, on average, 15% to aeration basin ammonia loads over the last three years. Nitrification performance changes seasonally (Figure E-17). During the winter months, little to no nitrification occurs. During summer ammonia and total nitrogen concentration levels are 2mg/L and less than 10 mg/L respectively.

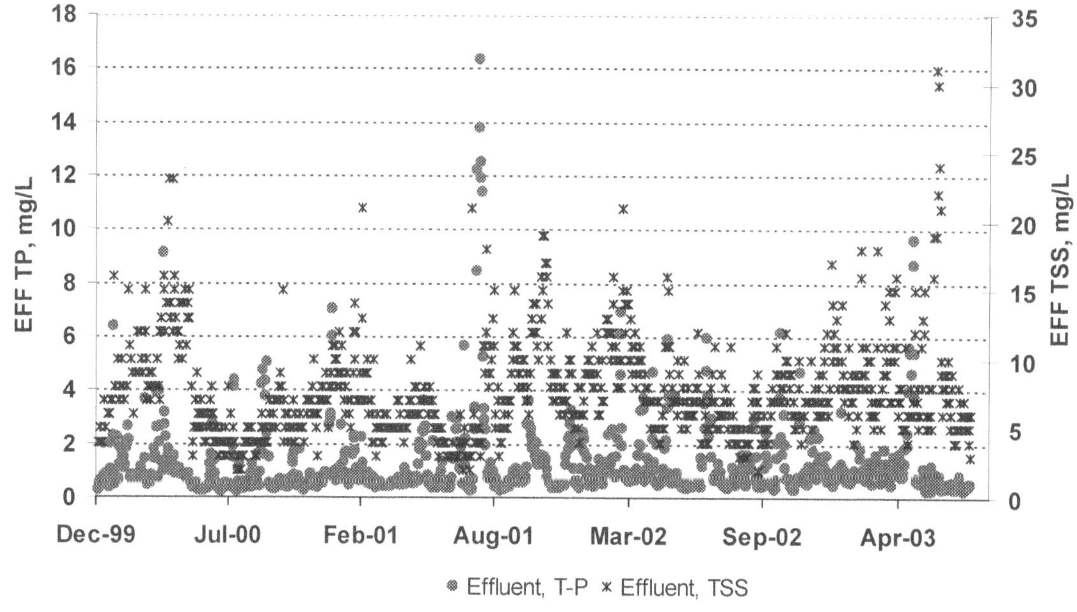


**Figure E-9. Effluent Total Phosphorus and Total Suspended Solids.**

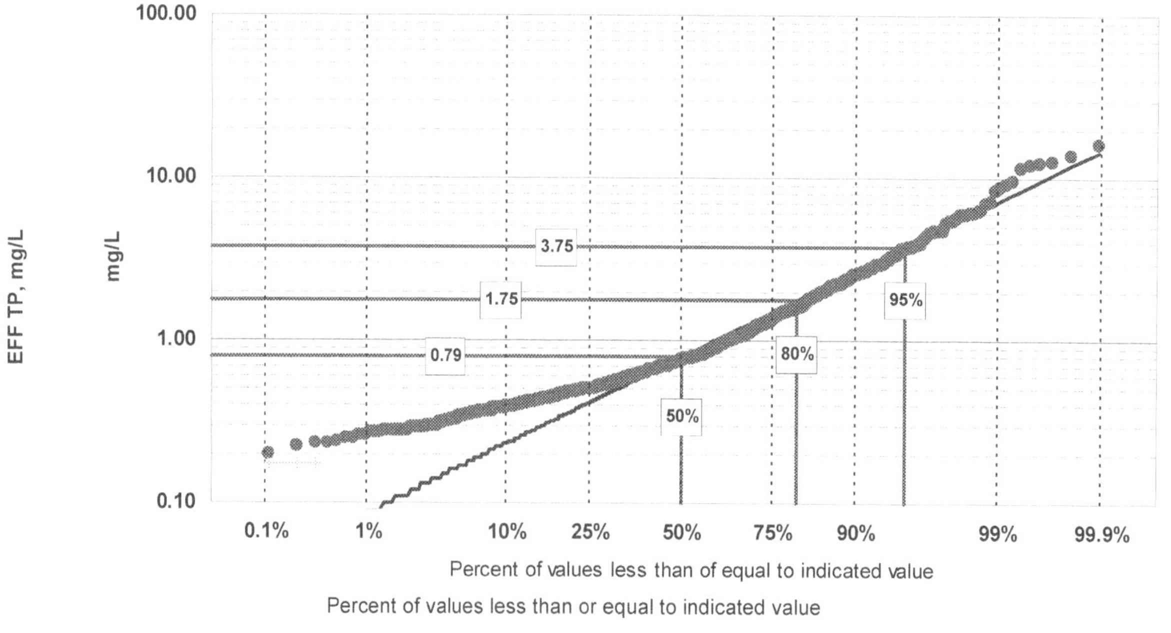


**Figure E-10. Effluent Total Phosphorus Probability Distribution.**



WERF

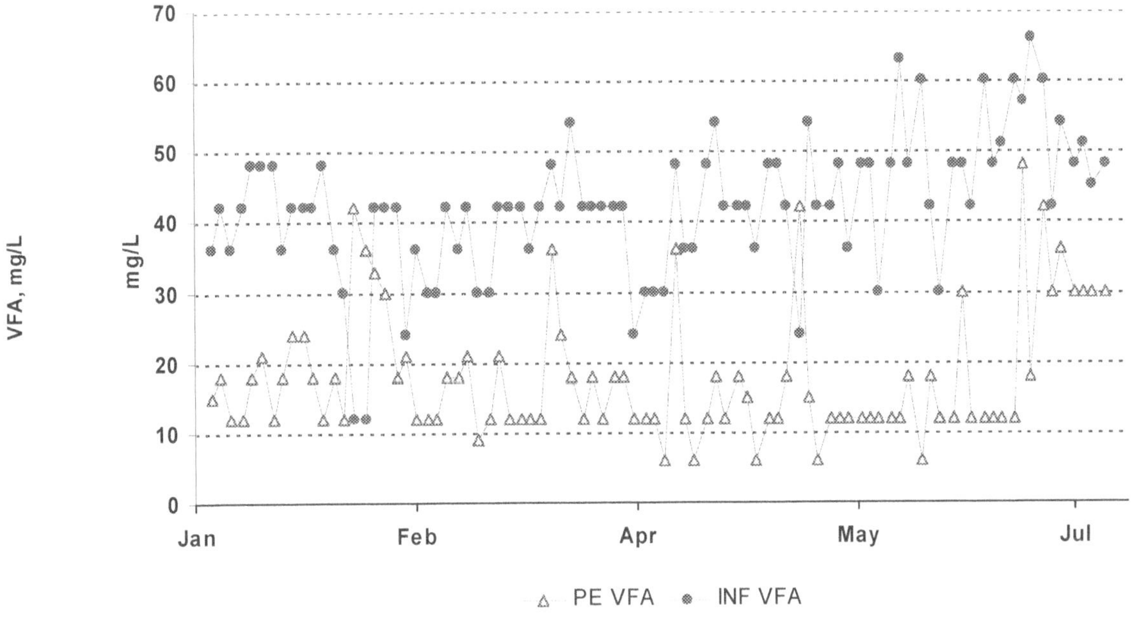


**Figure E-11. Influent and Primary Clarifier Effluent VFA During 2003.**

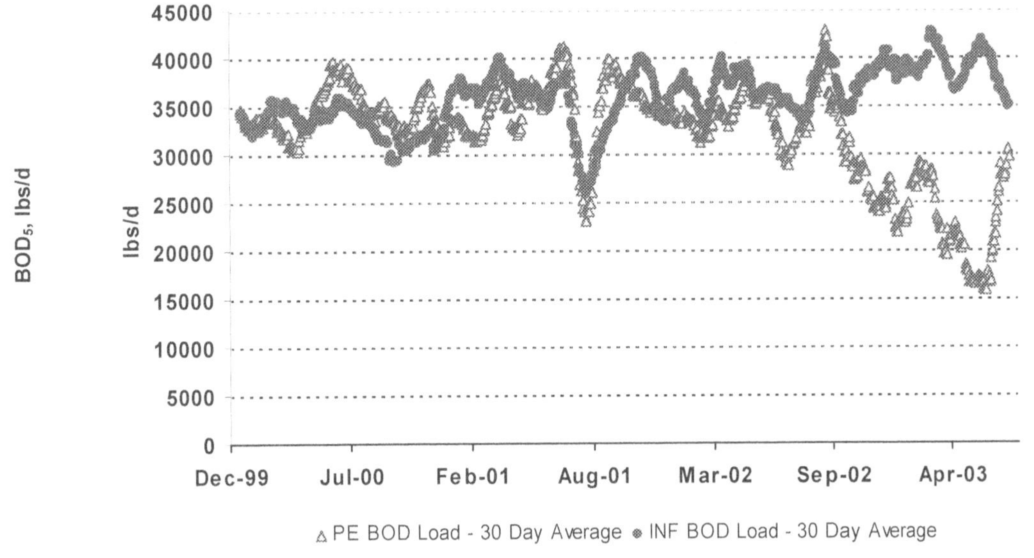


**Figure E-12. 30-day Averages for Raw Influent and Primary Effluent BOD Between 2000 and 2003.**

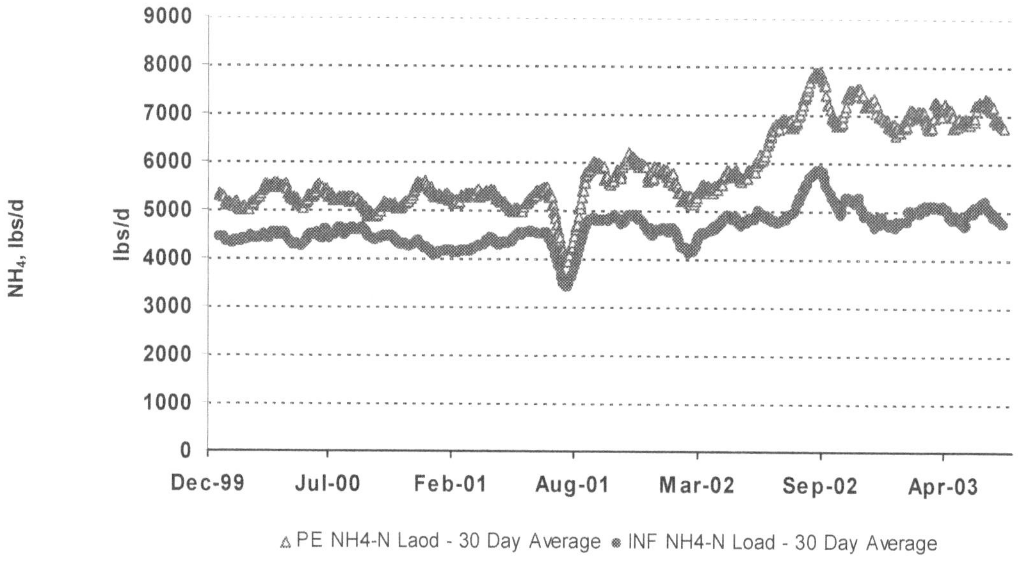


**Figure E-13. 30-day Averages of Raw Influent and Primary Effluent Ammonia from Between 2000 and 2003.**

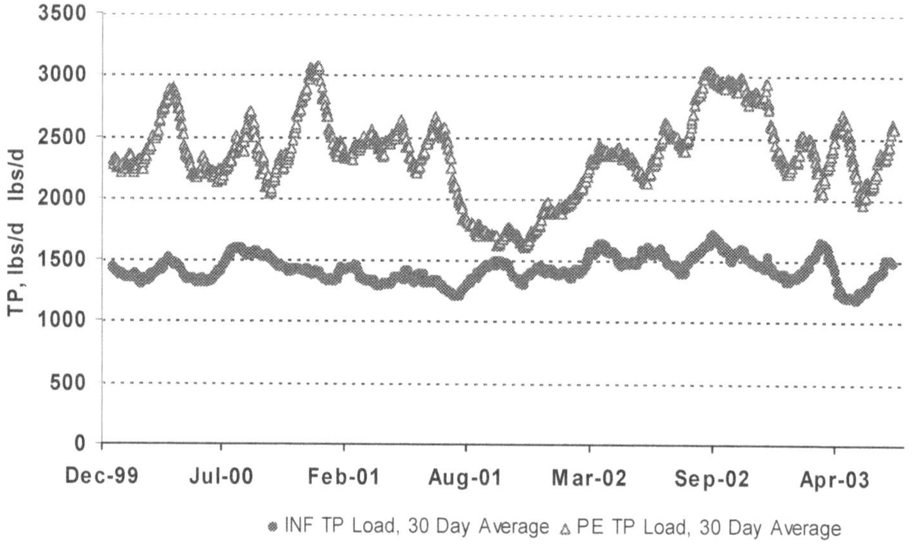


**Figure E-14. Influent and Primary Effluent Total Phosphorus Load.**

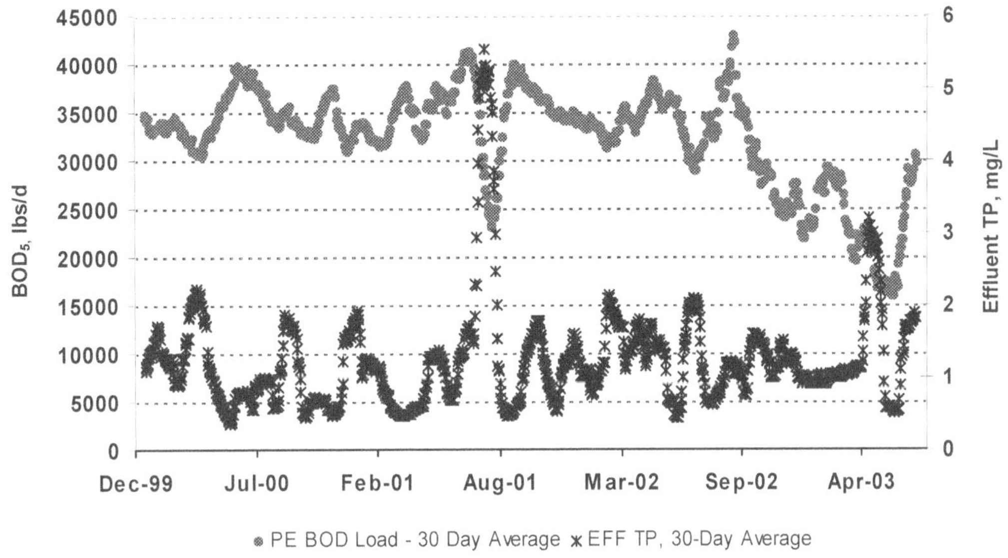


**Figure E-15. 30-day Averages of PE BOD and Effluent TP.**

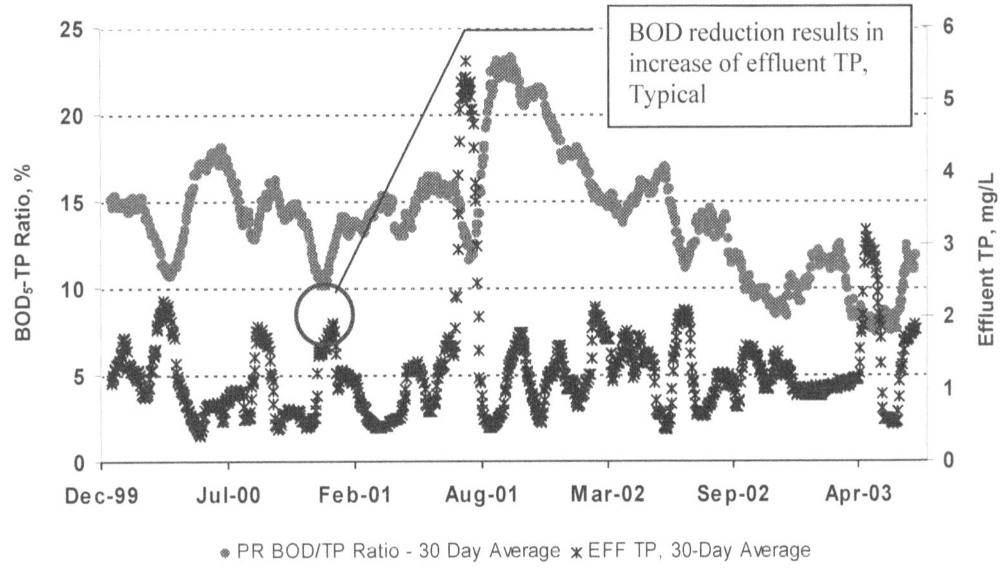


**Figure E-16. 30-day Averages of Primary Effluent BOD/TP Ratio and Effluent Total Phosphorus.**

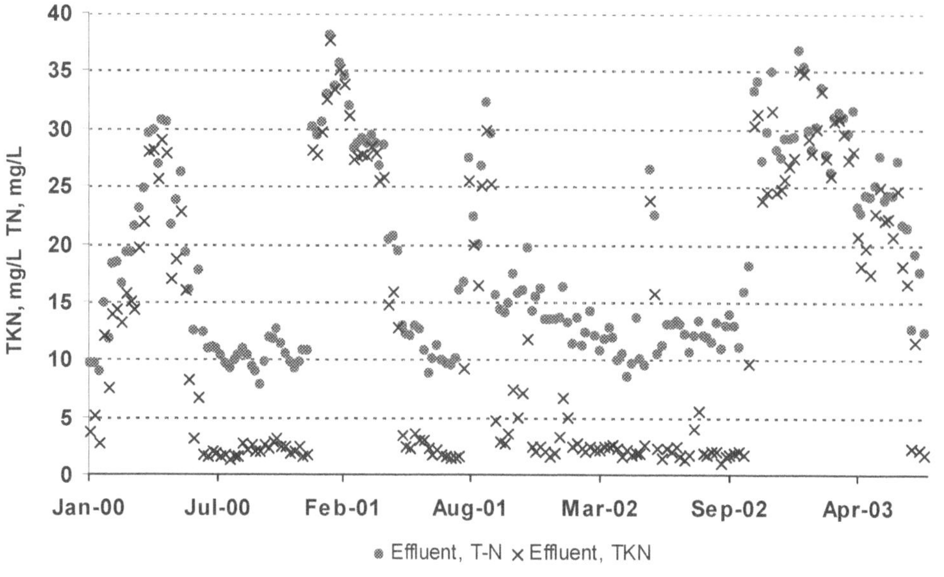


**Figure E-17. Effluent Total Nitrogen and TKN Concentrations.**

## E.3.3 Additional EBPR Process Evaluation

### E.3.3.1 Aeration Basin Phosphate Profiles

The aeration basin phosphate profile was measured during each sampling phase of the project. Charts of the profiles (Figure E-19, Figure E-22, and Figure E-23) show four lines:

- The individual daily profiles (thin shaded gray lines in background with diamond markers).
- Average phosphate profiles calculated from the daily profiles (thick line with circular markers).
- Flow corrected average phosphate profile (thick line with triangular markers). The flow corrected concentration is calculated as the theoretical phosphorus concentration based on the influent flow. This allows for a direct comparison of the concentration values without the influences of dilution from recycle flows.
- Phosphorus release rates (columns). The release rates (lb/d P / 1000 lb VSS) are calculated as daily phosphorus release compared to the aeration basin volatile solids content.

In order to understand the phosphate profile, the flow scheme must first be investigated. Flows are relatively complicated during VIP operation since three recycle streams must be accounted for (RAS, ANR, and ANX recycle). Figure E-18, Figure E-20, and Figure E-21 illustrate the internal recycle flows for the three periods when aeration basin profiles were measured.

#### *E.3.3.1.1 Phase 1 Profile*

Figure E-19 shows phosphate profiles measured during Phase 1 field tests. Two profiles were taken on the same day (morning and afternoon). The flow corrected profile shows phosphorus release continues in the anoxic zone. Additional release did not appear to impact phosphorus uptake. Phosphorus release amounts to 39 lb $PO_4$-P per 1000 lb VSS. This includes release in the anoxic zone. One good indicator for the PAO population is the initial release in the first anaerobic cell where VFA is more readily available. With a retention time of 0.4 hours and a release of 24.6 lb $PO_4$-P per 1000 lb VSS, the initial phosphorus release rate is 61.5 lb $PO_4$-P per 1000 lb VSS per hour.

The extent of phosphorus release is very surprising in light of low primary effluent VFA (17.9 mg/L). The 37.3 mg/L phosphorus release requires at minimum the equivalent of 74.6 mg/L aeration basin influent VFA. Assuming there was no VFA consumption by other organisms in the system, this suggests a minimum of 55 mg/L of VFA were generated in the anaerobic zone for consumption by PAOs.

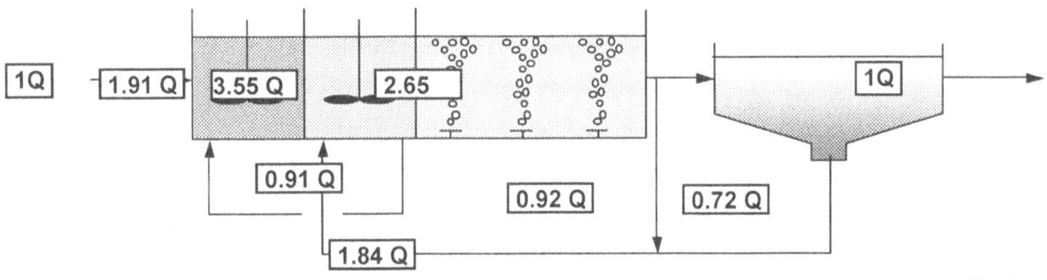


Figure E-18. RAS and Internal Recycle Flows at Nansemond during the Phase 1 Field Testing.

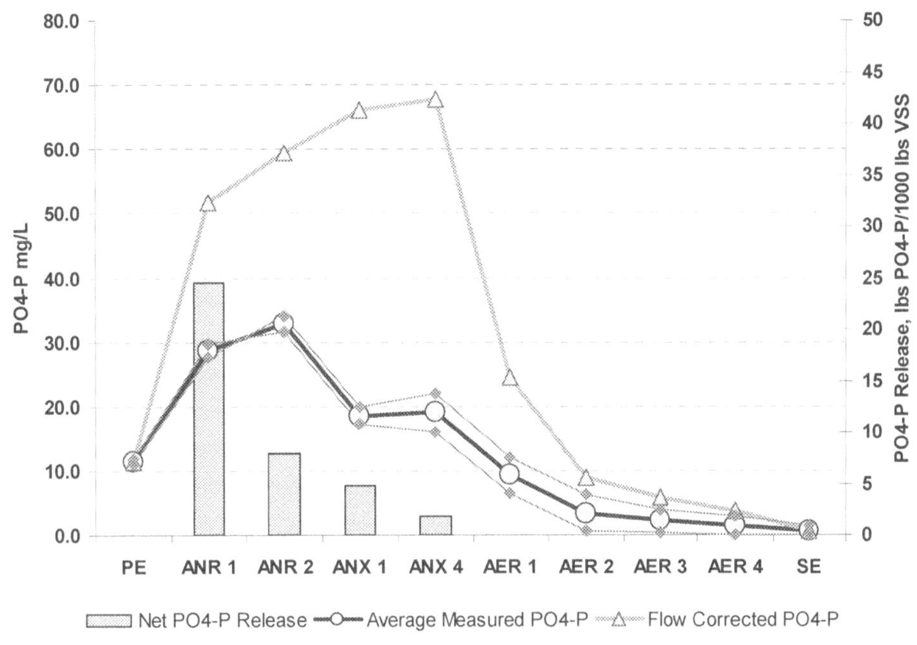


Figure E-19. Phosphate Profiles During Phase 1.

#### *E.3.3.1.2 Phase 2 Profile*

During Phase 2A testing the Nansemond plant operated under upset conditions (Figure E-22). Total phosphorus release was only 22.6 lb $PO_4$-P per 1000 lb AB VSS with an initial phosphorus release rate of 19.0 lb $PO_4$-P per 1000 lb AB VSS per hour.

Six weeks after Phase 2A testing, the process recovered from the upset, and the phosphorous release rate measured during Phase 2B intensive testing (Figure E-23) was three times higher than during Phase 2A testing (Figure E-22). Increased primary effluent VFA (now

48 mg/L) is reflected in the increased initial phosphorus release rates of 83 lb $PO_4$-P per 1000 lb AB VSS per hour). While primary effluent VFA increased, the majority (approximately 130 mg/L – assuming 0.5 mg P released/mg VFA consumed) of the VFA consumed during the phosphorus release was generated in the first anaerobic zone.

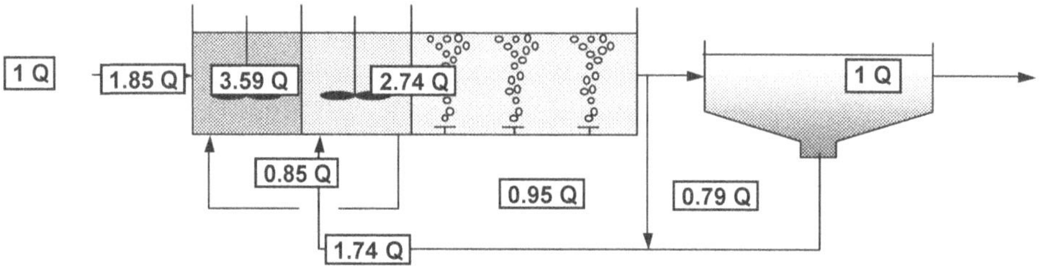


Figure E-20. RAS and Internal Recycle Flows at Nansemond During Phase 2A Testing.

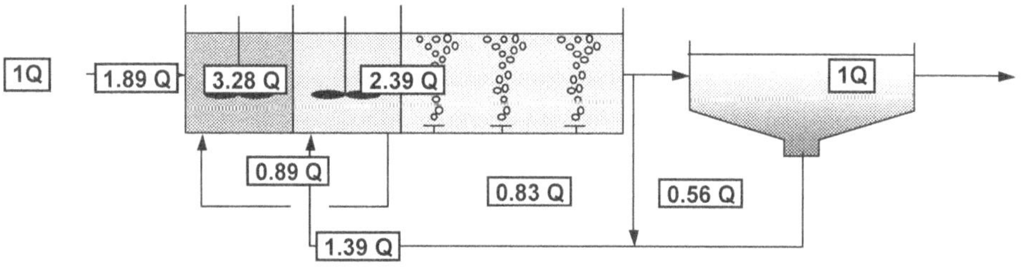


Figure E-21. RAS and Internal Recycle Flows at Nansemond During Phase 2B Testing.

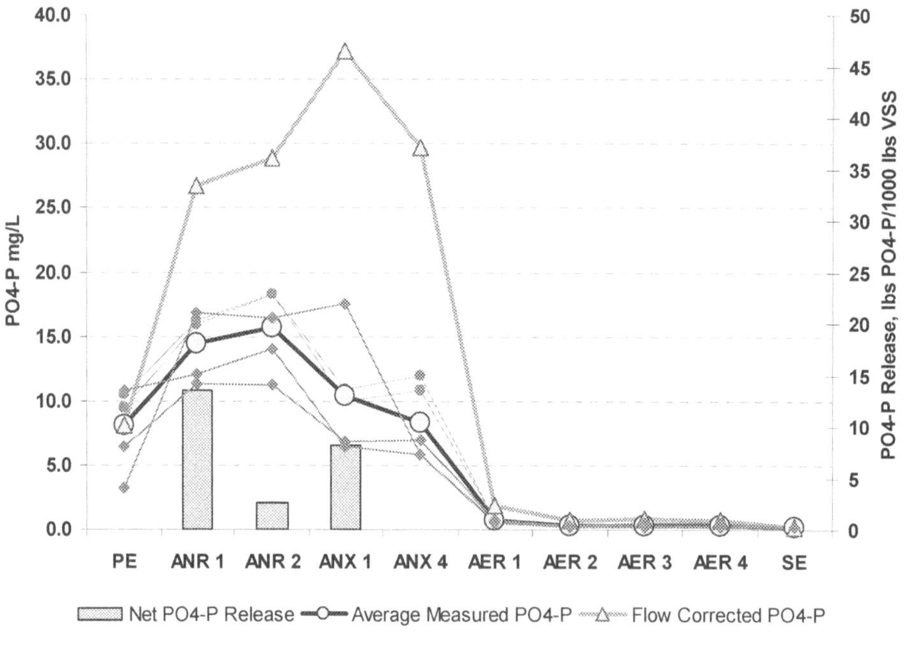


Figure E-22. Phosphate Profiles Measured During Phase 2A Field Testing.

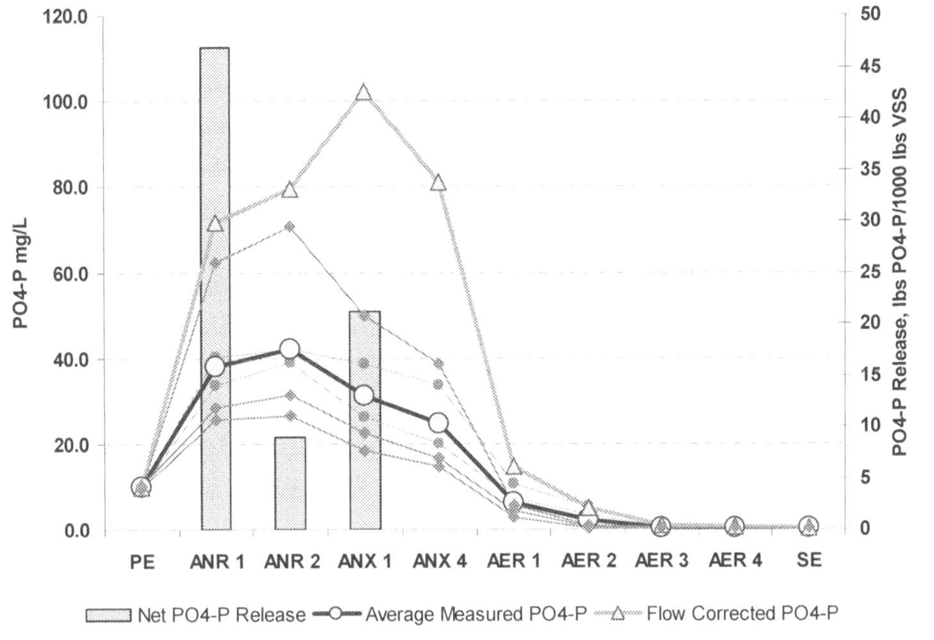


**Figure E-23. Phosphate Profiles Measured During Phase 2B Testing.**

**Table E-7. Process Control and Performance Measurement Results During Intensive Testing (two-week average, flow normalized).**

| Parameter | Unit | Phase 1 | Phase 2A | Phase 2B |
|---|---|---|---|---|
| Total Phosphorus Release | mg/L | 37 | 20 | 65 |
| Anoxic Phosphorus Uptake | mg/L | - | 7.4 | 21 |
| Phosphorus Release Rate | lb P per 1000 lb VSS per day | 39 | 25 | 77 |
| Phosphorus Uptake Rate | lb P per 1000 lb VSS per day | 55 | 39 | 95 |
| Total Aeration Basin Phosphorus Inventory | Lb P | 9,000 | 7,100 | 11,000 |
| Release of P Content | Lb/d | 6,400 | 3,400 | 13,000 |
| Initial Phosphorus Release Rate | lb/lb VSS/hr | 61.4 | 34 | 120 |

### E3.3.2 Uptake and Release Tests (URT)

URTs provide a quantification of the PAO population in the basin. By adding VFA to mixed liquor, the measured phosphorus release is limited by the PAO population in the basin and by the amount of stored phosphorus. URTs can be used for:

- Determining the amount of PAO activity from phosphorus release rate. The release rate (mg P/g VSS/h) is indicative of PAO activity.
- Determining the potential competition for VFA in the anaerobic zone. The phosphorus release ratio (mg P released/mg VFA removed) indicates PAO activity. A phosphorus release ratio of 0.5 mg P released/mg VFA removed indicates PAO

activity. Lower ratios are indicative of competition from other VFA consuming organisms such as GAOs.

- Determining key kinetic parameters for process simulation model calibration, such as ΔP/Acetic acid and magnesium and potassium uptake and release rates.
- Provides information regarding the phosphorus uptake kinetics, which represents the time required in the aerobic zone to remove the phosphorus.

Table E-8 summarizes URT results at Nansemond during Phase 2A (Figure E-24) and Phase 2B (Figure E-25) testing. The testing procedure is outlined in Appendix J. The test was conducted three days prior to the start of the intensive sampling and can be used to represent the same activated sludge captured by the intensive sampling. An additional test was completed on September 3, 2003 in conjunction with the GAO analysis described in Section 3.4.

Comparing URT results with the phosphorus release and uptake profiles measured during the intense sampling show the process was VFA-limited. A competing process consumed a fraction of the available VFA. URT also showed that when there is sufficient VFA, phosphorus release took place beyond rates measured in the field.

URT also shows a very low ΔP/acetic acid ratio of 0.2 mg P/mg VFA. This indicates that VFA are consumed by a competing process, potentially from denitrification or uptake by GAOs. In Nansemond's case, a third interpretation is possible: aeration basin phosphorus profiles show a significant amount of the consumed VFA is generated in the anaerobic zone. Being VFA-limited could also mean VFA generation in the anaerobic zone was inhibited.

**Table E-8. Uptake and Release Tests Results.**

| Test Parameter | Unit | Phase 2A | Phase 2B |
|---|---|---|---|
| Phosphorus Release Rate | mg P /g VSS /hr | 5.0 | 7.7 |
| Phosphorus Uptake Rate | mg P /g VSS /hr | 3.4 | 2.6 |
| Release Rate/Uptake Rate | -- | 1.5 | 3.0 |
| ΔP/Acetic acid | mg $PO_4$-P/ mg COD | 0.2 | 0.24 |
| ΔP/ΔMg - Release | mg $PO_4$-P / mg Mg | 26 | 2.8 |
| ΔP/ΔMg - Uptake | mg $PO_4$-P / mg Mg | 8.4 | 3.0 |
| MLVSS | mg/L | 2290 | 1740 |

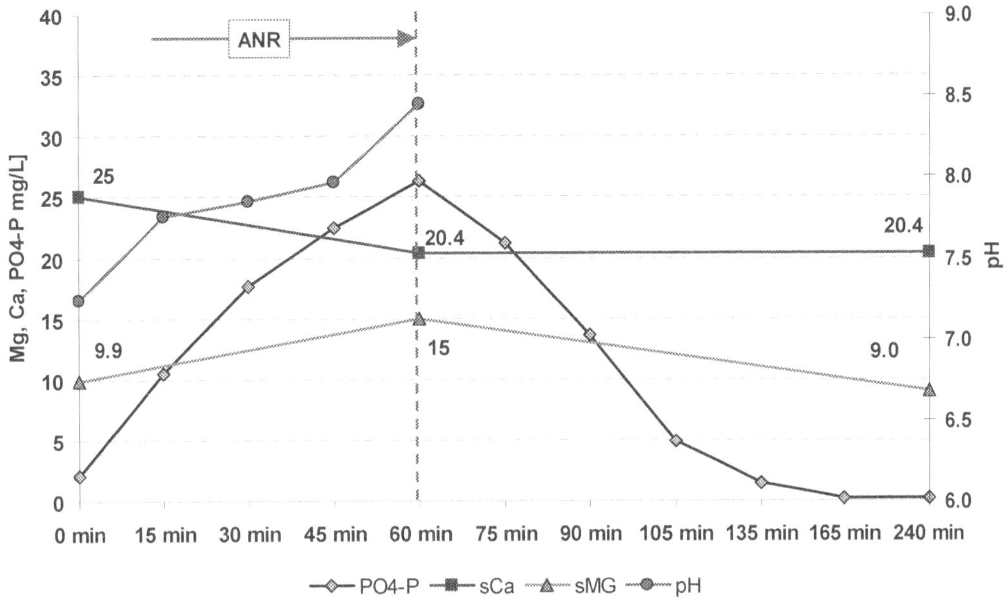


Figure E-24. Phosphorus Uptake and Release Test Results During Phase 2A.

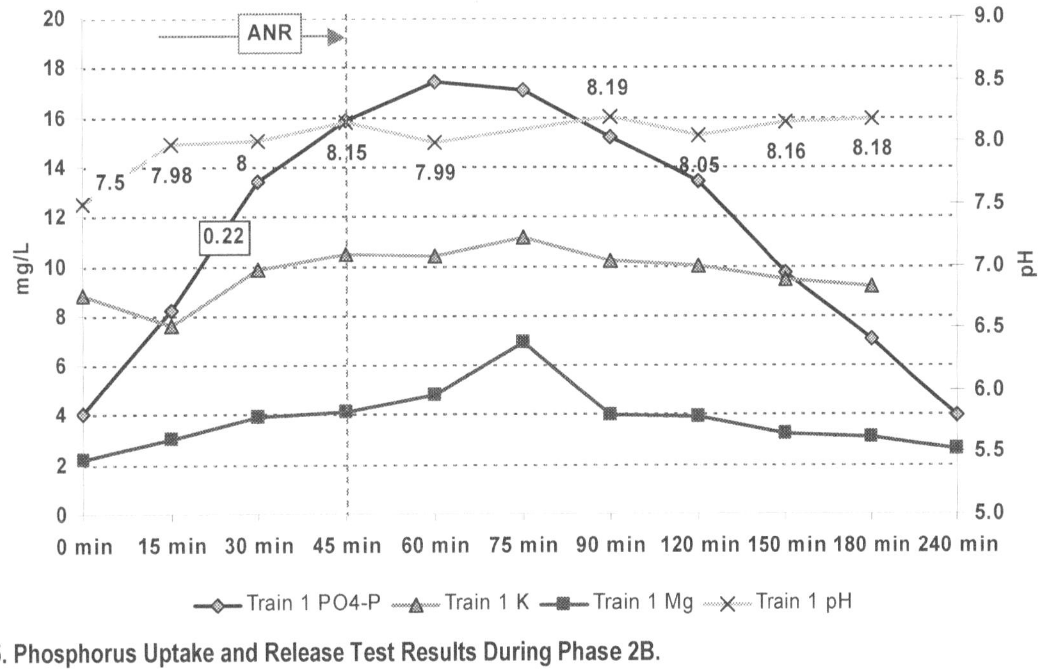


Figure E-25. Phosphorus Uptake and Release Test Results During Phase 2B.

### E3.3.3 Phosphorus and Metal Mass Balance

A phosphorus mass balance was constructed to determine the magnitude of recycle loads and their impact on the EBPR process. Values presented in the mass balance are either 2002 annual averages or grab sample measurements from the Phase 1 testing period. While the grab samples provide only a snapshot of this information, the samples are either WAS or digested biosolids. These samples (especially the biosolids samples) provide a good measurement due to the long digester retention times.

The most significant aspect of attaining phosphorus mass balance at Nansemond is the high recycle load (86%) from anaerobic digestion and dewatering. Figure E-26 shows phosphorus mass balance based on the 2000 average data. Even after primary clarification and removal of recycle particulate phosphorus, the primary effluent phosphorus is still just 60% higher than the raw influent. 70% of recycled phosphorus in the primary effluent is in the form of orthophosphophate.

Metals play an important role in the phosphorus mass balance. Digested biosolids have a phosphorus content of 5.1% (dry weight). The key metals in the digested biosolids (see Table E-9) are iron (9,900 mg/kg) and calcium (3,000 mg/kg) of wet cake. Phosphorous is present at 12,000 mg P/kg SS. Table E-9 and Table E-10 also show metal removal through uptake of precipitation during secondary treatment. Phosphorus and calcium levels significantly decrease between primary and secondary effluents; magnesium and potassium levels also decrease. When these compounds are expressed in equivalent fractions, calcium becomes the dominant cation removed across the process. Note that there is a discrepancy in the phosphorus balance – the WAB phosphorus content should be the same as the secondary removal, but the data shows a significant imbalance.

The metal mass balance across the anaerobic zone only shows a different picture. Table E-11 shows magnesium and potassium are released simultaneously with phosphorus in the anaerobic zone. Some calcium release was also detected. The total cation release (4,731 eq/sec) nearly matches the phosphate release (5,008 eq/sec).

Results of these four tests suggest calcium plays a very important role in phosphorus removal during the solids handling processes. Calcium phosphate precipitants could form in the digester (during release of phosphorus and metals in the digester) and potentially in the solids handling process. The anaerobic zone balance indicates that the key metabolic reaction appears to relies on magnesium and potassium as counter ions to the phosphate release.

It is important to note the data set is from very limited sampling.

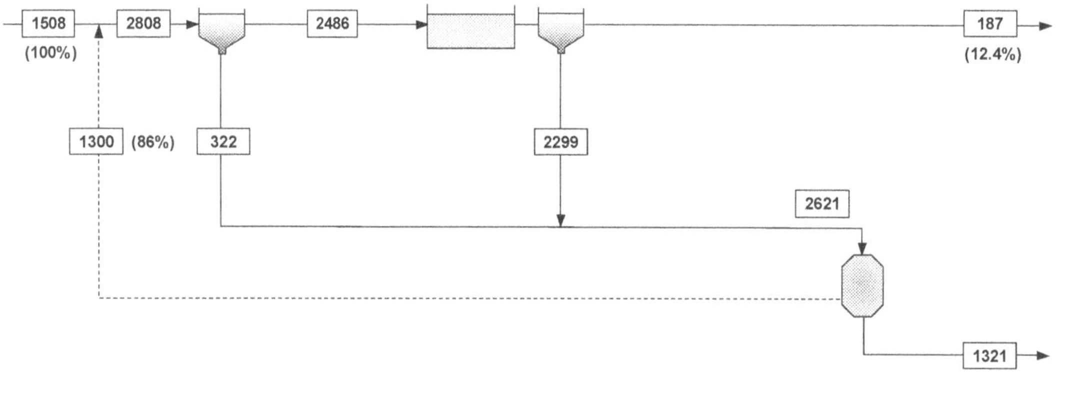


**Figure E-26. Phosphorus Mass Balance for Nansemond WWTP (2002 annual average).**

**Table E-9. Results of Phase 1 Phosphorus and Metals Testing from September 2002.**

| Analysis | Aeration Basin Influent mg/L | Anaerobic Zone[a] mg/L | Dewatered Solid Cake mg/kg[b] | Secondary Effluent mg/L |
|---|---|---|---|---|
| TP | | | 12,000 | |
| Mg | 7.2 | 16 | 2,300 | 4.7 |
| Ca | 28 | 17 | 3,300 | 17 |
| K | 25 | 40 | 900 | 22 |
| Fe | | | 9,900 | |
| Al | | | 2,200 | |

Note:
a. Filtered sample
b. Wet mass basis. Cake at 23.5 percent solids.

**Table E-10. Phosphorus and Metal Mass Balances Based on Phase 1 Field Test Results.**

| | | Primary Effluent | Anaerobic Zone[a] | Secondary Effluent | WAB | PriEff – SecEff | PriEff – SecEff | Sec + Cake |
|---|---|---|---|---|---|---|---|---|
| TP | lb/d | 3,350 | 13,200[b] | 160 | 2,220 | 3,190 | 1,620[c] | |
| Mg | lb/d | 1,190 | 5,280 | 750 | | 440 | 190[c] | 960 |
| Ca | lb/d | 4,620 | 5,610 | 2,730 | | 1,900 | 1,660[c] | 3,030 |
| K | lb/d | 4,130 | 13,200 | 3,530 | | 600 | 140[c] | 3,610 |

a. Soluble leaving anaerobic zone – filtered sample
b. Based on 40 mg/L orthophosphate measure in profile testing.
c. Decrease between primary and secondary effluent in equivalents per second

**Table E-11. Phosphorus and Metal Mass Balance across the Anaerobic Zone during the Phase 1 Field Testing.**

| | Primary Effluent lb/d | Anaerobic Zone[a] lb/d | Increase lb/d | Equivalent Eq/sec |
|---|---|---|---|---|
| TP | 3,350 | 13,200[b] | 9,860 | 5,000 |
| Mg | 1,190 | 5,280 | 4,100 | 1,790 |
| Ca | 4,620 | 5,610 | 990 | 870 |
| K | 4,130 | 13,200 | 9,080 | 2,070 |

Note:
a. Soluble – filtered sample
b. Based on 40 mg/L orthophosphate measure in profile testing.

### E.3.3.4 Microscopic Examination of Digested Sludge

A digital microscope was used to inspect dewatered biosolids at 200x magnification. 200x magnification is typically sufficient to detect struvite and other crystals that might form during solids stabilization and the dewatering processes. The pictures in Figure E-27 show some crystals. Only a few of these crystals were observed in the solids stream during Phase 1 testing. At some plants an abundance of crystals can be seen. Figure E-28 shows two images at the same magnification for a visual comparison of samples from other facilities. These photos are consistent with the operation of EBPR, showing that some of the compounds are crystallizing during the solids handling process.

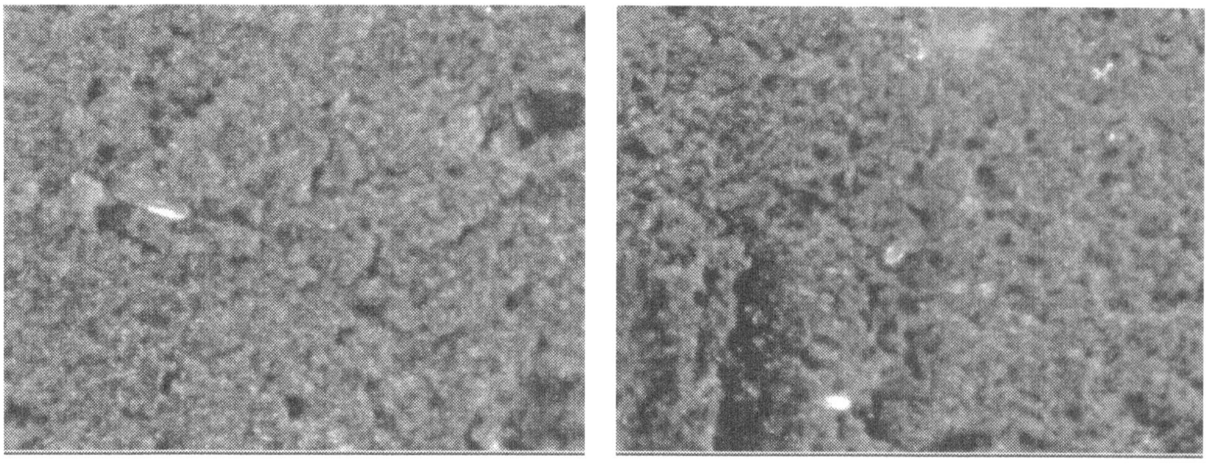

**Figure E-27. Digital Microscope Pictures of Struvite Crystals in Dewatered Biosolids (200x magnification).**

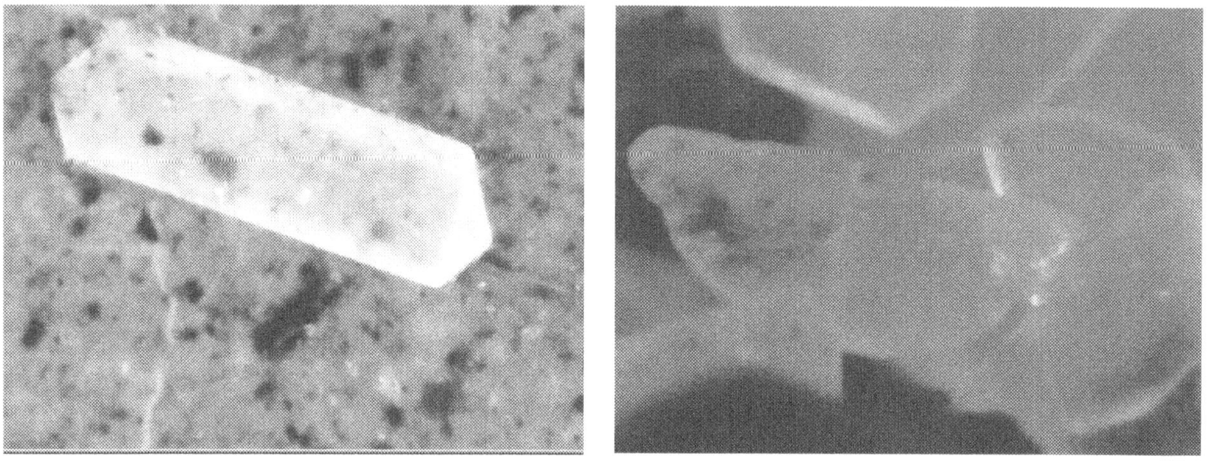

**Figure E-28. Digital Microscope Pictures of Struvite Crystals in Dewatered Biosolids from Eugene-Springfield WPCF and Washed Out Struvite Crystals from Durham AWWTP (200x magnification).**

## E.3.4 PAO and GAO Investigations

Mixed liquor samples from the Nansemond plant were examined to observe the presence of polyphosphate granules (indicative of EBPR), and to look for the presence of GAOs, which typically consist of large cocci (2-3 μm) arranged in tetrad clusters within the floc. Samples were collected on July 3, 2003 and September 3, 2003. Performance during these times was described in Section E-3.2.2 and E-3.3.2.

Both microscopic study and molecular techniques were used to examine the presence of GAOs in the mixed liquor from the plant. Light microscopy observation was used to observe the tetrad-forming GAOs, and molecular probing using specific oligonucleotide probes targeting GAOs was completed to identify both tetrad-forming and non-tetrad-forming GAOs. In addition Neisser and DAPI staining were applied to observe the presence and relative abundance of PAOs.

Figure E-29 and Figure E-30 show the observation of PAOs and GAOs using a Neisser stain. Figure E-31 shows the results of DAPI stain. The GAOs have Neisser positive stained cell walls, but are not positive for cell inclusions as are PAOs. The Neisser positive stained PAOs are dark blue to purple in color, and usually group in clusters within the flocs. The DAPI-stained

PAOs are yellow in color, which is characteristic of polyphosphate granules observed under the fluorescence microscope. Other non-polyphosphate-accumulating bacteria including GAOs show up in blue with DAPI stain.

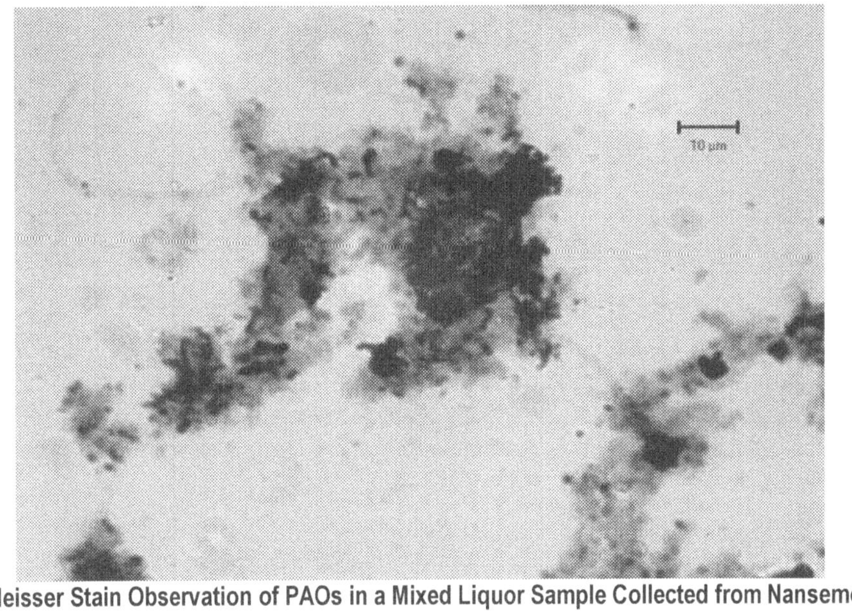


**Figure E-29. Neisser Stain Observation of PAOs in a Mixed Liquor Sample Collected from Nansemond on July 3, 2003.**

**Figure E-30. Neisser Stain Observation of PAOs in a Mixed Liquor Sample Collected from Nansemond on September 3, 2003.**

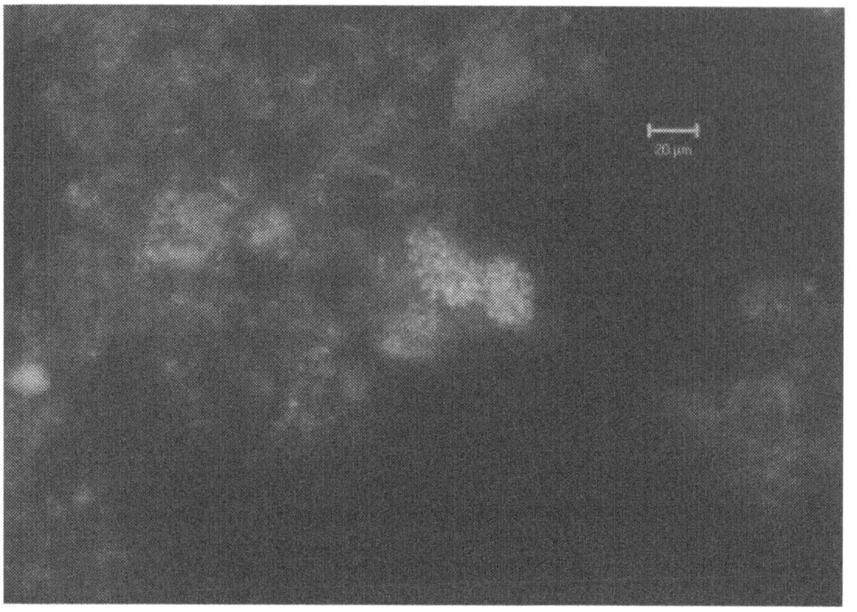

**Figure E-31. DAPI Stain Observation of PAOs and Tetrad-forming GAOs in a Mixed Liquor Sample Collected from on September 3, 2003.**

Table E-12 summarizes the observation of PAOs and GAOs for mixed liquor samples at Nansemond that were taken at two different times. No tetrad-forming GAOs were observed with the samples taken on 7-3-03 and 9-3-03. PAOs were found to be abundant in both samples studied.

Examination of the mixed liquor samples via molecular methods (fluorescence in situ hybridization (FISH))using 16S rDNA-based probes targeting *Accumulibacter* (represent PAOs) (Crocetti et al. 00) and some non tetrad-forming GAOs-*Competibacter* (GAOQ989 andGB-G2, Crocetti et al. 00 and Kong *et al*. 00) showed that 10-20% of the Eubacteria populations was GAOs and about 5-15% was PAOs (see Table E-13).

**Table E-12. Summary of PAOs and Tetrad-forming GAOs Observation for Nansemond WWTP Using DAPI and Neisser Stain.**

| Date of Sample | GAO observation | PAO observation | Comments |
|---|---|---|---|
| 7-3-03 | not observed | Abundant, cocci shaped cells group together in big flocs | PAO dominant, no GAOs was observed |
| 9-3-03 | not observed | Abundant, cocci shaped cells group together in big flocs | PAO dominant, no GAOs was observed |

**Table E-13. Summary of representative PAOs and Non-tetrad-forming GAOs Observation for Nansemond WWTP Using Molecular Methods.**

| Date of Sample | GAO observation | PAO observation | Comments |
|---|---|---|---|
| 9-3-03 | 10-20% of total Eubacteria | 5-15% of total Eubacteria | Both PAO and GAO had similar level of abundance. |

The phosphorus removal performance at the plant was evaluated for the time period when the samples were taken for the GAOs study. The results of phosphorus removal performance evaluation data are shown in Table E-14. The phosphorus removal performance at the plant was relatively stable during the study period as indicated by the phosphorus release rate and phosphorus uptake rate values. It is interesting that the FISH results suggested similar population abundance of GAOs as PAOs in the sludge. The presence of non-tetrad-forming GAOs as shown by the FISH results did not seem to greatly hamper the phosphorus removal. It should be noted that FISH methods only targeted certain PAO and GAO species for which the probes are available. DAPI and Neisser stain were targeting at all cells that are capable of accumulating poly-P, therefore should reveal more PAO populations than FISH method.

**Table E-14. Summary of P removal Performance at the Nansemond WWTP.**

| Date of Sample | GAOs abundance | PAOs abundance | P Release Rate gP/gVSS.hr | P Uptake Rate gP/gVSS.hr | P Release Ratio gΔ P/gΔAcetic Acid | Release Rate/ Uptake Rate |
|---|---|---|---|---|---|---|
| 6-03 | NA | NA | 5 | 3.4 | 0.2 | 1.5 |
| 7-3-03 | None | Abundant | | | | |
| 9-3-03 | None | Abundant | 7.7 | 2.6 | 0.24 | 3 |

## E.4 Conclusions

The study showed that while the plant meets the effluent TP target of 2 mg/L with 85% reliability, effluent TP concentrations are highly variable. This could be due in part to a high aeration basin influent phosphorus concentration (over 14 mg/L), which is strongly influenced by the recycle stream phosphorus concentration. Aeration influent VFA:P ratios range from 0.9 to 2.6, but periods of poor plant performance appear more closely linked to a change in influent organic concentration than the value of the ratio.

Anaerobic phosphorus release and low effluent phosphorus during intensive sampling periods (with no chemical addition) demonstrate that PAO growth is occurring. Microscopic analysis showed an abundance of PAOs and no GAOs, however molecular techniques suggested similar population abundance of GAOs as PAOs in the activated sludge. The uptake and release test also showed a lower ΔP/acetic acid ratio than anticipated, possibly indicating the consumption of VFAs by GAOs. However, the presence of GAOs did not appear to hamper phosphorus removal.

# APPENDIX F

# VIP Enhanced Biological Phosphorus Removal (EBPR) Evaluation Report

## F.1 Summary of Findings

The Virginia Initiative Plant (VIP) located in Norfolk, VA, is operated by the Hampton Roads Sanitation District (HRSD). The plant operates in an EBPR mode pioneered at and named for the plant (VIP process). Currently flows average around 30 million gallons per day (MGD), 75% of the 40 MGD design capacity. The plant has the ability to add alum to the secondary clarifiers for additional phosphorus removal, but rarely uses this capability. Polymer is occasionally added to the secondary clarifiers to aid TSS capture. Primary and waste activated sludge and alum solids from a water treatment plant are dewatered and incinerated, with centrate returned to the plant headworks.

The aeration basin influent at VIP shows a consistent wastewater composition with favorable nutrient ratios for BNR and EBPR. The BOD/TP ratio of 29 is higher than three times that measured at the nearby Nansemond facility, and is above the 20:1 ratio typically required for EBPR. Aeration basin influent total phosphorus concentrations averaged 5.3 mg/L during 2000-2003. While aeration basin influent volatile fatty acid (VFA) concentrations were not routinely measured, one would expect high VFA concentrations due to the primary sludge thickening in the primary clarifiers. Total phosphorus concentrations do increase approximately 8% across the primary clarifiers, indicating that phosphorus is released during the mixing of primary and waste activated sludge, and returned in the recycled centrate stream.

VIP has a year-round phosphorus limit of 2 mg/L. Operating data since 2000 indicate that the target is met the majority of the time. Between 2000 and 2003, effluent total phosphorus was under 1 mg/L 82% of the time, and under 0.5 mg/L 70% of the time. The plant did experience periods of upset with effluent total phosphorus concentrations of over 2 mg/L during two of the three summers in this period of record (2001 and 2002). The loss of EBPR coincided with the loss of nitrification, and could be due to a toxic industrial discharge caused by an annual event (i.e., cleaning) at one of the industrial facilities in the VIP service area.

Nutrient profiles collected during field testing indicate that nitrate is depleted prior to the final anoxic zone, and it appears that secondary phosphorus release occurs. This release, however, does not necessarily cause high effluent soluble phosphorus concentrations. The Phase 2 uptake and release tests (URT) showed very good rates during the first intensive sampling period in June 2003. Unlike the previous three summers VIP did not loose EBPR completely in 2003 but a reduction in phosphorus removal performance was evident from the effluent phosphorus concentration and August 2003 URT.

Microscopic observations of mixed liquor from July through September 2003 indicate a steady increase in the population of glycogen accumulating organisms (GAOs) and decrease in the population of phosphorus accumulating organisms (PAOs). The cause for this population shift is not known.

Table F-1 summarizes the key findings from the data review, field tests and intensive sampling at VIP.

**Table F-1. Summery of Key Results from Facility Investigation at VIP.**

| | |
|---|---|
| Reliability | EBPR is generally reliable, with effluent soluble phosphorus averaging 0.6 mg/L during the study period without any chemical addition. Effluent total phosphorus is under 1 mg/L 82% of the time and under 0.5 mg/L 70% of the time. Plant experienced upsets in EBPR in August 2001 and August/September 2002 for unknown reasons. |
| Operational Criteria | Solids retention time (SRT) ranged from 4 to 26 days, but was typically 6-10 days. Plant is designed for anaerobic hydraulic retention time (HRT) of 0.8 hours, anoxic HRT of 1.5 hours, and aerobic HRT of 4 hours. |
| Influent Organic Characterization | Aeration basin influent characteristics are consistent, and favorable for EBPR. Influent BOD/TP ratio averages 29:1. Thickening in primary clarifiers likely adds VFAs to the primary effluent. |
| Measures Showing PAO Growth | Plant experiences periods of high anaerobic P release and low effluent phosphorus, indicating PAO growth. This was confirmed by microscopic investigations in July 2003. |
| Factors Hindering PAO Growth | Periods of upset (low anaerobic release, low aerobic uptake, high effluent P) indicate some factors inhibiting PAO growth. Operating conditions do not point to PAO inhibition; possibly toxic release from an industrial discharger. |
| Phosphorus Mass Balance | Based on average 2002 values (including summer upset), plant removes 80% of influent total phosphorus. Some phosphorus is released when primary and waste activated sludge are mixed and dewatered, contributing about 9% of the influent total P load. Half of the recycle total P load is removed in the primary clarifier. |
| Secondary Release | Significant phosphorus release was observed in the final anoxic cell in all field testing profiles. Nitrate concentrations were depleted, supporting the observation that secondary release occurs. |

## F.2 Introduction

The analysis of this facility was completed as part of a project sponsored by the Water Environment Research Foundation (WERF). The project objective is to develop information for evaluating the ability of municipal wastewater treatment plants to efficiently and cost-effectively remove phosphorus to low levels through processes that involve EBPR. This was completed in large part through detailed studies at several treatment facilities that employ EBPR. The goals of the study were to:

- Determine process reliability.
- Determine the biological phosphorus removal efficiency.
- Determine the biological phosphorus removal capacity.
- Determine how chemical addition affects EBPR.

Definitions of terms used in this analysis are described in Appendix P.

## F.2.1 Facility Introduction

Figure F-1. Virginia Initiative Plant, Hampton Roads Sanitation District. Norfolk, VA.

The Virginia Initiative Plant (VIP) is located in Norfolk, Virginia, and is operated by the Hampton Roads Sanitation District (HRSD). VIP has a 40 MGD design treatment capacity with a current annual average flow of approximately 31.4 MGD. Plant influent is composed of 80% domestic wastewater and 20% industrial wastewater. HRSD has a strong pretreatment program and industrial dischargers must meet discharge permit requirements before releasing wastewater into the sewer system. Although the VIP service area has separate sewers, plant staff report that infiltration and inflow (I&I) can be substantial. The VIP plant was designed for biological nutrient removal, including enhanced biological phosphorus removal (EBPR) and has been in operation over 10 years. Each of the four activated sludge process trains at VIP include sequential anaerobic, anoxic, and aerobic basins.

VIP has a year-round phosphorus limit of 2 mg/L. No volatile fatty acid (VFA) but, on occasion, some chemical addition for phosphorus removal has been required to meet this permit. VIP experienced a significant decrease in EBPR during summer 2002, which has required the plant to add ferric chloride to meet effluent phosphorus requirements. VIP is also operated for nitrogen removal. The plant has no effluent nitrogen permit limits, but has a goal of 1 mg/L ammonia in the secondary clarifier effluent, 9 mg/L total nitrogen in the plant effluent, and 65 percent overall nitrogen removal. VIP adds polymer to the secondary clarifier influent to minimize effluent total suspended solids, as needed. VIP effluent is discharged into the Elizabeth River.

VIP dewaters (centrifuge) and incinerates primary and secondary solids, and alum solids from a water plant.

## F.2.2 Facility Process Design

### F.2.2.1 Liquid Stream

The biological nutrient removal (BNR) process follows screening, grit removal, and primary clarification. Secondary effluent (SE) is disinfected and discharged to the Elizabeth River. The BNR process configuration is that of the VIP design, named after this facility where it was developed. Figure F-2 shows the overall plant process schematic, and Figure F-3 shows the VIP process schematic.

The key feature of the VIP process is that return activated sludge (RAS) enters the process train at the head of the anoxic zone (ANX), where it combines with the mixed liquor recycle from the aerobic zone (the nitrified recycle or NRCY stream). A second mixed liquor recycle (the anoxic recycle or ARCY stream) returns biomass from the anoxic zone to the anaerobic zone (ANR) to achieve EBPR. This configuration limits the transfer of nitrate into the anaerobic zone to maximize VFA availability for phosphorus release. Primary effluent and ARCY mix at the head of the anaerobic zone. In the first anoxic cell the combined RAS/NRCY mixes with the anaerobic zone effluent. Operators also have the option of sending the RAS/NRCY stream to one of the two downstream cells if a longer anaerobic retention time is desired.

VIP has four parallel trains, each with two anaerobic zones, four anoxic zones, and four aerobic zones (Figure F-3). Effluent from the unaerated zones is split between the available aerobic basins. This configuration allows the operator to place any number of unaerated and aerated zones in operation.

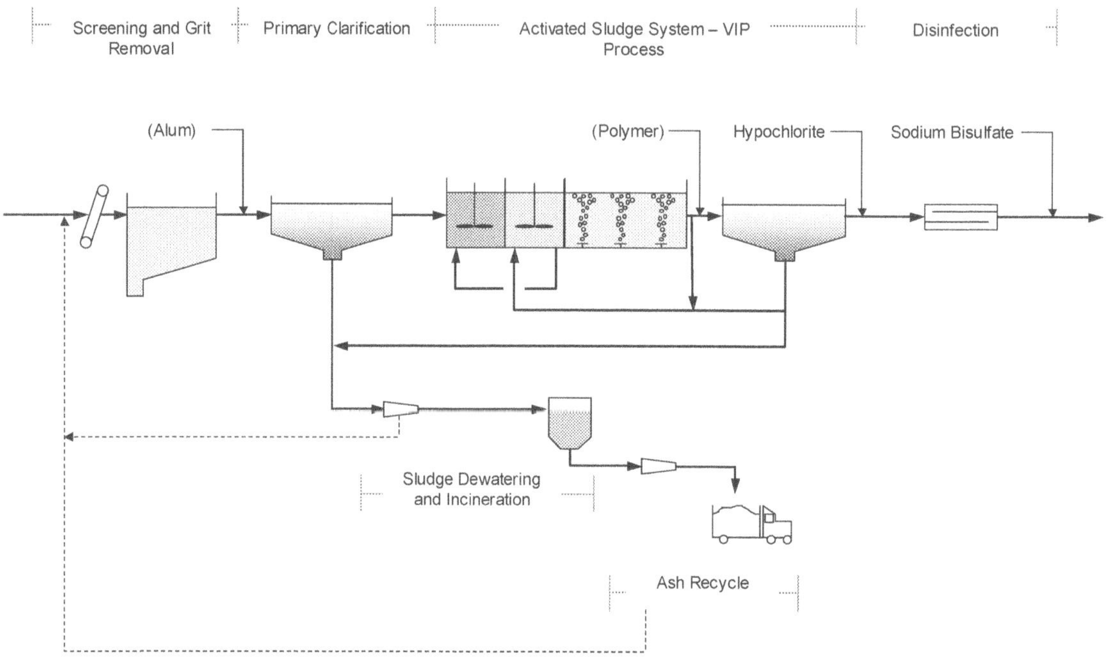


**Figure F-2. Process Schematic for VIP Liquid and Solids Treatment.**

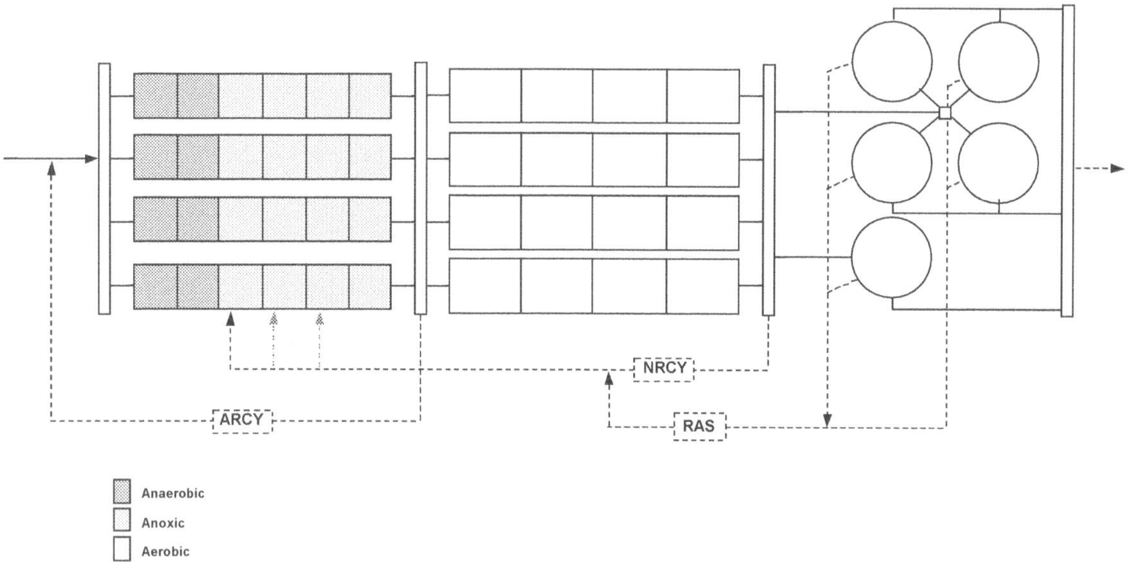


**Figure F-3. Secondary Treatment Layout at VIP.**

Table F-2. Key Process Parameters.

| Component | Unit | Design | Average |
|---|---|---|---|
| ANR HRT[1] | hrs | 0.78 | 0.87 |
| ANX HRT | hrs | 1.56 | 1.8 |
| AER HRT | hrs | 4.1 | 4.6 |
| Total HRT | hrs | 6.44 | 7.2 |
| RAS Rate | % | | 31 |
| ARCY Recycle Rate | % | | 73 |
| NRCY Recycle Rate | % | | 77 |
| SRT | days | | 9.5 |
| MLSS | mg/L | | 3090 |
| Solids Recycle Return | | Upstream of primary clarifier (no anaerobic digestion) | |
| Flow | MGD | 40 | 28.4 |

### F.2.2.2 Solids Stream

Primary sludge is thickened in primary clarifiers to about 4% total solids. Primary and waste activated sludges are mixed with alum solids from a water treatment plant and sent directly to dewatering centrifuges, then the dewatered cake is incinerated. Centrate is recycled to the plant headworks. Centrifuges operate continuously (24 hours per day, seven days per week). Scrubber water from the incinerator, which contains cyanide from the incineration process, is treated in a separate nitrification process before being recycled to the plant headworks. Table F-3 summarizes the solids processing at VIP.

Table F-3. Solids Processing Summary.

| Component | Technology | Comments |
|---|---|---|
| Primary Sludge Thickening | Gravity | Thickened in primary clarifier to 3% – 5% Total Solids |
| WAS Thickening | None | WAS sent directly to dewatering |
| Dewatering | Centrifuge | Operates continuously with no storage |
| Primary Sludge Stabilization<br>WAS Stabilization | Incineration | |

### F.2.2.3 Chemical Addition

VIP can feed alum to the secondary as EBPR backup but rarely uses this capability. Polymer is occasionally added to the secondary clarifiers to aid TSS capture when necessary. Alum has been added occasionally to the primary clarifier to improve BOD removal. Plant experience indicates that the best operation occurs when the primary effluent BOD loading to the EBPR system is below 40,000 lb/d. No chemicals are added to supplement alkalinity or control pH.

[1] Retention time is calculated based on aeration basin influent flow and does not include the recycle volume

### F.2.2.4 Process Control

The operating strategy is to maintain the solids retention time (SRT) just high enough to achieve nitrification even though the plant is not required to nitrify. The operator reduces the SRT to as low as possible while maintaining nitrification. The SRT averages approximately 9.5 days, but is operated as low as four days during the summer.

Dissolved oxygen (DO) is controlled to maintain about 2.5 mg/L in the aerated zone.

Operators at VIP monitor DO, MLSS, solids volume index (SVI), and effluent parameters. No hard and fast response strategy exists for dealing with process upsets. Appropriate actions are based on operator experience.

## F.3 Process Performance Evaluation

### F.3.1 Approach

The process evaluation entailed two data collection activities: a review of historical plant operating data, and the conduct of special sampling to supplement the routinely recorded water quality and process measurements. The following data sets were collected (see Table F-4):

**Routine Data Review**

The reviewed data represent all available water quality and process monitoring data collected from January 1, 2000 through August 31, 2003. Data were analyzed to determine general process performance, reliability of meeting effluent quality, and evaluation of the interaction between process operating parameters and performance. This data set is referred to as the **2000-2003 data set**.

**Special Data Collection**

During each phase of the project (Table F-4) project team members visited the VIP facility to conduct field tests and to interview plant staff. During Phase 1 the focus of the site visit was on supplementing the detailed survey information and data. Aeration basin phosphate and ammonia profiles were measured, and phosphorus uptake and release tests conducted. The results of the Phase 1 uptake and release tests are not presented in this report because the methodology was modified for Phase 2 to permit better interpretation of the results. Wastewater characteristics were collected for one day during Phase 1. During Phase 2, the data collection for each period extended over two weeks and generally included the collection of six samples for each water quality parameter.

During the intensive sampling period, the plant staff collected additional samples that are infrequently taken during routine sampling. The sampling plan included influent VFA, COD, readily biodegradable COD as measured by the flocculated filtered COD test (FFCOD), and soluble $BOD_5$ ($sBOD_5$) as well as the phosphorus and nitrogen species. Influent, effluent and solids processing recycle phosphorus and metals analyses were also included in the sampling plan to allow performance of P and metals mass balances.

The results of the special data collection efforts are included in the following data sets:

- **Phase 1 data set** This data set was collected in Year 1 of the project during August and September 2002.
- **Phase 2A data set** This data set was collected in Year 2 of the project during the first intensive sampling period in June 2003.

- **Phase 2B data set** This data set was collected in Year 2 of the project during the second intensive sampling period in August 2003.
- **Phase 2F data set** This data set was collected in Year 2 of the project during the field visit in June 2003.

**Table F-4. Summary of Field Tests and Sampling.**

| Field Testing Segment | Month | Data Set ID | Information Included |
|---|---|---|---|
| Historical Data Review | January 2000 – August 2003 | 2000 - 2003 | Influent, Effluent Averages<br>Process Parameter Averages |
| Phase 1 – Field Tests | September 2002 (One day) | Phase 1 | Aeration Basin Profiles<br>Uptake and Release Test<br>Operator Interviews |
| Phase 2 – Field Tests | June 2003 | Phase 2F | Uptake and Release Test<br>Operator Interviews<br>Data Collection |
| Phase 2 – Intensive Sampling Period 1 | June 2003<br>Two weeks | Phase 2A | Influent Characterization<br>Aeration Basin Profiles<br>Phosphorus Mass Balance<br>Metal Mass Balance |
| Phase 2 – Intensive Sampling Period 2 | August 2003<br>Two weeks | Phase 2B | Influent Characterization<br>Aeration Basin Profiles<br>Phosphorus Mass Balance<br>Metal Mass Balance |

## F.3.2 Overall Process Characterization and Performance

This section of the report summarizes the aeration basin influent characteristics and secondary treatment process performance based on the data collection described in Section F-3.1. It also discusses relationships between primary effluent loading and process performance.

### F3.2.1 Aeration Basin Influent Characterization

The aeration basin influent had a relatively consistent wastewater composition over the period of January 2000 to August 2003. Wastewater influent temperatures ranged between 12°C and 35°C. Similar to the nearby Nansemond facility, plant influent flows have returned to year 2000 levels after a steady decline between 2000 and 2001. The average influent flow for the period January 200 to August 2003 was 30 MGD (Figure F-4), which is 75% of the design flow. The diurnal flow has a typical pattern for a service area of this size (Figure F-5) with peaking factors for minimum flow of 0.73 and maximum flow of 1.14 during dry weather conditions.

**Table F-5. Aeration Basin Influent Wastewater Characterization.**

| ABI Wastewater Parameter | Unit | 2000 - 2003 | Phase 1 | Phase 2A | Phase 2B |
|---|---|---|---|---|---|
| Flow | MGD | 30 | 22 | 32 | 29 |
| $BOD_5$ | mg/L | 150 | 160 | 140 | 14 |
| $sBOD_5$ | mg/L | 92 | 89 | 97 | 86 |
| COD | mg/L | - | 260 | 280 | 270 |

| | | | | | |
|---|---|---|---|---|---|
| ffCOD | mg/L | - | 140 | 275 | 268 |
| VFA | mg acetic acid/L | - | 78.5 | 26 | - |
| TP | mg P /L | 5.3 | 4.3 | 4.3 | 4.8 |
| PO4-P | mg P /L | 4.1 | 3.4 | 2.9 | 3.8 |
| Total Alkalinity (as $CaCO_3$) | mg/L | 170 | 190 | 140 | 160 |
| pH | | 6.8 | 6.9 | 6.9 | 6.8 |
| TKN | mg N /L | 25 | 25 | 20 | 22 |
| $NH_4$-N | mg N /L | 18 | 22 | 18 | 17 |
| TSS | mg/L | 62 | 53 | 59 | 50 |
| Temperature | °C | 21 | 27 | 24 | 25 |
| VFA : P Ratio | - | - | 18 | 6.0 | - |
| ffCOD : P Ratio | - | - | 33 | 31 | 29 |
| COD : P Ratio | - | - | 58 | 64 | 56 |
| $BOD_5$:P Ratio | - | 29 | 37 | 33 | 29 |
| ffCOD : COD Ratio | - | - | 0.53 | 0.49 | 0.52 |
| VFA : ffCOD Ratio | - | - | 0.56 | 0.19 | - |

Both primary effluent ammonia and total phosphorus loads decreased between 2000 and 2003; primary effluent ammonia load decreased by approximately 3%, while total phosphorus load decreased by 15% (Figure F-6). Primary effluent $BOD_5$ load (Figure F-7) increased 19%, due largely to a steady increase from Fall 2000 through Spring 2001. Primary effluent soluble $BOD_5$ has remained constant. Raw influent flow has also remained almost constant.

Table F-5 summarizes the aeration basin influent composition and COD fractions for all testing phases. The aeration basin influent shows favorable nutrient ratios for BNR and EBPR. The $BOD_5$/TP ratio of 29 is more than three times the ratio measured at the nearby Nansemond facility, and is above the typical 20:1 ratio required for EBPR. Recycle from solids processing contributes approximately 10% of the aeration basin influent phosphorus on an average basis.

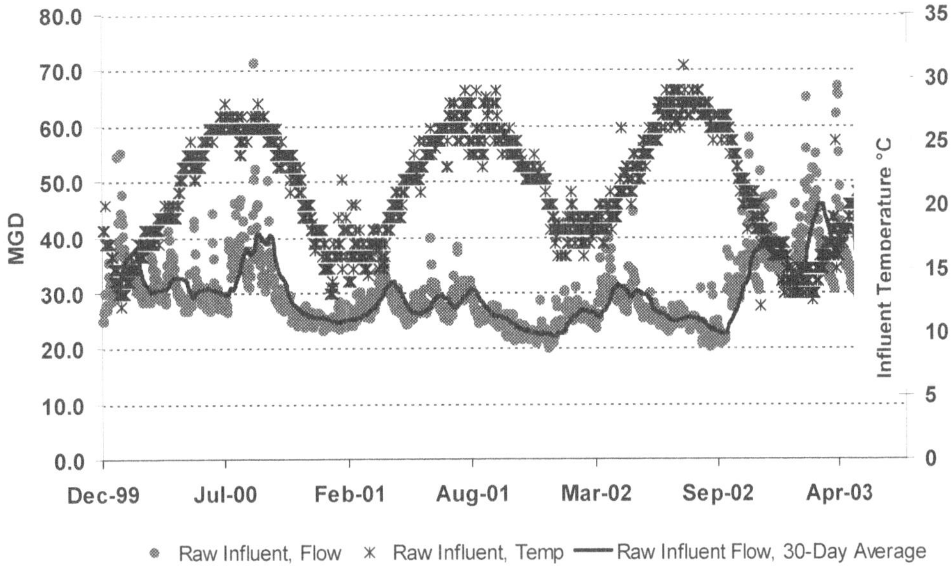


**Figure F-4. Influent Flow, Influent Temperature, and 30-day Average Influent Flow Between 2000 - 2003.**

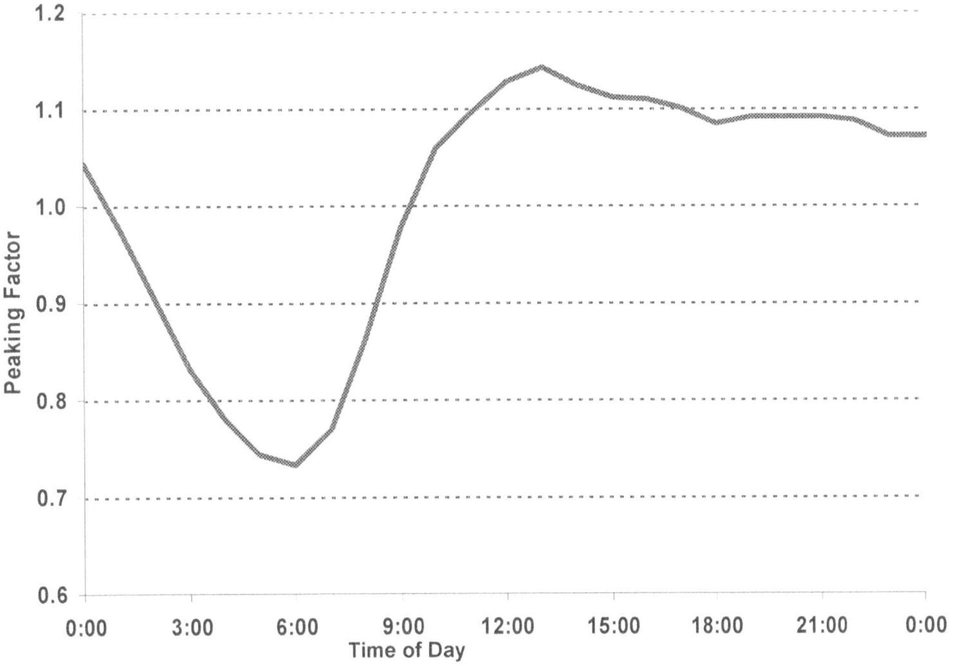


**Figure F-5. Dry Weather Diurnal Pattern of Peaking Factor at VIP.**

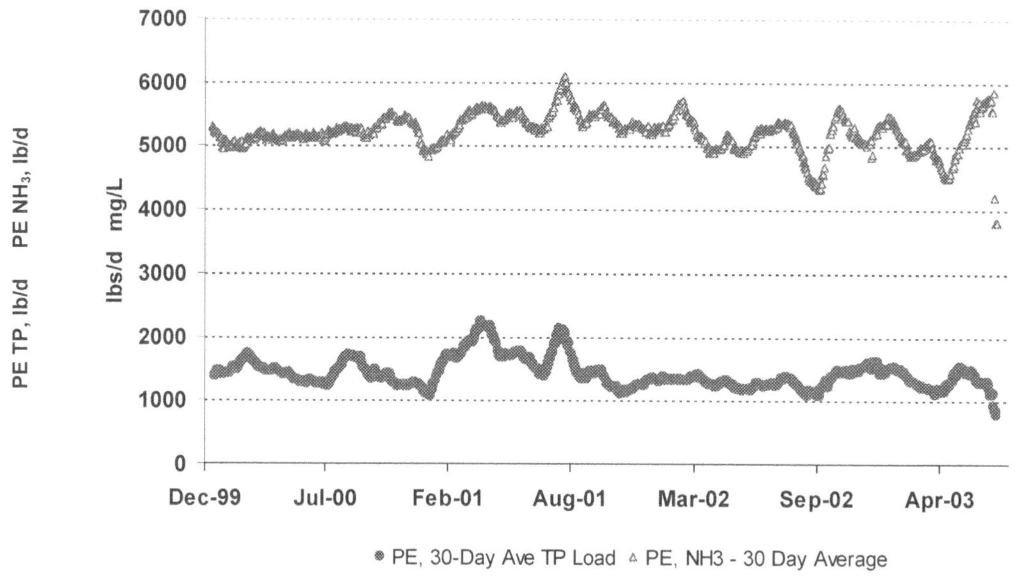


**Figure F-6. Primary Effluent Phosphorus and Ammonia Load.**

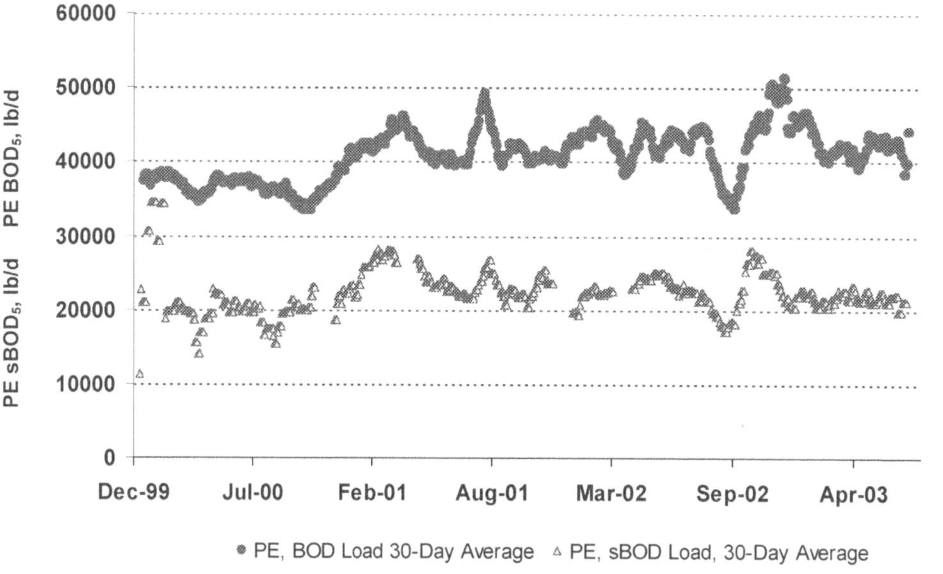


**Figure F-7. Primary Effluent BOD and Soluble $BOD_5$[2] Load.**

The aeration basin influent (ABI) shows a large fluctuation in flow and concentration. Concentrations of aeration basin influent $BOD_5$, TSS, and nutrients can vary by –75% to +200% (Table F-6). Even the 30-day average loads show large fluctuations, indicating that individual daily fluctuations can be large. Such large fluctuations make it very difficult to maintain process control and a stable biological inventory.

---

[2] Primary effluent values based on assumption that $sBOD_5$ does not change significantly between influent and primary effluent. Only influent $sBOD_5$ data were available

Table F-6. Selected Water Quality Measurement and Process Parameters from 2000 - 2003.

| Parameter | Unit | Min | Ave | Max |
|---|---|---|---|---|
| Influent Flow | MGD | 22 | 30 | 46 |
| ABI Temperature | °C | 12 | 21 | 35 |
| ABI $BOD_5$ | mg/L | 54 | 150 | 240 |
| TKN | mg N /L | 8.8 | 25 | 54 |
| ABI $NH_4$-N | mg N /L | 7.8 | 19 | 31 |
| ABI TP | mg P /L | 1.9 | 5.3 | 12 |
| ABI $PO_4$-P | mg/L | 1.0 | 4.1 | 10 |
| ABI TSS | mg/L | 16 | 62 | 190 |

### F3.2.2 EBPR Performance

Table F-7 summarizes the effluent water quality measurements and process parameters for the four data sets. The influent BOD, TSS, and nutrient loadings to the activated sludge process show large fluctuations, as to the SRT value (ranging from 4 to 26 days) and the MLSS concentration (ranging from 1,500 to 4,330 mg/L). The nitrate concentration in the first anoxic zone was 0.2 mg/L during Phase 2 testing (see discussion in Section F-3.2.2.2) at a time when the plant was nitrifying. This low nitrate concentration can be attributed to the high dilution rate through internal recycle and the hydraulic retention time of about 25 min. The net result is that the anoxic zone can become completely depleted of nitrate, thus creating an extended anaerobic zone at the end of the anoxic zone.

Table F-7. Process Control and Water Quality Measurements.

| Raw Wastewater Parameter | Unit | 2000 – 2003 | Phase 1 | Phase 2A | Phase 2B |
|---|---|---|---|---|---|
| Plant Flow (Q) | MGD | 30 | 22 | 33 | 29 |
| Aeration Basin Effluent $PO_4$-P | mg/L | - | - | < 0.2 | 0.42 |
| Secondary Effluent (SE) $PO_4$-P | mg/L | 0.55 | 1.8 | < 0.2 | 0.46 |
| SE TP | mg/L | 0.64 | 1.8 | 0.23 | 0.63 |
| SE TN | mg/L | 9.4 | 123 | 9.6 | 8.4 |
| SE $NH_4$-N | mg/L | 2.4 | 5.9 | 3.4 | 1.1 |
| SE $NO_3$-N | mg/L | 6.5 | 6.1 | 5.8 | 6.7 |
| SE BOD | mg/L | 6.2 | 5.0 | 80 | 2.0 |
| SE TSS | mg/L | 3.9 | 2.0 | 4.0 | 2.0 |
| SE Total Alk (as $CaCO_3$) | mg/L | 76 | 110 | 64 | 64 |
| Secondary P-Removal | % | 89 | 63 | 95 | 90 |
| Secondary N-Removal | % | 65 | 54 | 54 | 70 |
| RAS $PO_4$-P | mg P/L | - | - | 3.3 | 8.1 |
| RAS Nitrate | mg N/L | - | - | 0.8 | 0.49 |
| RAS DO | mg$O_2$/L | - | - | 0.3 | - |

Table F-8. Process Control and Water Quality Measurements (Continued).

| Raw Wastewater Parameter | Unit | 2000 – 2003 | Phase 1 | Phase 2A | Phase 2B |
|---|---|---|---|---|---|
| SRT | days | 9.2 | 8.1 | 6.0 | 5.9 |
| RAS Rate | % Q | 30 | 27 | 28 | 29 |
| ML Recycle to ANR | % Q | 72 | 68 | 72 | 71 |
| ML Recycle to ANX | % Q | 77 | 83 | 71 | 82 |
| ANR HRT | hrs | 0.87 | 1.0 | 0.72 | 0.80 |
| ANX HRT | hrs | 1.7 | 2.0 | 1.4 | 1.6 |
| AER HRT | hrs | 4.6 | 5.5 | 3.8 | 4.2 |
| MLSS | mg/L | 3010 | 2400 | 2850 | 2340 |
| MLVSS | mg/L | 2320 | 1944 | 2190 | 1800 |
| MLSS/MLVSS | % | 77 | 81 | 77 | 77 |

The plant rarely experiences problems with filamentous growth and consistently shows low SVIs, which are reflected in the average effluent TSS of 3.9 mg/L (Table F-7).

At VIP, EBPR contributes to the majority of phosphorus removal. Only at the time of process upsets polymer is added to the secondary clarifiers to aid solids retention. The plant operated without alum addition to the primary from Dec 2002 through the end of this study in September 2003 producing an average effluent soluble phosphorus concentration of 0.6 mg/L.

#### *F.3.2.2.1 Phase 1 Testing*

Before 2001, the VIP plant has operated in EBPR mode without significant problems. Chemical supplementation of EBPR was not required until 2001 (although polymer is occasionally added to the secondary clarifier influent to enhance settling). During the first field visit in 2002, the VIP EBPR process experienced upset conditions similar to the previous year. While the operators' perception was that this was the first upset in a very long time, the data review revealed that similar upsets had occurred during the same timeframe over the last three years, increasing in magnitude each year (Figure F-12). A cause and effect relationship with monitored parameters has not been identified.

Figure F-8 shows the effluent total phosphorus, TKN, and TSS concentrations during the summer of 2001. The EBPR upset lasted several months and the plant did not fully recover until December 2002. The cause and reasons for the long recovery period are unknown.

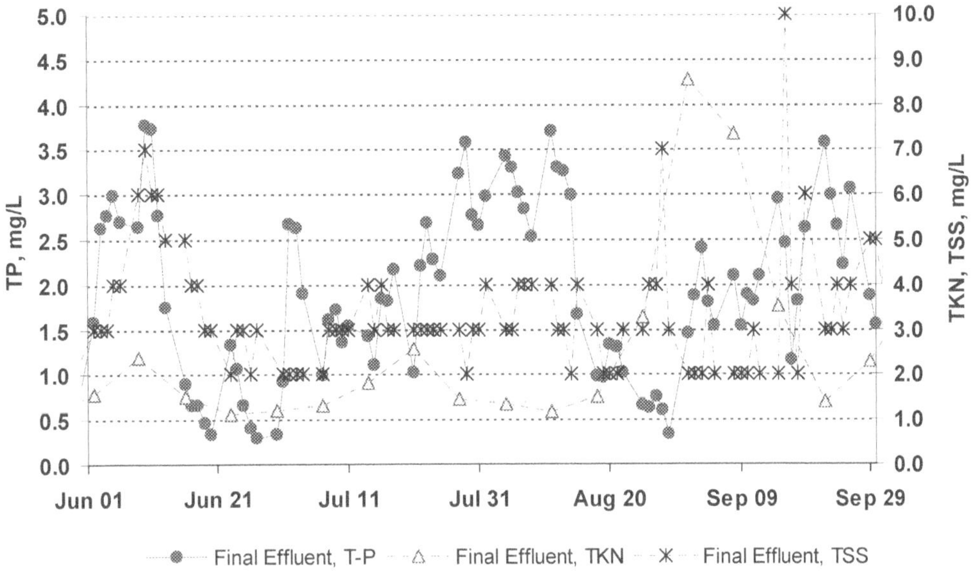


**Figure F-1. Secondary Effluent Phosphorus, TKN, and Suspended Solids (Summer, 2001).**

### *F.3.2.2.2 Phase 2 Testing*

During the second field visit in June 2003 (Phase 2A), EBPR operation was stable and effluent phosphorus concentrations consistently low. Despite the low RAS rate it appears that there was little phosphorus re-release in the secondary clarifier blanket (Figure F-9). The SRT was relatively low during the first intensive sampling period (averaging seven days), and the net aerobic retention time was only 3.9 hours. During the second intensive testing the aerobic retention time was slightly higher.

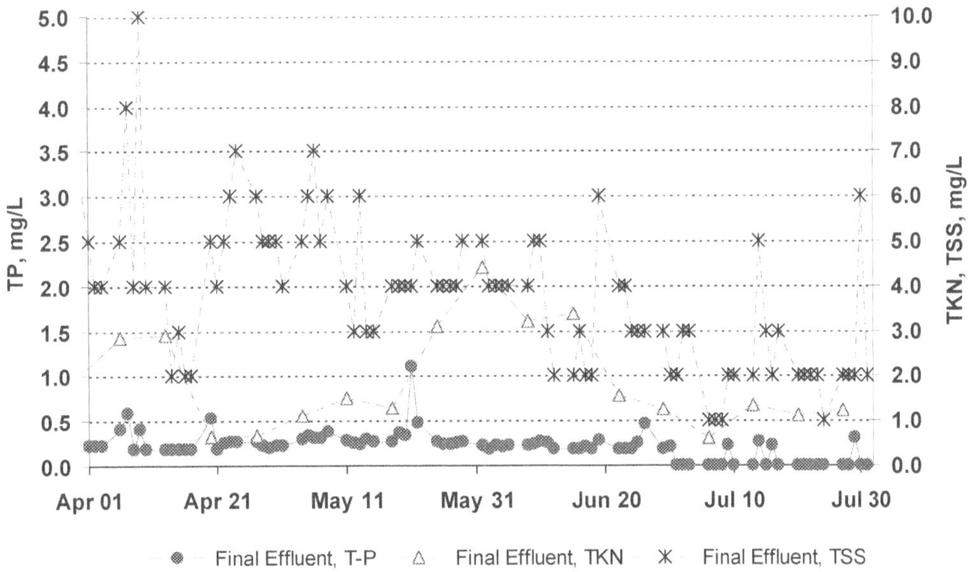


**Figure F-9. Phase 2 Secondary Effluent Phosphorus, TKN, and Suspended Solids (2002).**

### F3.2.3 Relationship Between Performance and Process Loading

Based on the primary effluent wastewater characteristics one would expect good EBPR performance. However, data review for the period 2000-2003 revealed periodic EBPR performance problems. Figure F-10 shows effluent total phosphorus and suspended solids. The plant achieved less than 1 mg/L effluent total phosphorus during 82% of the year and less than 0.5 mg/L 70% of the time (Figure F-11). Occasionally it seemed to go through phases of complete EBPR loss.

In the summer of 2000, 2001 and 2002, EBPR was lost due to an unknown reason. The summer of 2003 did not see an EBPR upset. Because the loss of EBPR coincided with the loss of nitrification (Figure F-12) one could suspect toxic industrial discharge caused by an annual event (i.e. cleaning) at one of the industrial facilities in VIP's service area. Since January 1$^{st}$ 2003, effluent TP has stayed below 0.3 mg/L without chemical addition.

Changes in primary effluent and recycle loads at many plants have a significant impact on the aeration basin influent loading and BOD:P and BOD:N ratios. In the case of VIP this impact is much smaller than normal. The primary clarifier removals average only 15% of the influent BOD (Figure F-13) and due to biosolids incineration there are no significant recycle loads. However the phosphorus load increase across the primary clarifier (Figure F-14) by about 8%. Though much lower than at facilities with anaerobic digestion, where recycle typically ranges between 30 and 100%, the recycle load is higher than one would expect considering that primary and waste activated sludge are directly dewatered and incinerated. It is possible that during the mixing of primary and secondary solids, stored phosphorus is released and recycled to the plant. Since the primary sludge is thickened in the primary clarifier, one could expect VFA concentrations in excess of 100 mg/L in the primary sludge. Currently, the additional recycle load does not appear to pose a particular problem for VIP and no relationship between the recycle load and the EBPR upsets exist. The recycle ammonia loads (Figure F-15) are small and not expected to have an impact on the overall secondary process performance.

No relationship between aeration basin influent BOD/TP ratio and effluent phosphorus is evident (Figure F-16). This is contrarary to the relationship found at the Nansemond WWTP. The excursions at VIP occurred seasonally and appear to be triggered by some upset event.

Analysis by Rabinowitz et. al. (2004) of data from three full-scale warm-weather BNR facilities showed that EBPR was significantly compromised or failed at temperatures over 30°C. The researchers hypothesized that the high temperatures reduced the PAO growth rate, resulting in washout. This condition can have the secondary impact of allowing readily biodegradable substrate normally consumed by PAOs to enter the aerobic zone, promoting the growth of filamentous bacteria. Figure F-17 shows the relationship between temperature and effluent TP for VIP plant. The results (also reported in Rabinowitz et al., 2004) supports this hypothesis: performance of the EBPR process deteriorates when the temperature approaches and exceeds 30 C.

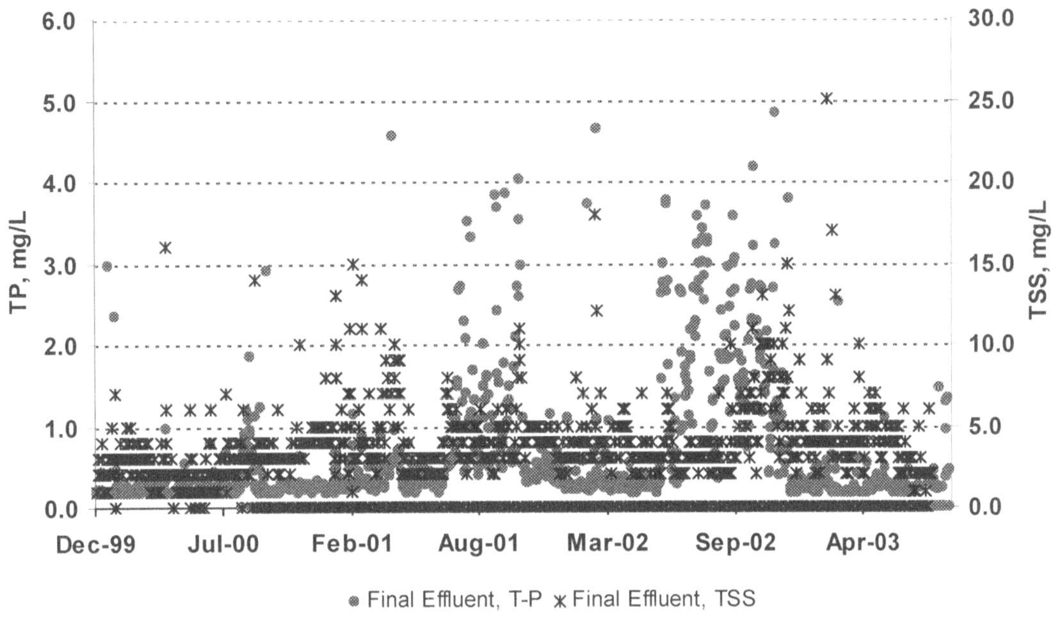


**Figure F-2. Effluent Total Phosphorus and Total Suspended Solids.**

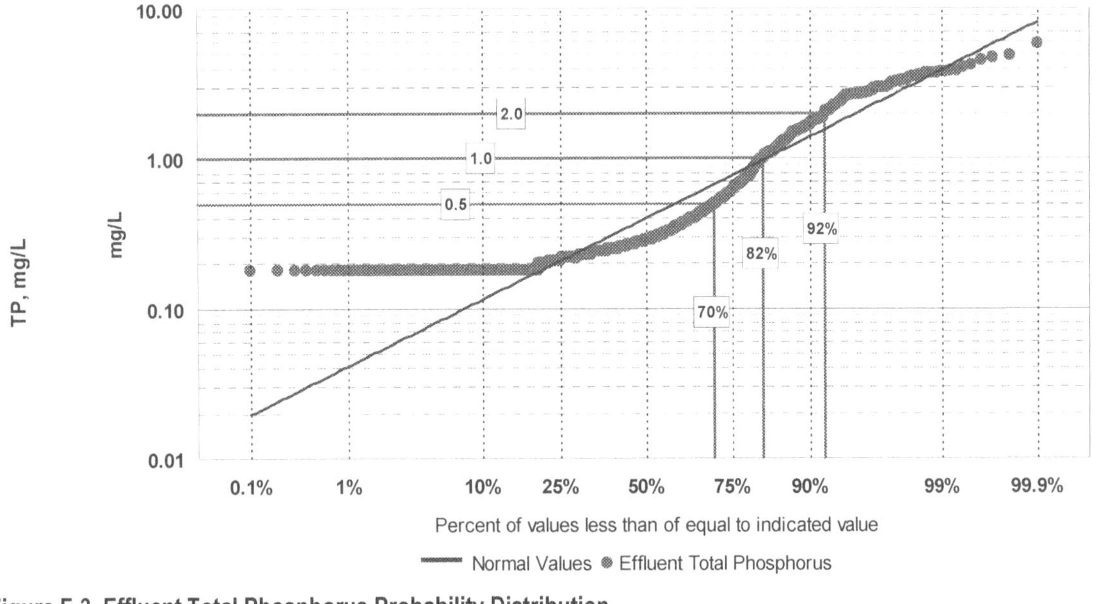


**Figure F-3. Effluent Total Phosphorus Probability Distribution.**

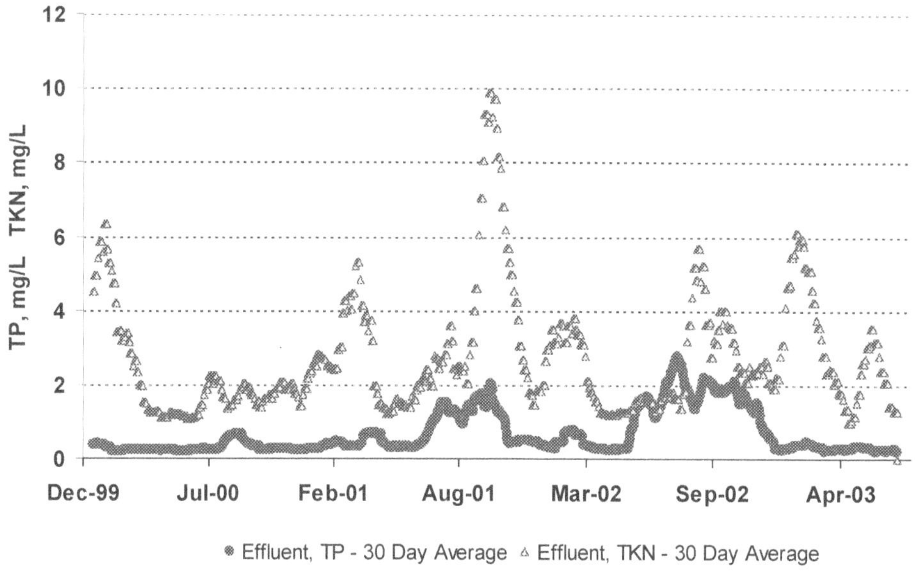


**Figure F-4. 30-day Average for Effluent Total Phosphorus and TKN[3].**

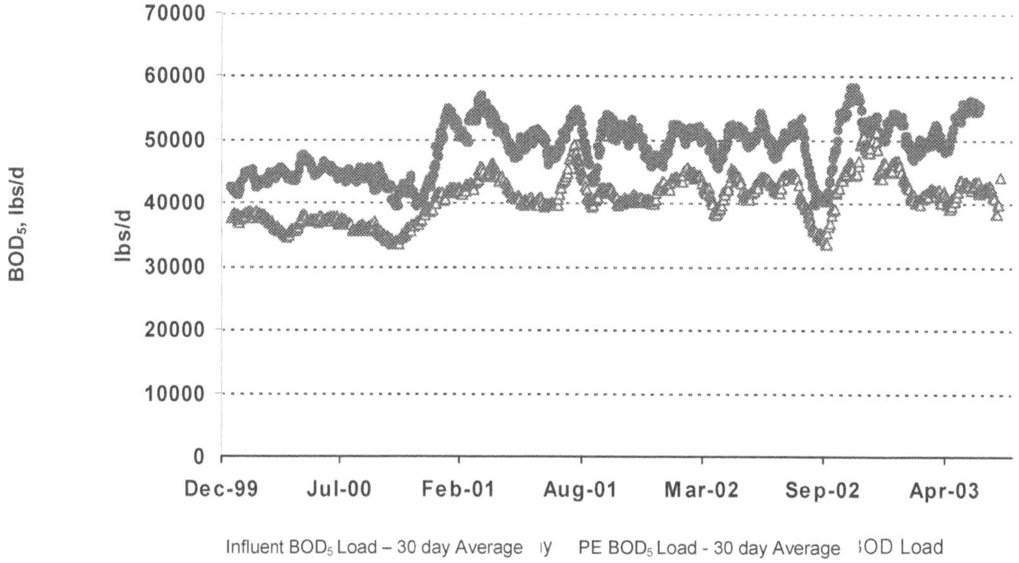


**Figure F-5. 30-day Averages for Raw Influent and Primary Effluent $BOD_5$ Between 2000 and 2003.**

[3] No effluent ammonia data available.

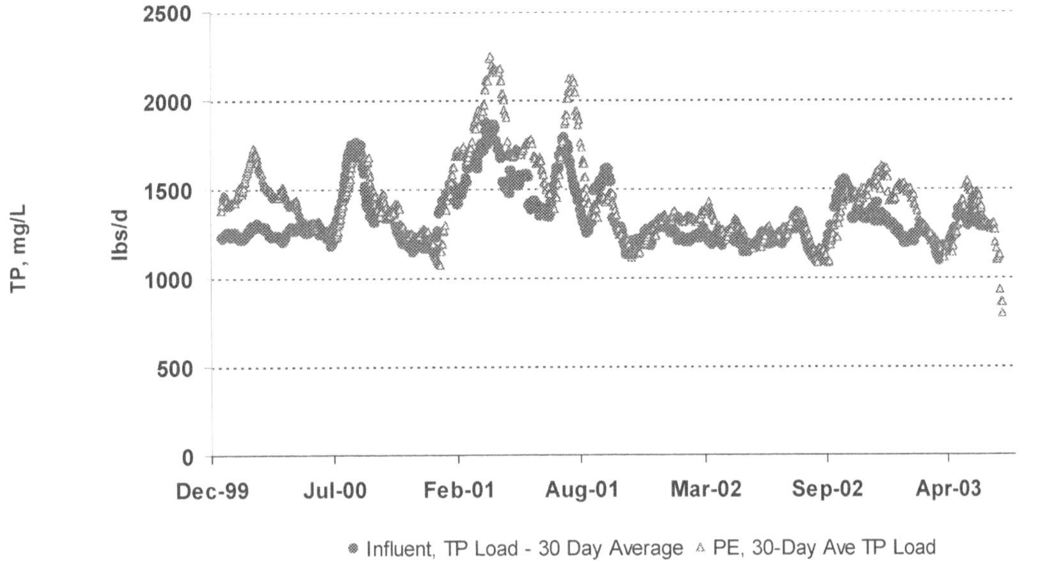


**Figure F-6. Influent and Primary Effluent Total Phosphorus Load.**

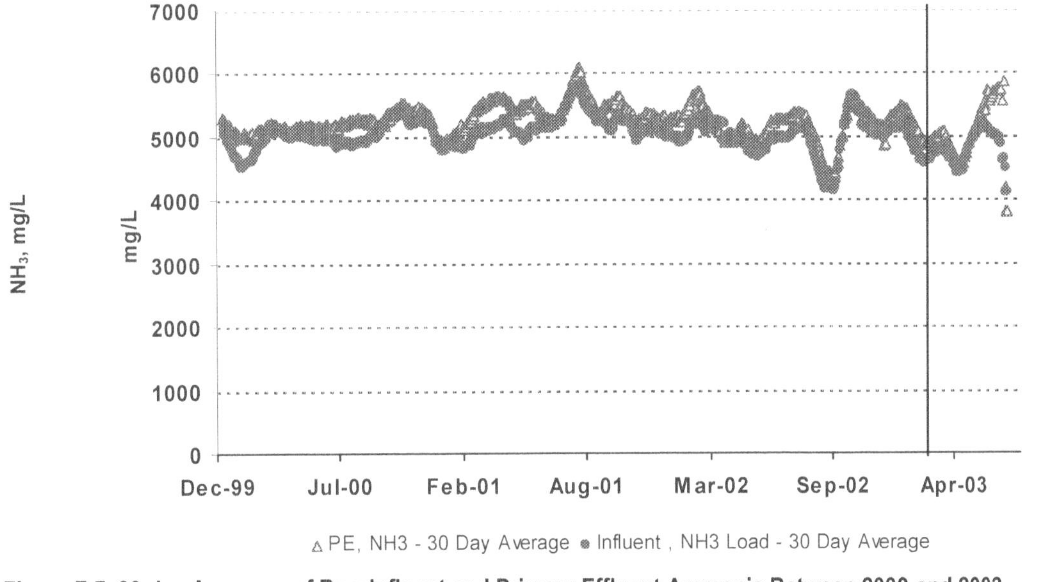


**Figure F-7. 30-day Averages of Raw Influent and Primary Effluent Ammonia Between 2000 and 2003.**

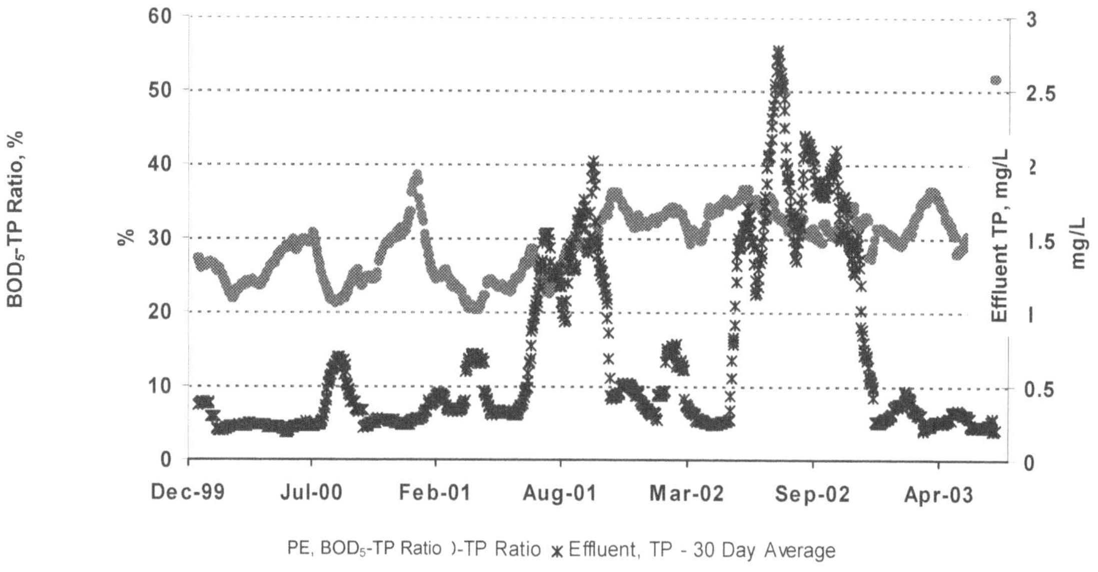


Figure F-8. 30-day Averages of Primary Effluent BOD5/TP Ratio and Effluent Total Phosphorus.

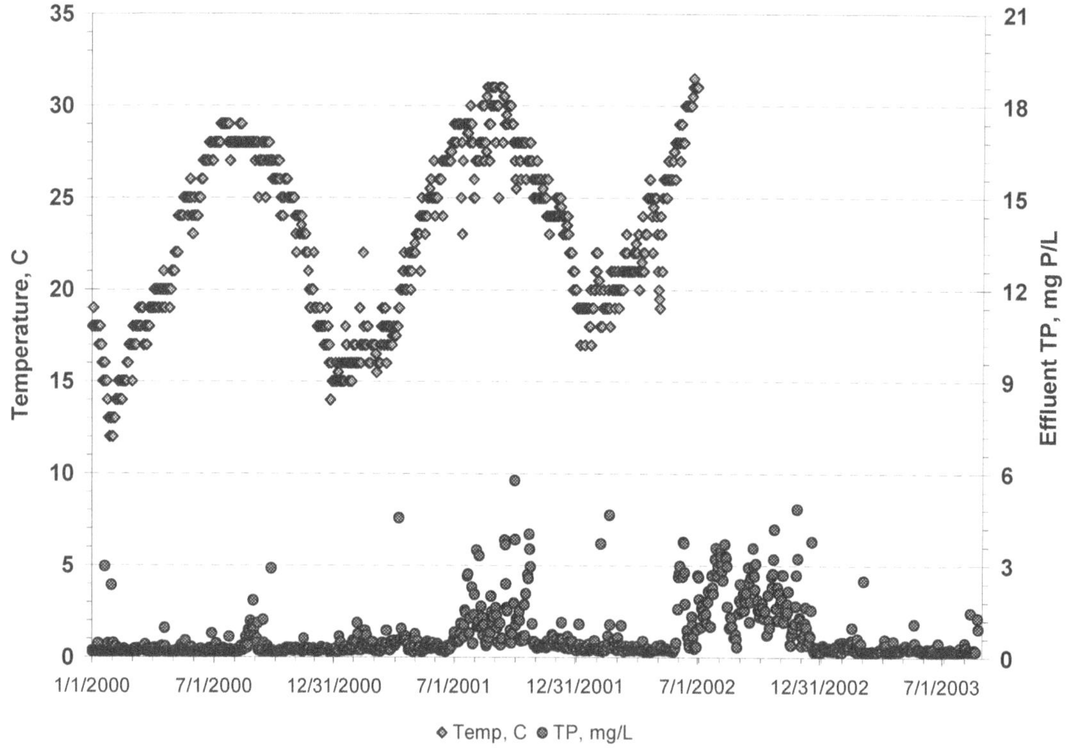


Figure F-17. Trend of Temperature and Effluent Total Phosphorus for VIP.

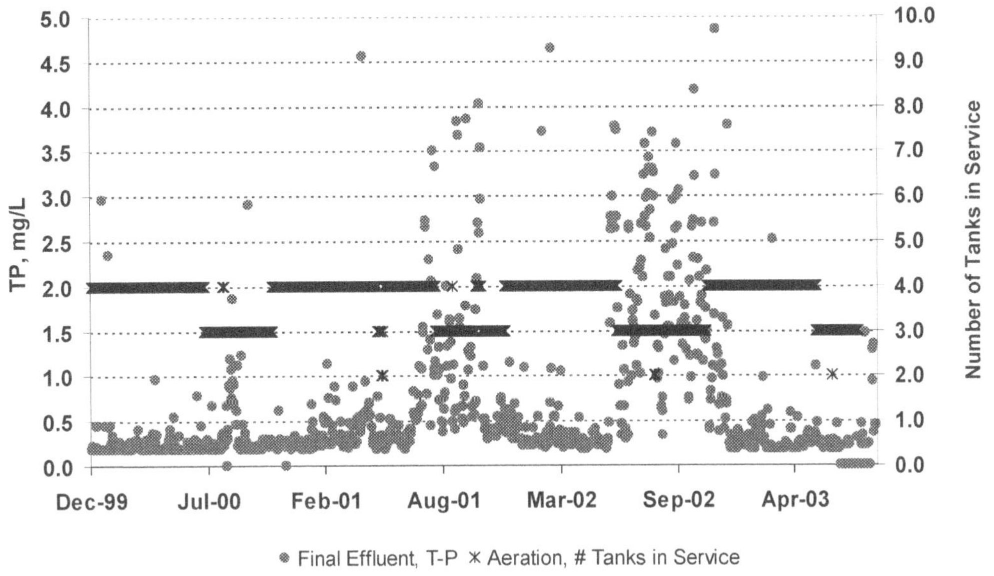


**Figure F-18. Effluent Total Phosphorus and Number of Aeration Basins Online.**

Figure F-17 shows the effluent TP and number of aeration basins online. There appears to be a relationship between the number of aeration basins online and effluent TP. The number of online aerobic basins is reduced in summer, and coincides with the periods of elevated effluent TP. This could be attributed to aerobic SRT, or simply to slow phosphorus uptake kinetics.

## F.3.3 Additional EBPR Process Evaluation

### F.3.3.1 Aeration Basin Phosphate Profiles

The aeration basin phosphate profile was measured during each sampling phase. Charts of the profiles Figure-20, Figure F-23, and Figure F-24) show four lines:

- The individual daily profiles (thin "shaded grey" lines in background with diamond points)
- Average phosphate profiles calculated from the daily profiles (thick line with circular points)
- Flow corrected average phosphate profile (thick line with triangular points). The flow corrected concentration is calculated as the theoretical phosphorus concentration based on the influent flow. This allows for a direct comparison of the concentration values without the influence of dilution from recycle flows.
- Phosphorus release rates (columns). The release rate (lb/d P/1000 lb VSS) is calculated as phosphorus released per day normalized by the total aeration basin volatile solids content.

In order to understand the phosphate profiles, the flow scheme must first be investigated. The flows are relatively complicated for the VIP mode of operation since three recycle streams must be accounted for: RAS, NRCY, and ARCY (described previously in Section F-2.2.1). Figure F-19, Figure F-21, and Figure F-22 illustrate the internal recycle flows for the three periods when the aeration basin profiles where measured.

#### *F.3.3.1.1 Phase 1 Profiles*

Figure F-20 shows the phosphate profiles measured at VIP during the Phase 1 field tests. The two profiles were taken on the same day, one in the morning and one in the afternoon. No release was measured in the first anaerobic zone, which is difficult to explain in light of the 80 mg/L primary effluent VFA measured on the same day. The concentration of VFAs in the second anaerobic zone was 12 mg/L, suggesting a total phosphorus release of approximately 35 mg/L. However, the flow corrected phosphorus release during in the anaerobic zones only reached about 12 mg/L. The uptake and release ratio of 0.1 (mg P released per mg of acetate removed) measured during the field test confirms what can be inferred from the lack of release measured—another organism type was out-competing phosphorus accumulating organisms (PAOs) and consuming the available VFA.

The plant also had trouble nitrifying and achieving EBPR during Phase 1 testing. This is evident in the profiles, as the phosphorus concentration continues to increase through the anoxic zones. Due to the lack of nitrate in the ARCY, the anoxic zones became anaerobic, allowing phosphorus release to continue.

With a mixed liquor concentration of 1650 mg/L and a volatile solids content of 81%, the normalized phosphorus release during the anaerobic retention time amounts to 5.3 lb $PO_4$-P per 1000 lb AB VSS. Including the release during the "anoxic" retention time, the value increases to 19.2 lb $PO_4$-P per 1000 lb AB VSS. Under good EBPR operation with a healthy PAO population and adequate VFAs, one would expect high release in the first anaerobic cell and no release (or some phosphorus uptake) in the anoxic cells. The situation observed at VIP during the first field visit was reversed—there was no release in the first anaerobic zone, and limited release through the subsequent anaerobic and anoxic zones. The phosphorus release was marginal considering the high aeration basin influent VFA concentration of 80 mg/L. All these measures indicated that EBPR was not active during this visit. The EBPR process took several months to fully recovery and reach phosphorus removal performance at levels experienced prior to the upset.

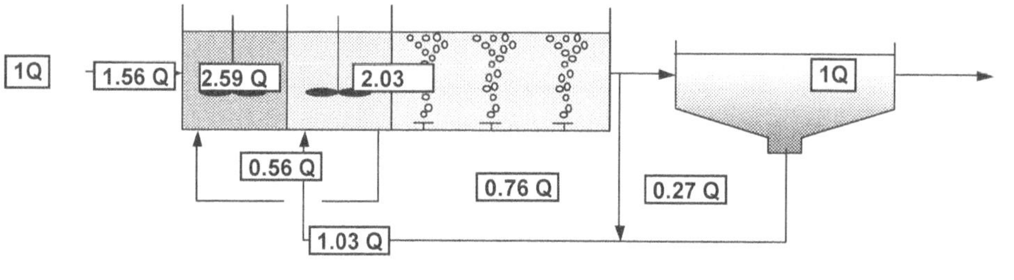


**Figure F-19. RAS and Internal Recycle Flows at VIP During the Phase One Field Testing.**

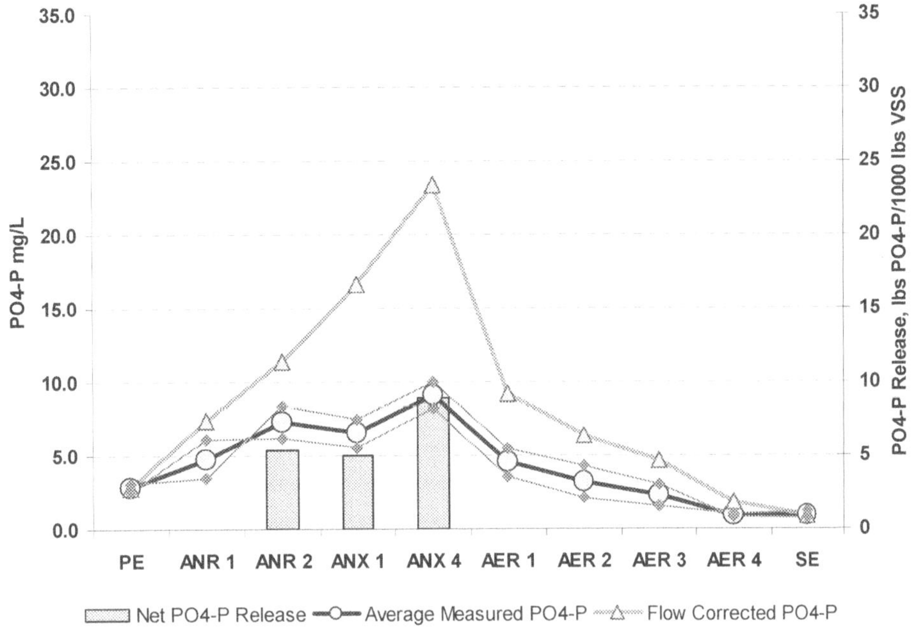


**Figure F-20. Phosphate Profiles Measured During Phase 1 Field Testing.**

### *F.3.3.1.2 Phase 2 Profiles*

Figure F-21 and Figure F-22 show the process configuration during Phase 2A and 2B, respectively. During the intensive testing periods in June and August 2003, the VIP EBPR process achieved secondary effluent total phosphorus of 0.23 mg/L and 0.63 mg/L respectively (Figure F-23 and Figure F-24). During Phase 2A, the total phosphorus release was 46.8 lb/d $PO_4$-P per 1000 lb VSS, including release during anoxic retention time. The initial release rate (Phase 2A) of 66.4 lb $PO_4$-P per hour and 1000 lb AB VSS (76.5 lb $PO_4$-P per hour and 1000 lb AB VSS) is an indication for a well-established PAO population. The aeration basins influent VFA of 26 mg/L suggests a significant VFA generation in the anaerobic and anoxic zone. The measured phosphorus release of 24 mg/L would require 48 mg/L of VFA, indicating that close to 50% of the VFA taken up in the EBPR process was generated in the anaerobic zone.

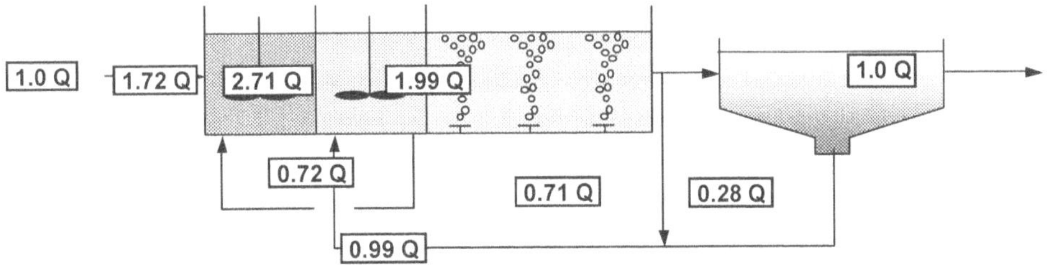


**Figure F-21. RAS and Internal Recycle Flows at VIP During Phase 2A Testing.**

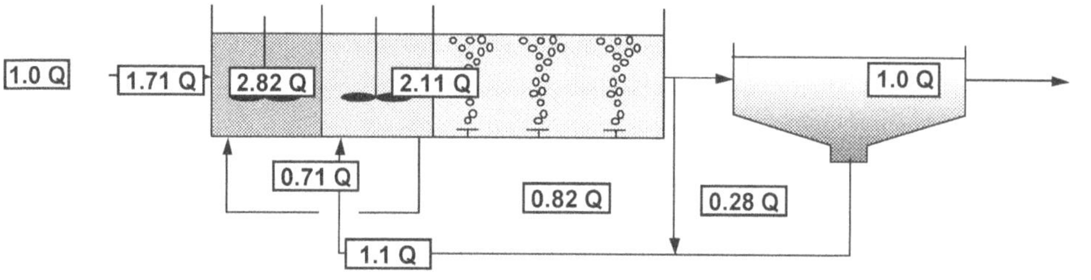


Figure F-22. RAS and Internal Recycle Flows at VIP the Phase 2B Testing.

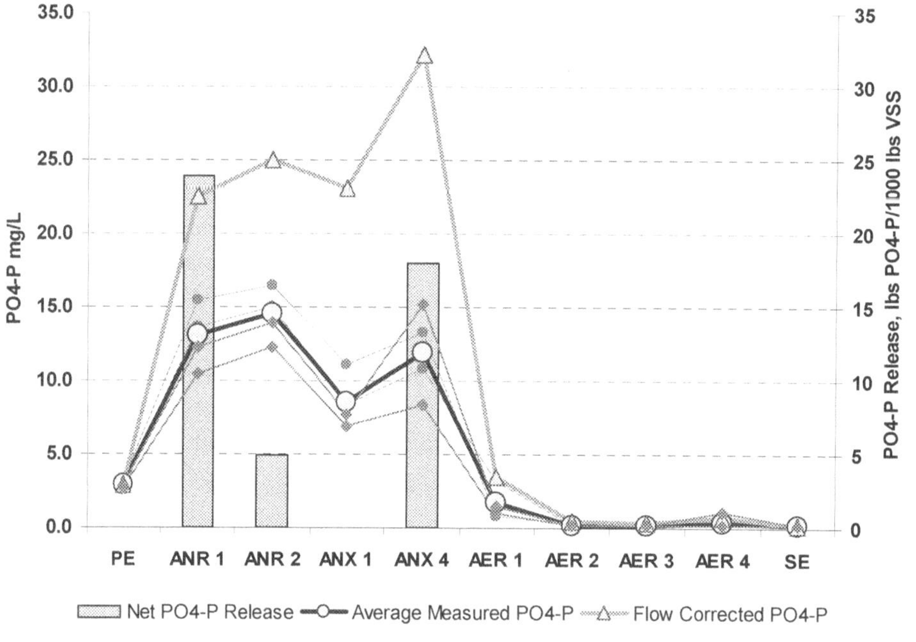


Figure F-23. Phosphate Profiles During First Intensive Sampling Period, Phase 2A.

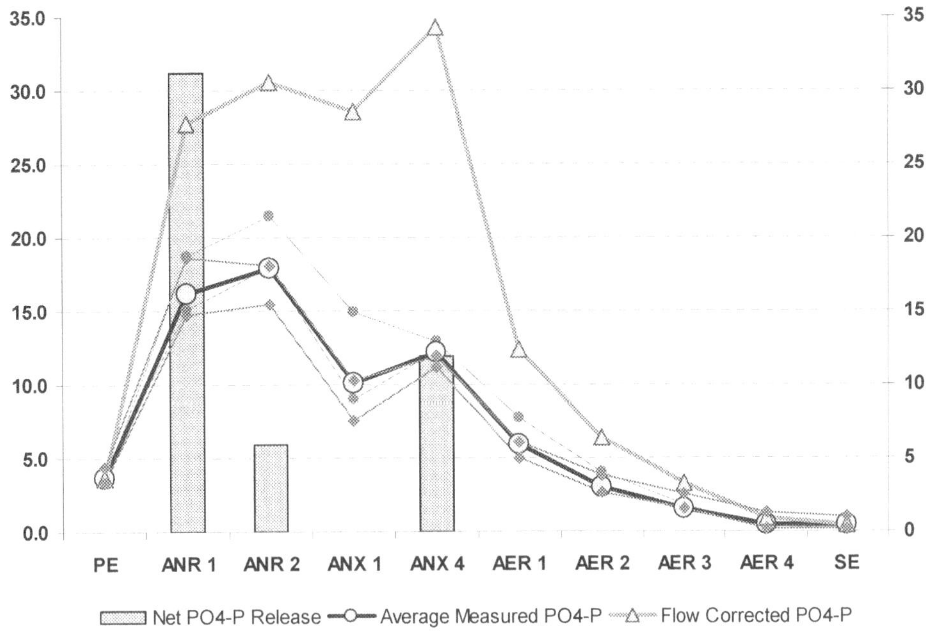


**Figure F-24. Phosphate Profiles During Second Intensive Sampling Period, Phase 2B.**

A second release under in the anoxic zone was measured during both intensive sampling periods. This release can be explained based on the low nitrate concentration in the anoxic zone (ANX 1 nitrate = 0.8 mg N/L). The low nitrate concentration is likely due to a combination of low mixed liquor (nitrate) recycle rate, high dilution with RAS and PE (2.8Q), and long anoxic retention time. Table F-8 summarizes the EBPR performance parameters based on the phosphorus profiles measured during intensive testing.

**Table F-8. Process Control and Water Quality Measurement Results During Intensive Testing (two-week average).**

| Raw Wastewater Parameter | Unit | Phase 1 | Phase 2a | Phase 2B |
|---|---|---|---|---|
| Total Phosphorus Release | mg/L | 14.5 | 23.6 | 23.6 |
| Anoxic Phosphorus Uptake | mg/L | 0.0 | 3.1 | 4.7 |
| Phosphorus Release Rate | Lb P per 1000 lb VSS per day | 19.2 | 46.8 | 49.0 |
| Phosphorus Uptake Rate | Lb P per 1000 lb VSS per day | 26.1 | 54.4 | 63.8 |
| Total Aeration Basin Phosphorus Inventory | Lb P | 5812 | 6050 | 5842 |
| Release of P Content | Lb P/d | 62% | 106% | 96% |
| Initial Phosphorus Release Rate | Lb P per 1000 lb VSS per hour | 0.0 | 66.4 | 76.5 |

### F3.3.2 Uptake and Release Tests (URT)

URTs provide a quantification of the PAO population and its activity in the basin. By adding VFA to the influent, phosphorus release is limited by the PAO population in the basin and by the amount of stored phosphorus. Theoretically, the phosphorus release under anaerobic conditions is 0.5 mg P/mg VFA. Lower measure values could indicate competition from other bacteria in the system such as GAOs.

URTs can be used for:

- Determining the amount of PAO activity from phosphorus release rate. The release rate (mg P/g VSS/h) is indicative of PAO activity.
- Determining the potential for competition in the anaerobic zone for VFA. The phosphorus release ratio (mg P released/mg VFA removed) indicates PAO activity. A phosphorus release ratio of 0.5 mg P released/mg VFA removed indicates PAO activity. Lower ratios are indicative of competition from other VFA consuming organisms such as GAOs.
- Determining key kinetic parameters for process simulation model calibration, such as ΔP/Acetic acid and magnesium and potassium uptake and release rates.

Table F-9 summarizes the results from the URT at VIP during Phase 2A and 2B (figure F-25). The testing procedure is outlined in Appendix J. The test was conducted three days before the intensive sampling to represent the activated sludge biomass during intensive sampling. An additional test was completed on September 3, 2003 in conjunction with the GAO analysis described in Section F-3.4

The URT confirm the observed trend of reduced EBPR performance and are also consistent with the increased GAO population (Section F-F.3.5). Similar to the trends in the recent years EBPR performance seems to be deteriorating during the summer. While there is some evidence pointing to increased GAO population as a cause, Figure F-18 suggests a relationship between HRT and EBPR performance. More than likely however, a combination of seasonally changing factors like temperature, retention time, SRT, etc are responsible.

**Table F-9. Uptake and Release Tests Results.**

| Test Parameter | Unit | Phase 2A | Phase 2B |
|---|---|---|---|
| Phosphorus Release Rate | mg P /g VSS /hr | 12.4 | 11.6 |
| Phosphorus Uptake Rate | mg P /g VSS /hr | 7.6 | 3.0 |
| Release Rate/Uptake Rate | | 1.6 | 3.8 |
| ΔP/Acetic acid | mg $PO_4$-P/ mg COD | 0.39 | 0.28 |
| ΔP/ΔMg - Release | mg $PO_4$-P / mg Mg | 4.71 | 6.38 |
| ΔP/ΔMg - Uptake | mg $PO_4$-P / mg Mg | 4.38 | 4.89 |
| ΔP/ΔK - Release | mg $PO_4$-P / mg K | - | 4.36 |
| ΔP/ΔK - Uptake | mg $PO_4$-P / mg K | - | 2.49 |
| MLVSS | mg/L | 1953 | 2180 |

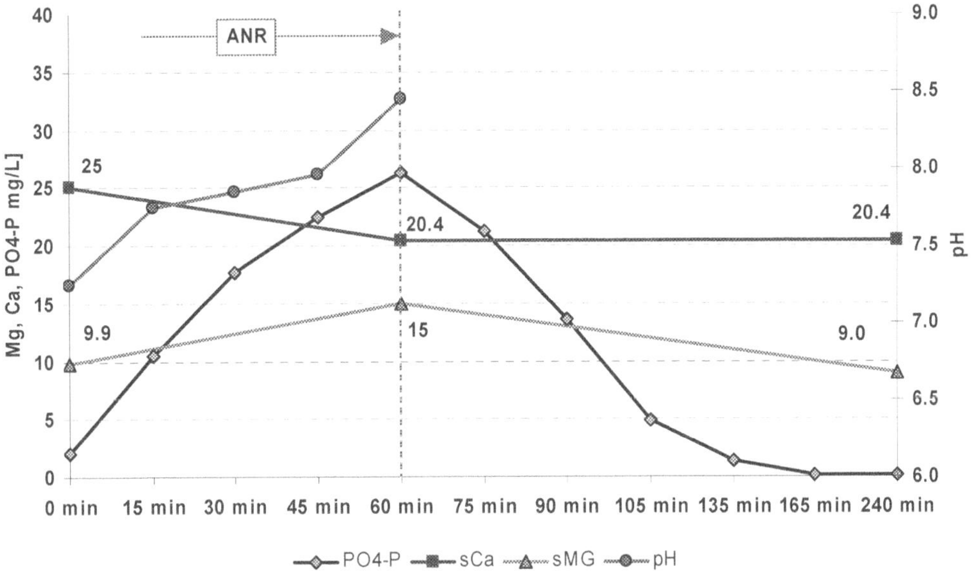


Figure F-25. Phosphorus Uptake and Release Test Results from Phase 2A.

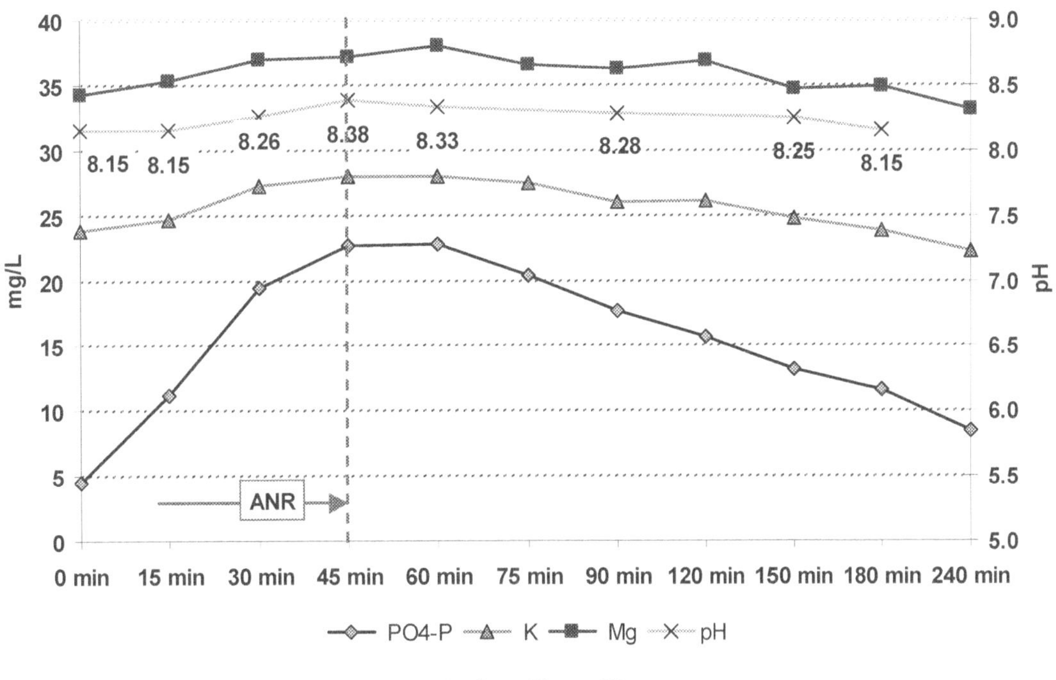


Figure F-26. Phosphorus Uptake and Release Test Results from Phase 2B.

### F3.3.3 Phosphorus Mass Balance

Balancing phosphorus mass allows tracking of phosphorus mass flow throughout the system. Values presented in this section are either 2002 annual averages or grab sample measurements from the Phase 1 testing period. While the grab samples provide only a snapshot of information, WAS or digested solids samples provide a solid measurement due to the long digester retention times. Year 2002 annual averages reflect periods of process upset as shown in Figure F-10 and described in Section F-3.2.

Figure F-27 shows the VIP phosphorus mass balance. Primary and waste activated sludge are dewatered and incinerated. The plant removes 81% of the influent phosphorus during primary and secondary treatment. The recycle load from primary and waste activated sludge dewatering amounts to 9% of the influent phosphorus load. About half of the recycle load is removed in the primary clarifier. Approximately 85% of the influent phosphorus is removed from the liquid stream with the waste activated sludge.

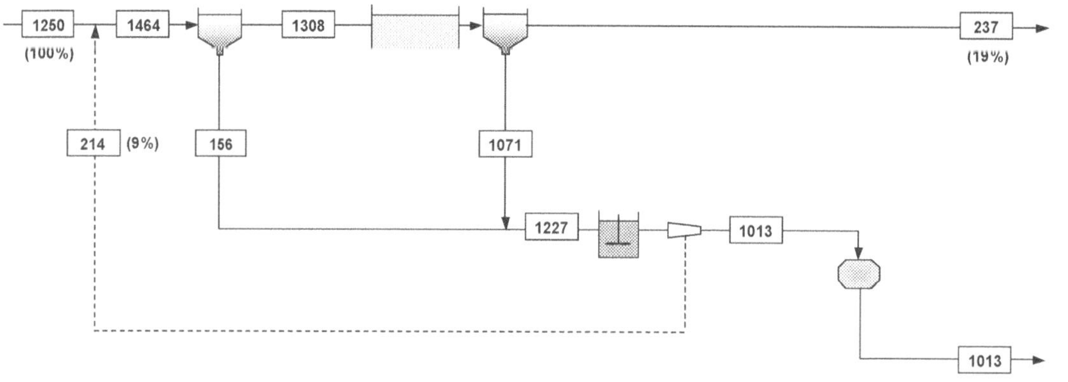


**Figure F-27. Phosphorus Mass Balance for VIP WWTP (2002 annual average mass flow lb P/d).**

### F.3.4 Microscopic Examination of Digested Sludge

At facilities that operate anaerobic digesters (especially those that digest waste activated sludge), struvite (magnesium ammonium phosphate) can often be found in the digested biosolids and can be observed under magnification. A digital microscope was used to inspect dewatered solids at 200x magnification. Because VIP does not use anaerobic digesters, the concentrations of ammonia and phosphate would be expected to be too low to promote struvite crystal formation. The pictures in Figure F-28 do not show any crystals in the VIP biosolids.

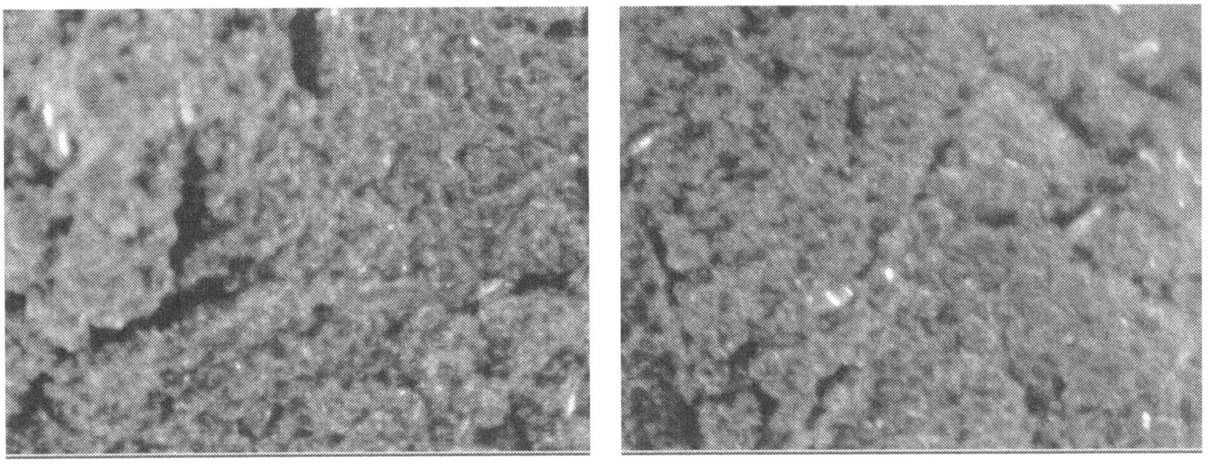

**Figure F-28. Digital Microscope Pictures of Dewatered Biosolids at 200x Magnification.**

### F.3.5 PAO and GAO Investigations

Mixed liquor samples from the plant were examined to observe the presence of polyphosphate granules (indicative of EBPR), and to look for the presence of GAOs, which typically consist of large cocci (2-3 μm) arranged in tetrad clusters within the floc. Three samples were collected in July and September of 2003, in close proximity with the Phase 2A and 2B Field Testing. Performance during these times was described in Sections F-3.2.2 and F-3.3.2.

Both microscopic study and molecular techniques were used to examine the presence of GAOs in the mixed liquor from the plant. Light microscopy observation was used to observe the tetrad-forming GAOs, and molecular probing using specific oligonucleotide probes targeting GAOs was completed to identify both tetrad-forming and non-tetrad-forming GAOs. In addition Neisser and DAPI staining were applied to observe the presence and relative abundance of PAOs.

Figure F-29 and Figure F-30 show the observation of PAOs and GAOs using a Neisser stain. Figure F-31 and Figure F-32 show the results of DAPI stain. The GAOs have Neisser positive stained cell walls, but are not positive for cell inclusions as are PAOs. The Neisser positive stained PAOs are dark blue to purple in color, and usually group in clusters within the flocs. The DAPI-stained PAOs are yellow in color, which is characteristic of polyphosphate granules observed under the fluorescence microscope. Other non-polyphosphate-accumulating bacteria including GAOs show up in blue with DAPI stain.

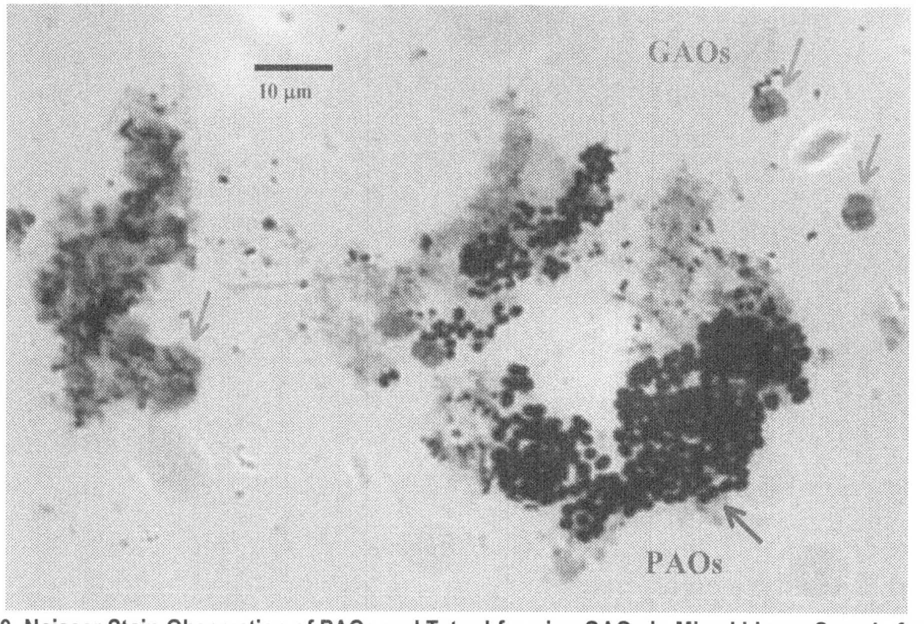


Figure F-29. Neisser Stain Observation of PAOs and Tetrad-forming GAOs in Mixed Liquor Sample from VIP Plant (July, 2003).

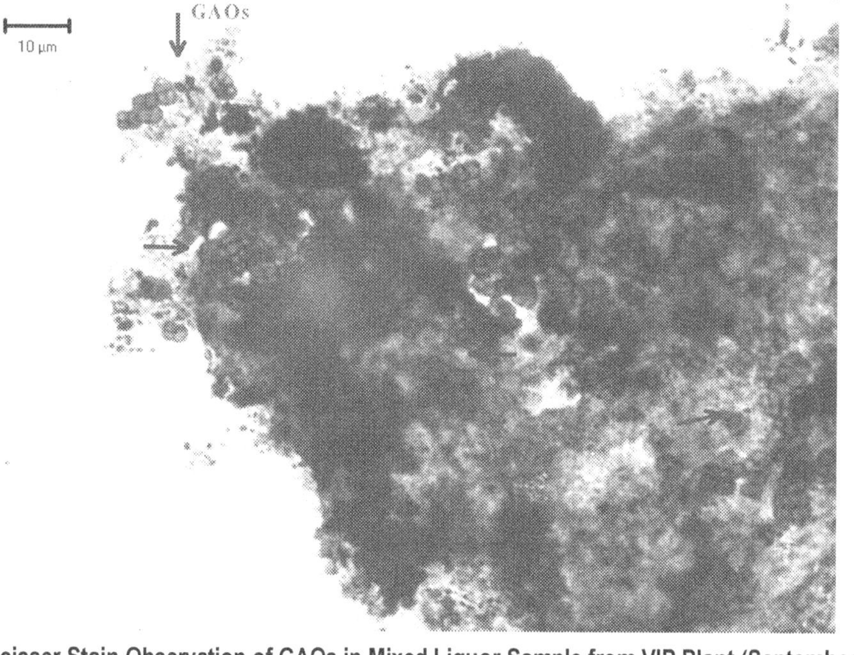


Figure F-30. Neisser Stain Observation of GAOs in Mixed Liquor Sample from VIP Plant (September 12, 2003).

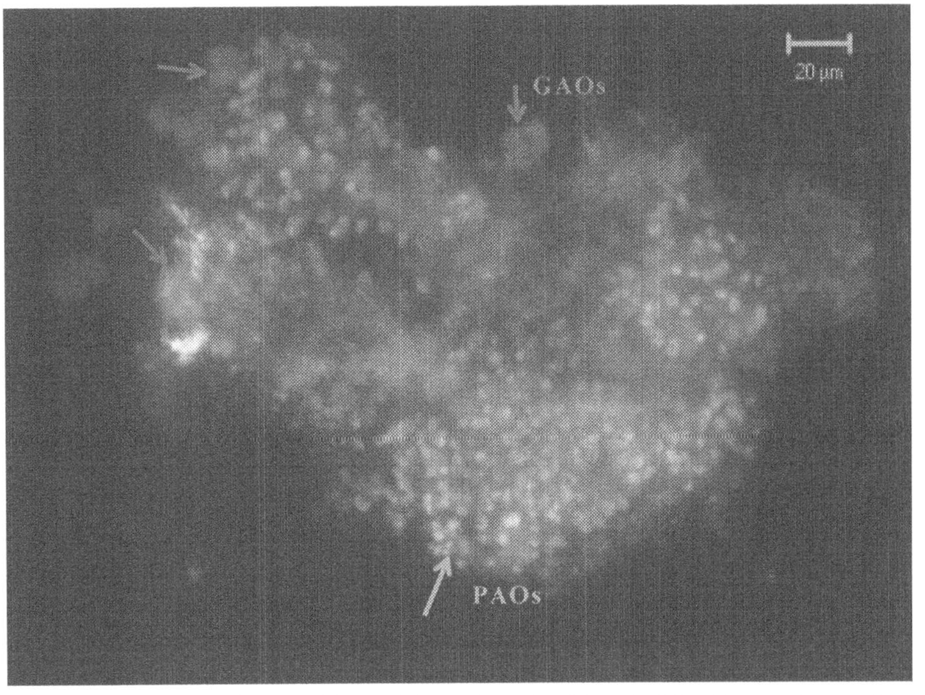


**Figure F-31. DAPI Stain Observation of PAOs and Tetrad-forming GAOs in Mixed Liquor Sample from VIP Plant (September 3, 2003).**

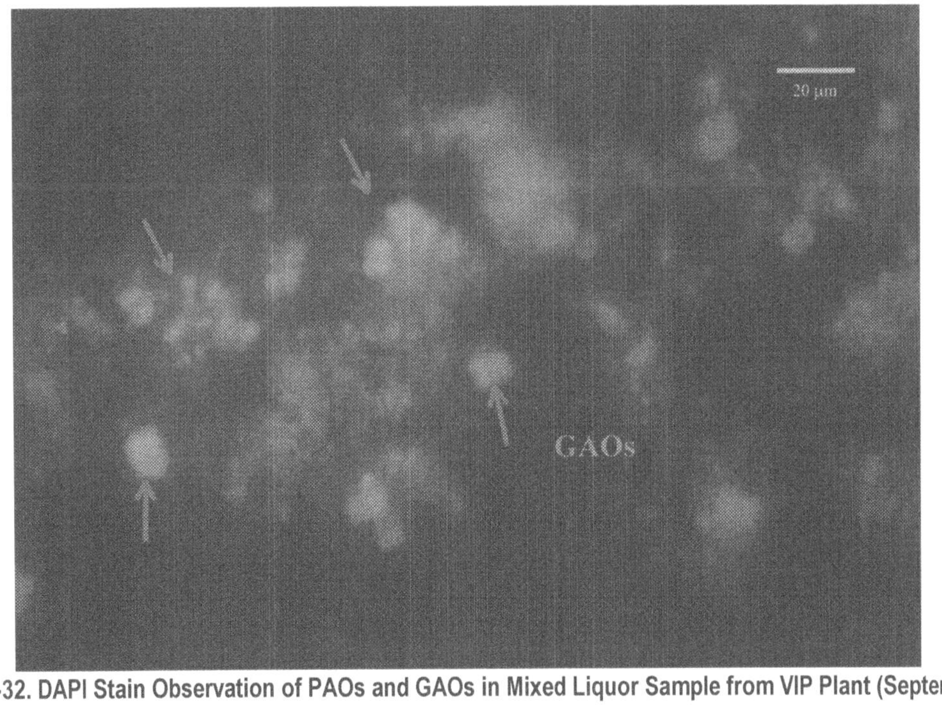


**Figure F-32. DAPI Stain Observation of PAOs and GAOs in Mixed Liquor Sample from VIP Plant (September 12, 2003).**

Table F-10 summarizes the observation of PAOs and GAOs for mixed liquor samples collected through the summer of 2003. During this time, the microbiology showed a shift from PAO to GAO organisms. In July, PAOs were predominant and only a small number of GAOs were found in the sludge flocs. The GAO population had increased by early September, and in mid-September the clusters of GAOs were almost as abundant as the clusters of PAOs. The sample taken in September, however, had abundant GAOs, especially in the sample taken on Sep. 12th,2003, in which clusters of GAOs were almost as abundant as the clusters of PAOs.

Examination of the mixed liquor samples using molecular methods (fluorescent in situ hybridization (FISH))using 16S rDNA-based probes targeting *Accumulibacter* (representing PAOs) (Crocetti et al. 00) and some non tetrad-forming GAOs-*Competibacter* (GAOQ989 and GB-G2, Crocetti et al. 00 and Kong *et al.* 00) showed that 10-20% of the Eubacteria populations was *Accumulibacter*-GAOs. About 1-10% of the Eubacteria population was found to be *Competibacter*-GAOs and their abundance remained relatively constant during the study period. It should be noted that FISH methods only targeted certain PAO and GAO species for which the probes are available. DAPI and Neisser stain were targeting at all cells that are capable of accumulating poly-P, therefore should reveal more PAO populations than FISH method. Additional GAO species could also be observed using the staining methods.

**Table F-10. Summary of PAOs and GAOs observation for VIP plant.**

| Date of Sample | GAO observation (Tetrad-forming) | GAO observation (*Competibacter*) | PAO observation | Comments |
|---|---|---|---|---|
| **VIP 7-3-03** | ***Some*** - 3 out of 10 fields observed GAOs, 2-27 tetrapods per field and spread, not in big flocs grouped together | 1-10% of total Eubacteria | ***Abundant***, cocci shaped cells group together in big flocs | PAO dominant, some tetra pod-shaped GAOs observed |
| **VIP 9-3-03** | ***Easily found***, some - 5 out of 10 fields observed GAOs, some group together | 1-10% of total Eubacteria | ***Abundant***, in big flocs | PAO dominant, tetra pod-shaped GAOs easily observed |
| **VIP 9-12-03** | ***Abundant***, 8 out of 10 fields observed GAOs, in big flocs, some fields had almost all GAOs | 1-10% of total Eubacteria | ***Easily found***, in flocs | Abundant GAOs, almost as abundant as PAOs. A lot filaments spread out in star-like shape |

The phosphorus removal performance at the plant was evaluated for the time period when the samples were taken for the GAO study. The results of phosphorus removal performance evaluation data are shown in Table F-11.

**Table F-11. Comparison of GAOs abundance and P Removal Performance.**

| Date of Sample | GAOs abundance | P Release Rate gP/gVSS.hr | P Uptake Rate gP/gVSS.hr | P Release Ratio gΔ P/gΔAcetic Acid | Release Rate/ Uptake Rate |
|---|---|---|---|---|---|
| VIP 6-03 | NA | 12.4 | 7.6 | 0.36 | 1.6 |
| VIP 7-3-03 | Some | | | | |
| VIP 9-3-03 | NA | | | | |
| VIP 9-12-03 | Abundant | 11.6 | 3 | 0.26 | 3.8 |

## F.4 Conclusions

VIP has a year-round phosphorus limit of 2 mg/L. Operating data since 2000 indicate the target is met the majority of the time. Between 2000 and 2003, effluent total phosphorus was under 1 mg/L 82% of the time, and under 0.5 mg/L 70% of the time. The aeration basin influent has a high BOD:P ratio (30:1), and thickening in the primary clarifiers likely adds VFAs to the BOD-rich influent. The plant did experience periods of upset with effluent total phosphorus concentrations of over 2 mg/L during two of the three summers in this period of record (2001 and 2002). The loss of EBPR coincided with the loss of nitrification, and could be due to a toxic industrial discharge caused by an annual event (i.e., cleaning) at one of the industrial facilities in VIP's service area.

Nutrient profiles collected during field testing indicate that nitrates are depleted prior to the final anoxic zone, and it appears that secondary phosphorus release occurs. This release, however, does not necessarily translate to high effluent soluble phosphorus. The uptake and release tests showed high uptake and release rates during the first intensive sampling period in June 2003. The release rate was somewhat lower and uptake significantly lower in the uptake and release test conducted in August 2003.

Microscopic observations of mixed liquor taken from July through September 2003 indicate a steady increase in the population of GAOs and decrease in the population of PAOs. This observation is consistent with the change in EBPR performance seen in the uptake and release tests. The cause of a population shift is unknown.

# REFERENCES


Barnard, J.L. 1976. A review of biological phosphorus removal in the activated sludge process. *Water SA*. 2, 3: 136-144.

Barnard, J.L. 1984. Activated primary tanks for phosphate removal. *Water SA*. 10, 3: 121-126.

Barnard, J.L., Stevens, G.M., Leslie, P.J. 1985. Design strategies for nutrient removal plants. *Water Science & Technology*. 17, (11/12): 233-242.

Benisch, M., Neethling, J.B., Baur, R. 2001. Role of Metals in Phosphate Crystallization in Biological Phosphorus Removal. *Proceedings of Water Env. Federation Annual Conference*, Atlanta, GA.

Bond, P.L., Erhart R., Wagner, M., Keller, J., Blackall, L.L. 1999. Identification of some of the major groups of bacteria in efficient and nonefficient biological phosphorus removal activated sludge systems. *Applied and Environmental Microbiology*, September: 4077-4084.

Brdjanovic, D., Logemann, S., van Loosdrecht, M.C.M., Hooijmans, C.M., Alaerts, G.J., Heijnen, J.J. 1998. Influence of temperature on biological phosphorus removal: process and molecular ecological studies. *Water Research*. 32 (4): 1035-1048.

Cech, J.S., Hartman, P. 1990. Glucose induced break down of enhanced biological phosphate removal. *Environmental Technology*. 11: 651-656.

Cech, J.S., Hartman, P. 1993. Competition between polyphosphate and polysaccharide accumulating bacteria in enhanced biological phosphate removal systems. *Water Research*. 27, 1219-1225.

Carucci, A., Lindrea, K., Majone, M., Ramadori, R. 1999. Different mechanisms for the anaerobic storage of organic substrates and their effect on enhanced biological phosphate removal (EBPR). *Water Science & Technology*. 39, (6): 21-28.

Comeau, Y. 1989. The Role of Carbon Storage in Biological Phosphate Removal from Wastewater. PhD Dissertation. University of British Columbia.

Crocetti, G.R., Hugenholtz, P., Bond, P.L., Shuler, A., Keller, J., Jenkins, D., Blackall, L. 2000. Identification of Polyphosphate-Accumulating Organisms and the Design of 16S rRNA-directed Probes for their Detection and Quantification. *Applied and Environmental Microbiology*. 66: 1175-1182.

Crocetti, G.R., Banfield, J.F., Keller, J., Bond, P.L., Blackall, L.L. 2002. Glycogen Accumulating Organisms in Laboratory-scale and Full-scale Activated Sludge Processes. *Microbiology-SGM*. 148: 3353-3364.

Eikelboom, D.H. 2000. *Process Control of Activated Sludge Plants by Microscopic Investigation*. IWA Publishing, London.

Erdal, U.G., Erdal, Z.K., Randall, C.W. 2002. Effect of temperature on EBPR system performance and microbial community. *Proceedings of Water Env. Federation Annual Conference.* Chicago, IL.

Filipe, C.D.M., Daigger, G.T., Grady, Jr., C.P.L. 2001a. pH as a key factor in the competition between glycogen-accumulating organisms and phosphorus-accumulating organisms. *Water Environment Research* 73, (2): 223-232.

Filipe, C.D.M., Daigger, G.T., Grady, Jr., C.P.L. 2001b. Effects of pH on the rates of aerobic metabolism of phosphate-accumulating and glycogen-accumulating organisms. *Water Environment Research* 73, (2): 213-223.

Goins, P., Parker, D., deBarbadillo, C., Wallis-Lage, C. 2003. Building a Better Nutrient Trap. Water Environment & Technology. 15, (7): 54-57.

Griffiths, P.C., Stratton, H.M., Seviour, R.J. 2002. Environmental factors contributing to "G bacteria" population in full-scale EBPR plants. Water Science & Technology, 46, (4/5): 185-192.

Janssen, P.M.J, Meinema, K, van der Roest, H.F. 2002. *Biological Phosphorus Removal: Manual for Design and Operation.* IAW Publishing: London.

Kong, Y, Ong, S.L., Ng, W.J., Liu, W.T. 2002. Diversity and distribution of a deeply branched novel proteobacterial group found in anaerobic-aerobic activated sludge processes. *Environmental Microbiology.* 4: 753-757.

Levin, G.V., Shapiro, J. 1965. Metabolic uptake of phosphorus by wastewater organisms. *Journal of Water Pollution Control Federation.* 37: 800.

Liu, W.T., Mino, T., Nakamura, K., Matsuo, T. 1996. Glycogen accumulating population and its anaerobic substrate uptake abilities in anaerobic-aerobic activated sludge without biological phosphorus removal. *Water Research.* 30: 75-87.

Liu, W.T., Nakamura, K., Matsuo, T., Mino, T. 1997. Internal energy-based competition between polyphosphate and glycogen-accumulating bacteria in biological phosphorus removal reactors—effect of P/C feeding ratio. *Water Research.* 31, (6): 1430-1438.

Mamais, D., Jenkins, D., Pitt , P. 1993. A rapid physical-chemical method for the determination of readily biodegradable soluble COD in municipal wastewater. *Water Research.* 27, (1): 195-197.

Maszenan, A.M., Seviour, R.J., Patel, B.K.C., Rees, G.N., McDougall, B.M. 1998. The hunt for the G-bacteria in activated sludge biomass. *Water Science & Technology.* 37, (4/5): 65-69.

Maurer M, Gujer W, Hany R, Bachmann S. 1997 Intracellular carbon flow in phosphorus accumulating organisms in activated sludge systems. Wat. Res. 31:907-917.

Merseth, H.L. 1995. An Investigation of Factors Affecting Biological Phosphorus Removal: Primary Sludge Fermentation and System Operating Conditions. Masters Dissertation. University of Washington.

Mino, T., Arun, V., Tsuzuki, Y., Matsuo, T. 1987. Effect of phosphorus accumulation on acetate metabolism in the biological phosphorus removal process. *Proceedings of the Biological Phosphate Removal from Wastewater IAWPRC Specialized Conference.* Rome, Italy.

Mino, T., Satoh, H., Matsuo, T. 1994. Metabolism of different bacterial populations in enhanced biological phosphorus removal processes. *Water Science & Technology*. 29, (7): 67-70.

Mino, T., van Loosdrecht, M.C.M., Heijnen, J.J. 1998. Microbiology and biochemistry of the enhanced biological phosphate removal process. *Water Research*. 32: 3193-3207.

Rabinowitz, B., Daigger, G.T., Jenkins, D., Neethling, J.B. 2004. The Effect of High Temperatures On BNR Process Performance. *Proceedings, Water Environment Federation Annual Conference,* New Orleans, LA.

Satoh, H., Mino, T., Matsuo T. 1992. Uptake of organic substrates and accumulation of polyhydroxyalkanoates linked with glycolysis of intracellular carbohydrates under anaerobic conditions in the biological excess phosphate removal processes. *Water Science & Technology*. 26 (5/6): 933-942.

Schuler, A.J., Jenkins, D. 2003. Enhanced biological phosphorus removal from wastewater by biomass with different phosphorus contents, part 1: experimental results and comparison with metabolic models. *Water Environment Research*. 75: 485.

Serafim, L.S., Lemos, P.C., Reis, M.A.M. 2002. Effect of pH control on EBPR stability and efficiency. *Water Science and Technology*. 46, No. 4-5: 179-184.

Seviour, R.J., Maszenan, A.M., Tandoi, V., Patel, B.K.C., Kong, Y., Schumann, P. 2000. Microbiology of the ‘G-bacteria” in activated sludge. *Environmental Microbiology*. 2, (6): 581-593.

Stensel, H.D. 1991. *Principles of Biological Phosphorus Removal. In Phosphorus and Nitrogen Removal from Municipal Wastewater: Principles and Practice, Second Edition.* Sedlak, R., Editor. Lewis Publishers.

Stensel, H.D., Barnard, J.L. 1992. *Principles of Biological Nutrient Removal. In Design and Retrofit of Wastewater Treatment Plants for Biological Nutrient Removal, Volume 5.* Randall, C.W., Barnard, J.L., Stensel, H.D., editors. Technomic Publishing Company. Lancaster, PA.

Stephens, H.M., Neethling, J.B., Benisch, M., Baake, B., Stensel, H.D., Moore, R. Reliability of biological phosphorus removal. 2003.*Proceedings of the Water Env. Federation Conference,* Los Angeles.

Stephens, H., Stensel, H.D. 1998. Effect of operating conditions on biological phosphorus removal. *Water Environment Research*. 70, (3): 362-369.

Sudiana, I.M., Mino, T., Satoh, H., Matsuo, T. 1998. Morphology, in-situ characterization with rRNA targeted probes and respiratory Quinone profiles of enhanced biological phosphorus removal sludge. *Water Science & Technology*. 38, (8/9): 69-76.

Tsai, C.S. and Liu, W.T. 2002. Phylogenetic and physiological diversity of tetrad-forming organisms in deteriorated biological phosphorus removal systems. *Water Sci. and Technol.* 46(1-2):179-184.

Wang, J.C., Park, J.K., Whang, L.M. 2000. Comparison of fatty acid composition and kinetics of phosphorus-accumulating organisms and glycogen-accumulating organisms. *Proceedings of the Water Env. Federation Annual Conference*. New Orleans, LA.

Wanner, J., Ruzickova, I., Krhutkova, O., Pribyl, M. 2000. Activated sludge population dynamics and wastewater treatment plant design and operation. *Water Science & Technology*. 41, (9): 217-225.

Wentzel, M.C., Lotter, L.H., Loewenthal, R.E., Marais, G.V.R. 1986. Metabolic behaviour of Acinetobacter spp. in enhanced biological phosphorus removal — a biochemical model. *Water SA*. 12, 4: 209-223.

Whang, L-M., Park, J.K. 2001. Competition between polyphosphate- and glycogen-accumulating organisms in enhanced biological phosphorus removal systems; effect of temperature and sludge age. *Proceedings of the Water Env. Federation Annual Conference*. Atlanta, GA.

Whang, L-M., Park, J.K. 2002. Competition between polyphosphate- and glycogen-accumulating organisms in biological phosphorus removal systems; effect of temperature. *Water Science & Technology*. 46, (1/2): 191-194.

Whang, L-M., Filipe, C.D.M., Park, J.K. 2002. Model-based evaluation of competition between polyphosphate- and glycogen-accumulating organisms. *Proceedings of the Water Env. Federation Annual Conference*, Chicago.

WERF SUBSCRIBERS

## WASTEWATER UTILITY

### Alabama
Montgomery Water Works & Sanitary Sewer Board

### Alaska
Anchorage Water & Wastewater Utility

### Arizona
Gila Resources
Glendale, City of, Utilities Department
Mesa, City of
Peoria, City of
Pima County Wastewater Management

### Arkansas
Little Rock Wastewater Utility

### California
Calaveras County Water District
Central Contra Costa Sanitary District
Crestline Sanitation District
Delta Diablo Sanitation District
Dublin San Ramon Services District
East Bay Dischargers Authority
East Bay Municipal Utility District
Eastern Municipal Water District
El Dorado Irrigation District
Fairfield-Suisun Sewer District
Fort Lauderdale, City of
Irvine Ranch Water District
Las Virgenes Municipal Water District
Livermore, City of
Lodi, City of
Los Angeles, City of
Los Angeles County, Sanitation Districts of
Napa Sanitation District
Orange County Sanitation District
Palo Alto, City of
Riverside, City of
Sacramento Regional County Sanitation District
San Diego Metropolitan Wastewater Department, City of
San Francisco, City & County of
San Jose, City of
Santa Barbara, City of
Santa Cruz, City of
Santa Rosa, City of
South Bayside System Authority
South Coast Water District
South Orange County Wastewater Authority
Sunnyvale, City of
Union Sanitary District
West Valley Sanitation District

### Colorado
Boulder, City of
Colorado Springs Utilities
Greeley, City of
Littleton/Englewood Water Pollution Control Plant
Metro Wastewater Reclamation District, Denver

### Connecticut
The Mattabassett District
New Haven, City of, WPCA

### District of Columbia
District of Columbia Water & Sewer Authority

### Florida
Broward, County of
Fort Lauderdale, City of
Gainesville Regional Utilities
JEA
Miami-Dade Water & Sewer Authority
Orange County Utilities Department
Reedy Creek Improvement District
Seminole County Environmental Services
St. Petersburg, City of
Stuart Public Utilities
Tallahassee, City of
Tampa, City of
Toho Water Authority
West Palm Beach, City of

### Georgia
Atlanta Department of Watershed Management
Augusta, City of
Clayton County Water Authority
Cobb County Water System
Columbus Water Works
Fulton County
Gwinnett County Department of Public Utilities
Savannah, City of

### Hawaii
Honolulu, City & County of

### Idaho
Boise, City of

### Illinois
American Bottoms Wastewater Treatment Plant
Greater Peoria Sanitary District
Kankakee River Metropolitan Agency
Metropolitan Water Reclamation District of Greater Chicago
Wheaton Sanitary District

### Iowa
Ames, City of
Des Moines Metro Wastewater Reclamation Authority
Iowa City

### Kansas
Johnson County Unified Wastewater Districts
Unified Government of Wyandotte County/ Kansas City, City of

### Kentucky
Louisville & Jefferson County Metropolitan Sewer District

### Louisiana
Sewerage & Water Board of New Orleans

### Maine
Bangor, City of
Portland Water District

### Maryland
Anne Arundel County Bureau of Utility Operations
Howard County Department of Public Works
Washington Suburban Sanitary Commission

### Massachusetts
Boston Water & Sewer Commission
Upper Blackstone Water Pollution Abatement District

### Michigan
Ann Arbor, City of
Detroit, City of
Holland Board of Public Works
Lansing, City of
Saginaw, City of
Wayne County Department of Environment
Wyoming, City of

### Minnesota
Rochester, City of
Western Lake Superior Sanitary District

### Missouri
Independence, City of
Kansas City Missouri Water Services Department
Little Blue Valley Sewer District
Metropolitan St. Louis Sewer District

### Nebraska
Lincoln Wastewater System

### Nevada
Henderson, City of

### New Jersey
Bergen County Utilities Authority
Passaic Valley Sewerage Commissioners

### New York
New York City Department of Environmental Protection

### North Carolina
Charlotte/Mecklenburg Utilities
Durham, City of
Metropolitan Sewerage District of Buncombe County
Orange Water & Sewer Authority

### Ohio
Akron, City of
Butler County Department of Environmental Services
Columbus, City of
Metropolitan Sewer District of Greater Cincinnati
Northeast Ohio Regional Sewer District
Summit, County of

### Oklahoma
Trinity River Authority
Tulsa, City of

### Oregon
Clean Water Services
Eugene/Springfield Water Pollution Control
Water Environment Services

### Pennsylvania
Philadelphia, City of
University Area Joint Authority

### South Carolina
Charleston Commissioners of Public Works
Mount Pleasant Waterworks & Sewer Commission
Spartanburg Sanitary Sewer District

### Tennessee
Cleveland, City of
Knoxville Utilities Board
Murfreesboro Water & Sewer Department
Nashville Metro Water Services

### Texas
Austin, City of
Dallas Water Utilities
Denton, City of
El Paso Water Utilities
Fort Worth, City of
Gulf Coast Waste Disposal Authority
Houston, City of
San Antonio Water System
Trinity River Authority

### Utah
Salt Lake City Corporation

### Virginia
Alexandria Sanitation Authority
Arlington, County of
Fairfax County
Hampton Roads Sanitation District
Henrico, County of

Hopewell Regional Wastewater Treatment Facility
Loudoun County Sanitation Authority
Lynchburg Regional WWTP
Prince William County Service Authority
Richmond, City of
Rivanna Water & Sewer Authority

**Washington**
Bellevue Utilities Department
Edmonds, City of
Everett, City of
King County Department of Natural Resources
Seattle Public Utilities
Sunnyside, Port of
Yakima, City of

**Wisconsin**
Green Bay Metro Sewerage District
Madison Metropolitan Sewerage District
Milwaukee Metropolitan Sewerage District
Racine, City of
Sheboygan Regional Wastewater Treatment
Wausau Water Works

**Australia**
South Australian Water Corporation
Sydney Water Corporation
Water Corporation of Western Australia

**Canada**
Greater Vancouver Regional District
Regina, City of, Saskatchewan
Toronto, City of, Ontario
Winnipeg, City of, Manitoba

**Mexico**
Servicios de Agua y Drenaje de Monterrey, I.P.D.

**New Zealand**
Watercare Services Limited

**United Kingdom**
Yorkshire Water Services Limited

## STORMWATER UTILITY

**California**
Los Angeles, City of, Department of Public Works
Monterey, City of
San Francisco, City & County of
Santa Rosa, City of

**Colorado**
Boulder, City of

**Georgia**
Griffin, City of

**Iowa**
Cedar Rapids Wastewater Facility
Des Moines Metro Wastewater Reclamation Authority

**Kansas**
Overland Park, City of

**Maine**
Portland Water District

**North Carolina**
Charlotte, City of, Stormwater Services

**Pennsylvania**
Philadelphia, City of

**Tennessee**
Chattanooga Stormwater Management

**Washington**
Seattle Public Utilities

## STATE

Arkansas Department of Environmental Quality
Fresno Department of Public Utilities
Kansas Department of Health & Environment
Urban Drainage & Flood Control District, CO

## CORPORATE

ADS Environmental Services
The ADVENT Group Inc.
Alan Plummer & Associates
Alden Research Laboratory Inc.
Aqua-Aerobic Systems Inc.
Aquateam–Norwegian Water Technology Centre A/S
ARCADIS
Associated Engineering
BaySaver Inc.
Black & Veatch
Blasland, Bouck & Lee Inc.
Boyle Engineering Corporation
Brown & Caldwell
Burns & McDonnell
CABE Associates Inc.
The Cadmus Group
Camp Dresser & McKee Inc.
Carollo Engineers Inc.
Carpenter Environmental Associates Inc.
CDS Technologies Inc.
Chemtrac Systems Inc.
CH2M HILL
Damon S. Williams Associates, LLC
David L. Sheridan, P.C.
Dewling Associates Inc.
Earth Tech Inc.
Ecovation
EMA Inc.
The Eshelman Company Inc.
EWT Holdings Corporation
Fay, Spofford, & Thorndike Inc.
Freese & Nichols Inc.
ftn Associates Inc.
Fuss & O'Neill Inc.
Gannett Fleming Inc.
GE Betz
Greeley and Hansen LLC
Hazen & Sawyer, P.C.
HDR Engineering Inc.
HNTB Corporation
HydroQual Inc.
Infilco Degremont Inc.
Ingersoll-Rand Energy Systems
Insituform Technologies Inc.
Institute for Environmental Technology & Industry, Korea
Jacobson Helgoth Consultants Inc.
Jason Consultants LLC Inc.
Jordan, Jones, & Goulding Inc.
KCI Technologies Inc.
Kelly & Weaver, P.C.
Kennedy/Jenks Consultants
Komline Sanderson Engineering Corporation
Lawler, Matusky & Skelly Engineers, LLP
LTI Inc.
Lombardo Associates Inc.
Malcolm Pirnie Inc.
McKim & Creed
MEC Analytical Systems Inc.
Metcalf & Eddy Inc.
MWH
New England Organics
O'Brien & Gere Engineers Inc.
Odor & Corrosion Technology Consultants Inc.
Oswald Green, LLC
PA Government Services Inc.
Parametrix Inc.
Parsons
Post, Buckley, Schuh & Jernigan
R&D Engineering/Conestoga Rover & Associates
RMC Inc.
R.M. Towill Corporation
Ross & Associates Ltd.
Royce Technologies
SAIC Maritime Technical Group
Short Elliott Hendrickson Inc.
Stantec Consulting Inc.
Stearns & Wheler, LLC
Stone Environmental Inc.
Stormwater Management Inc.
Stratus Consulting Inc.
Synagro Technologies Inc.
Tetra Tech Inc.
Trojan Technologies Inc.
URS Corporation
USfilter
Veola Water North America
Wade-Trim Inc.
Westin Engineering Inc.
Weston Solutions Inc.
Woodard & Curran
WRc/D&B, LLC
WWETCO, LLC
Zenon Environmental Inc.
Zoeller Pump Company

## INDUSTRY

American Electric Power
ChevronTexaco Energy Research & Technology Company
The Coca-Cola Company
Dow Chemical Company
DuPont Company
Eastman Chemical company
Eastman Kodak Company
Eli Lilly & Company
Merck & Company Inc.
ONDEO Services
Procter & Gamble Company
PSEG Services Corporation
RWE Thames Water Plc
Severn Trent Services Inc.
United Water Services LLC

**Note: List as of 2/1/05**

## WERF Board of Directors



## WERF Research Council



## WERF Technical Advisory Committees

### Stormwater Technical Advisory Committee



### Decentralized Research Advisory Council

# WERF Product Order Form

a benefit of joining the Water Environment Research Foundation, subscribers are entitled to receive one complimentary copy of all final orts and other products.Additional copies are available at cost (usually $10).To order your complimentary copy of a report, please write e" in the unit price column.WERF keeps track of all orders.If the charge differs from what is shown here, we will call to confirm the total re processing.

ne Title

anization

ress

State Zip Code Country

ne Fax Email

| Stock # | Product | Quantity | Unit Price | Total |
|---|---|---|---|---|
| | | | | |
| | | | | |
| | | | | |
| | | | | |
| | | | | |
| | | | | |
| | | | | |
| | | | | |
| | | | Postage & Handling | |
| | | | VA Residents Add 4.5% Sales Tax | |
| | | | Canadian Residents Add 7% GST | |
| | | | TOTAL | |

## Method of Payment:

(All orders must be prepaid.)

Check or Money Order Enclosed

Visa ❑ Mastercard ❑ American Express

ount No. Exp. Date

ature

## Shipping & Handling:

| Amount of Order | United States | Canada & Mexico | All Others |
|---|---|---|---|
| Up to but not more than: | Add: | Add: | Add: |
| $20.00 | $5.00* | $8.00 | 50% of amount |
| 30.00 | 5.50 | 8.00 | 40% of amount |
| 40.00 | 6.00 | 8.00 | |
| 50.00 | 6.50 | 14.00 | |
| 60.00 | 7.00 | 14.00 | |
| 80.00 | 8.00 | 14.00 | |
| 100.00 | 10.00 | 21.00 | |
| 150.00 | 12.50 | 28.00 | |
| 200.00 | 15.00 | 35.00 | |
| More than $200.00 | Add 20% of order | Add 20% of order | |

*minimum amount for all orders

e: Please make checks payable to the Water Environment Research Foundation.

## To Order (Subscribers Only):

Log on to www.werf.org and click on the "Product Catalog."

Phone: (703) 684-2470
Fax: (703) 299-0742.

WERF
Attn: Subscriber Services
635 Slaters Lane
Alexandria, VA 22314-1177

## To Order (Non-Subscribers):

Non-subscribers may be able to order WERF publications either through WEF (www.wef.org) or IWAP (www.iwapublishing.com).Visit WERF's website at www.werf.org for details.

IWA Publishing's authorised EU representative for General Product Safety Regulations is Diane D'Arras, 15 rue Duret, 75116 Paris, France, e-mail: safety@iwap.co.uk.

Printed and bound by CPI Group (UK) Ltd, Croydon, CR0 4YY

12/05/2026

02108868-0009